The History of English Leicester Sheep in Australia

Brenton Heazlewood

The Rural Publishing Company

First published by The Rural Publishing Company 2025.

Copyright © Brenton Heazlewood 2025

Print (Paperback): 978-1-923008-32-8
eBook: 978-1-923008-33-5

This work is copyright. Apart from any use permitted under the Copyright Act 1968, no part of this publication may be reproduced, stored in a retrieval system or transmitted in any form or by any means, electronic, mechanical, photocopying, recording or otherwise, without the prior written permission of Brenton Heazlewood or The Rural Publishing Company.

Cover Design: The Rural Publishing Company
Typesetting & Design: The Rural Publishing Company

The Rural Publishing Company
Website: https://theruralpublishingcompany.com.au
Email: hello@theruralpublishingcompany.com.au

To my wife Anne,

You have supported and encouraged me for the past six years while I have researched and written this book.

You have read and re-read the manuscript many times offering your suggestions and correcting my poor spelling and punctuation. Above all you have patiently tolerated your kitchen table being constantly covered in reference books, magazines, flock books and my written manuscripts for many years, only asking for it to be tidied up when visitors came.

Without your encouragement and support this task would have been a lot more difficult.

Thank you.

BRENTON HEAZLEWOOD

Contents

Introduction	1
Timeline	7
1. History of English Leicester Sheep in Australia	17
2. Tasmania	31
3. New South Wales	83
4. Western Australia	131
5. South Australia	137
6. Victoria	169
7. Queensland	221
8. The Leicester and the Merino	233
9. Cross-breeding with Leicesters	247
10. Dual-Purpose Breeds	257
11. Boiling Down and the Frozen Meat Trade	261
12. Quarantine	279
13. England Imports	283

14.	New Zealand Imports	313
15.	Transport of Sheep	323
16.	Exports	355
17.	Flock Book Notes	369
18.	English Leicester Association of Australia	403
19.	Lustre Wool	429
20.	Melton Vale / Melton Park English Leicester Stud History	437
21.	How Has the English Leicester Changed, and Should it Have Modernised?	471
22.	Young Breeders	481
23.	Bakewell's Legacy in Australia	489
24.	Robert Bakewell (1725-1795)	501
References		555
Appendix 1: English Leicester Flock Numbers		585
Appendix 2: Sheep Numbers		605
About the Author		611

Introduction

I have been involved with English Leicesters all my life, and now, as I have got older they have become, according to my wife Anne, an obsession. I would not deny this as my family has a long history of involvement with the breed. My great grandfather, Robert George Heazlewood registered a stud in 1871 and each generation since has bred and shown English Leicesters. There is little doubt that Robert's father, Henry (my great, great grandfather) would have been running English Leicesters on his property, part of which I own today, from the time he purchased it in 1854.

I have written this history as a breeder of these sheep. I am not an academic, but am proud to perhaps classify myself on a similar level to Robert Bakewell, a yeoman farmer, and it is from this level that I look at this breed. These sheep have character, attitude, personality, good looks, beautiful long wool and of course that lively, bright eye.

My earliest memories of the English Leicester sheep would be when I was about 5 years old and I was allowed to hold the sheep for judging at the local Westbury Show. Dad was also showing Border Leicesters and Dorset Horns which we were not allowed to hold because they were not as quiet as the English Leicesters. History is now repeating itself, as I am now confident to allow my young grandchildren to handle the sheep, due to their quiet temperament.

For all my life the English Leicester part of our sheep operation has been a non-profit enterprise, but I am thankful that my father kept the stud going, if only in a small way, due to the long connection we have had with the breed. As with all heritage breeds, their shepherds keep them going because of the love they have for the breed and the valuable genetic material that they carry. Today the breed is being cared for by some 25 stud breeders in Australia, and although total numbers are relatively low the breed is in safe hands and its future is secure.

My research into the history of the English Leicester in Australia started several years ago, when I realised that we did not know exactly when the breed was first introduced into Australia, but I did think that it must be getting close to 200 years ago.

After some five years of research, I have still not been able to answer that question with certainty, and we may never be able to state with absolute certainty when that happened. My best estimate is that it was on 24 October 1825, when the *Mountaineer* arrived in Hobart carrying the *three fine Leicester rams and two young ewes* belonging to the Bryant Brothers, that we can say that they had arrived in the colony. This is the first documented arrival of English Leicesters that I can locate, and this timing fits well with the removal of the live sheep export ban in England. We do know that special permission had been granted for the export of some special sheep prior to 1824, but it was after the ban had been lifted that the steady stream of sheep started to arrive in the colony.

THE HISTORY OF ENGLISH LEICESTER SHEEP IN AUSTRALIA

The fact that they first arrived in Van Diemen's Land is not surprising. The sailing ships of the day would call at Cape Town for supplies before heading further south to take advantage of the Roaring Forties, which would bring them around the bottom of Van Diemen's Land, before they turned north to Sydney. Most ships would call into Hobart before going on to Sydney. Van Diemen's Land, and the opportunities available here were also getting good reports back in England. The climate being similar to that in England was also an advantage to the prospective new settler, as the farming system and livestock that they knew, was more easily transplanted here than into New South Wales.

The story of the English Leicester in this country is more about its association with other breeds, principally the Merino, than it is of the breed on its own. Before either the Merino or the Leicester had arrived in Australia, they had had a close relationship, but it was here that the Leicester's success as a sire to put over the Merino, the breed that it had played a strong part in making it so adaptable to the Australian conditions, meant that these two breeds very much depended on one another.

The Leicester's eventual decline to a very minor heritage breed is very much due to its ability to improve other breeds. It is in some ways a victim of its own success, particularly in relation to its main successor and direct descendant, the Border Leicester which replaced the English Leicester as the principal sire to mate with the Merino to produce the first cross ewe. The first cross ewe being the main dam for the country's prime lamb industry.

Virtually all the early importers of the English Leicester into Van Diemen's Land before 1870 were also importers of Merinos. The *Mountaineer* was carrying more Merinos than English Leicesters when it berthed in Hobart with the first three English Leicester rams. The dominant Merino breeders in Van Diemen's Land early days were also the dominant English Leicester breeders of the day, and they were mainly in the south of the state, an area today we do not associate with English Leicesters. The Bryant Brothers, W.J.T. Clarke, George Stokell, Robert Jones and Samuel Blackwell were all early importers of not only Merinos and English Leicesters, but also cattle and horses.

This same pattern of the English Leicester and Merino being introduced together is repeated in New South Wales, Victoria, Queensland and South Australia. While wool was the main object of production for export back to England, the Leicester was needed to help improve the wool and supply a better carcass. Both English Leicesters and Merinos were on the *Elizabeth* when it arrived in Sydney with the first English Leicesters into that state. English Leicesters and Merinos arrived in Western Australia during its first two years of settlement. The South Australian Company brought both English Leicesters and Merinos on the first voyage of settlement to South Australia. English Leicesters would have been taken across Bass Strait with the thousands of very early sheep from Van Diemen's Land to Port Phillip. Again, into Queensland, the English Leicester would have accompanied the Merino on the early overland walks from New South Wales into the Darling Downs.

While there was a steady stream of imports from England in the first half of the 1900's, the majority of stud imports came from New Zealand. This is most likely due to it being cheaper and easier to import from New Zealand than England. New Zealand breeders were also bringing sheep to the

THE HISTORY OF ENGLISH LEICESTER SHEEP IN AUSTRALIA

Sydney and Melbourne sales. This period also saw a change to importers being mainly those specialising in breeding stud sheep and producing large numbers of flock rams for sale to the Merino breeders. As Australia's sheep numbers increased and the export of frozen meat gave a ready market to the lamb and mutton that was being produced, the stud sheep industry became a big and specialised business for most of the British breeds of sheep that had been introduced.

While the Border Leicester gradually superseded the English Leicester during the first half of the 1900's, the English Leicester still maintained strong stud numbers. There was gradual increase in both registered studs and ewe numbers to a peak in 1949. In that year there were 136 registered studs running 10,368 ewes. From this peak there was a sharp decline. In 1952, just three years later there were just 75 studs with 4,263 ewes. From there ewe numbers slowly decreased until about 1974 from where they hovered about the 900 to 1,000 number until about the year 2000 when another slow decrease started. Registered studs reached their lowest point in 2007 and 2013 when only 15 were registered.

While Australia's English Leicester numbers may be small (the United States, England and Sweden have larger numbers of ewes), our reputation in the English Leicester world is strong. We have exported sheep to New Zealand, Canada and the United States, semen to England and the United States, and embryos to Sweden.

The name used to describe this breed of sheep has varied and changed since Bakewell's time, and it still varies today. From Bakewell's time it has been called the Dishley Leicester, the New Leicester, the Improved Leicester, the English Leicester (to differentiate it from the Border Leicester), just plain

Leicester, and more recently the Leicester Longwool. Today Australia and New Zealand are the only countries who refer to them as English Leicesters, although they were referred to as 'English Leicester' in England to differentiate them from the 'Border Leicester' up until about 1950. Today they are called Leicester Longwools in the rest of the world. Regardless of the name used, they are the same breed and in this book I have used all these names, trying to refer to them as they were, at the time and place in question.

This is the story of the English Leicester breed in Australia, and the wide influence it has had on this country's sheep industry. It is also the story of the people who have been closely associated with these sheep, from the early importers to those who have in latter years, been guardians of the valuable genetics that this breed possess, so they are not lost to future generations.

Timeline

1725 (May 23): Robert Bakewell born.

1788 (January 26): First Fleet arrives in Australia.

1788 (June 18): UK, the Act is passed preventing the export of sheep from Britain.

1793: Bengal sheep from India arrive in Australia.

1795 (October 1): Robert Bakewell died.

1796: Irish Longwool sheep (Leicester improved) imported.

1797: The first Merinos arrive from Cape Town.

1800: First Southdowns imported by Major George Johnstone.

1804: Major Johnston receives the first Teeswaters.

1804 (August 15): Macarthur purchases Spanish Merinos from the flock of King George III.

1821: First auction of Australian wool in London.

1822 (June 28): The Sydney Gazette mentions Leicester or Lincoln wool sent from VDL. This was most likely Teeswater wool.

1824 (April): The Australian Agricultural Company was formed in London.

1824 (May): The Van Diemen's Land Company was formed in London.

1824 (December 10): Repeal of British Wool Act came into effect allowing the export of live sheep.

1825: New South Wales & Van Diemen's Land Establishment (later known as The Cressy Company) formed in London.

1825 (October 24): *Mountaineer* arrives in Hobart with 3 Leicester rams and 2 ewes, the property of the Bryant brothers, the first documented Leicester's to arrive into the colony.

1825 (December 3): Van Diemen's Land proclaimed independent from New South Wales.

1826 (May 4): The Cressy Company imports Leicesters.

1826: First Cotswold's arrive in Australia.

1827 (April 7): Edward Sparke Jr returns to Sydney from England aboard the *Elizabeth* with 60 Leicester and Merino sheep, the first Leicesters to arrive in NSW.

1829 (June 18): Western Australia founded with Leicesters being imported sometime during the next 2 years.

THE HISTORY OF ENGLISH LEICESTER SHEEP IN AUSTRALIA

1829 (December 23): W.J.T. Clarke arrives in Hobart with Leicesters.

1830: Van Diemen's Land Company imports Leicesters.

1832: VDL Co. imports Cheviots, perhaps the first to Australia.

1835: Steam starting to replace sail in shipping.

1835: In separate expeditions, John Batman and John Pascoe Faulkner leave Launceston to launch first successful European settlements at Port Phillip, which developed into Melbourne.

1835 (October 9): The South Australian Company formed in London.

1836 (April 1): John Aitken lands sheep at Port Phillip, perhaps including Leicesters.

1836 (August 16): The *John Pirie* arrives at Kangaroo Island with the first Leicesters to South Australia for the South Australian Company.

1837: Sheep prices reach their highest point.

1837 (April): W.J.T. Clarke takes sheep from VDL to the Port Phillip district, so perhaps Leicesters were also taken then.

1837 (July 21): The *Achilles* arrives in Sydney with 3 Leicester rams and 12 ewes for the AA Co.

1838: First Lincolns arrive in Van Diemen's Land.

1839: Shetland Sheep imported by the Clyde Company, perhaps the first and only importation of this breed.

1839: First Romney Marsh importation by William Deene, NSW.

1840: Transportation to NSW ceased.

1840: Patrick Leslie takes the first sheep overland from NSW to the Darling Downs.

1840's: Introduction of iron wire for fencing.

1842: Transportation to Queensland ceased and free settlement allowed.

1843: The boiling down of sheep takes off due to low sheep prices.

1843: The first wool auction in Australia by Thomas Mort.

1843 (July): Possible first export of Leicesters from Australia to New Zealand.

1845 (approximately): Invention of the drafting race.

1851: The Ritchie 'Malton' English Leicester Stud, Flock No. 5, founded.

1851: Gold discovered in Victoria.

1853: Transportation to Van Diemen's Land ceased.

1855: First Shropshire sheep imported by Charles Price, SA.

1856: Van Diemen's Land name officially changed to Tasmania.

1857: Duncan MacGregor Sr arrives in Victoria.

1859 (December 10): Queensland becomes a self-governing colony.

THE HISTORY OF ENGLISH LEICESTER SHEEP IN AUSTRALIA

1861: Hampshire Downs in Australia

1864: William Field (TAS) sold 191 English Leicester rams to New Zealand.

1865: Border Leicesters in Victoria.

1870: 56 Leicesters from NSW sold to California.

1870's: The Corriedale breed established in Australia.

1871 (September 26): Formation of the Long-woolled Sheep Association of Victoria.

1877: First successful shearing machine.

1879 (December 6): Refrigerated meat shipments to Britain start.

1880: R.V. Dennis, VIC, fixed a type of comeback, the Polwarth.

1887: Holford Wettenhall, VIC, established a fixed comeback flock which he named 'Ideals', later to be also known as Polwarths.

1892: Sheep numbers in Australia at 106 million.

1892 (April 30): Formation of the Tasmanian Long-Wool Sheep Breeders' Association.

1893: The Improved Leicester Sheep Breeders Association (UK) formed.

1894: South Australian Shropshire Society publishes the first sheep flock book in Australia.

1895 (May 14): The first Dorset Horn sheep imported by John Melrose, SA.

1898: Flock Book for British Breeds of Sheep in Victoria, Vol 1 published.

1902: Sheep numbers reduced to 56 million due to severe drought.

1902: First Suffolk's arrive in Australia (G.R. Jackson, VIC). This may not be correct as there are references to Benjamin Boyd running Suffolk's on his properties near Eden in the 1840's.

1906 (November): First Wensleydale's imported by E. Carr, NSW.

1907: The Australian Longwool Sheepbreeders' Association was formed.

1909: The Australian Longwool Sheepbreeders' Association publish Vol. 1 of their Flock Book.

1909 (July 1): Commonwealth quarantine legislation came into force.

1909: First Roscommon sheep imported from Ireland by W. Killen, NSW.

1919: First Ryeland's imported by NSW DPI.

1925: Australian Society of Breeders of British Sheep formed.

1926: Volume 18 of the flock book is the first to include a stud prefix.

1934: The Australian Longwool Sheepbreeders' Association amalgamates with the Australian Society of Breeders of British Sheep.

1935: First combined flock book published, Vol. 17.

1937: Poll Dorset's developed from Dorset Horns.

THE HISTORY OF ENGLISH LEICESTER SHEEP IN AUSTRALIA

1939: First Dorset Down's imported by L.J. McMaster, NSW.

1941: Sheep numbers increased to 125 million.

1946: First South Suffolk's imported by Mr C. Kennedy, WA.

1946: The last English Leicester ewes arrive in Australia from the UK before live sheep imports are banned.

1947: The Australian English Leicester Breeders' Association was formed.

1951: The last English Leicester ram arrives in Australia from the UK before live sheep imports are banned.

1952: First and only Wiltshire Horn importation (10 ewes and 1 ram) from the UK by G.D. Crossthwaite, WA.

1952 (May): Live sheep imports into Australia banned, except from New Zealand.

1953: Zenith Sheep Association formed. This breed was developed over the proceeding years by L. Basset, Donald, VIC.

1960's: Perendale's introduced from New Zealand.

1960's (early): Cormo's developed in Tasmania by Corriedale x Saxon Merino.

1965: Gromark development commenced in NSW, breed society formed 1969.

1968: Coopworth's formally established in Australia.

1967-1976: Elliotdale's developed in Tasmania.

1970: Australia's sheep population reaches 180 million, the highest on record.

1970's: Drysdale's introduced from New Zealand.

1975: Tukidale's introduced from New Zealand.

1980's: First Finn sheep semen imported from North America by the University of New South Wales.

1981: Two English Leicester rams sent to New Zealand from Tasmania.

1982 (September 19): English Leicester Association of Australia re-formed.

1983: Awassi fat tail sheep embryos imported to WA.

1985: White Suffolk breed Association formed.

1987 (March): English Leicester Society of South Australia formed.

1987: Two English Leicester rams sent to New Zealand from Melton Vale, Tasmania.

1989: Nine ewes and one English Leicester ram sent from Tasmania to Colonial Williamsburg Foundation, USA, to re-introduce the breed back into the USA.

1990's (early): Aussiedown's developed.

1990: Koenarl stud sends English Leicester semen to Britain.

THE HISTORY OF ENGLISH LEICESTER SHEEP IN AUSTRALIA

1993 (September): First Texal's born in Australia.

1996: Prime SAMM's arrive in Australia.

1996: East Friesian's arrive in Australia.

1996: Damara genetics from South Africa.

1996: Wiltipoll's developed in Australia and breed association formed.

1998: Van Rooy first embryos imported.

1999: Afrikaner embryos imported (may have been on the first fleet).

1999: Persian's introduced via embryos.

2003: Ostlers Hill stud sends English Leicester semen to the USA.

2005: Poll Wiltshire's evolved.

2005: First Charollais embryos imported from the UK.

2007: First Ile de France embryos from South Africa.

2008: Melton Park stud sends English Leicester semen to the USA.

2009: 68 million sheep in Australia, the smallest flock size since 1905.

2011: Australian White breed launched.

2014: Melton Park stud sends English Leicester semen to the USA.

2015: Melton Park stud sends English Leicester semen to the USA.

2018: Langs Crossing and Melton Park send English Leicester embryos to Sweden.

2020: Valais Blacknose semen and embryos imported from the UK.

2020: Babydoll breed officially recognised.

2022: 70 million sheep in Australia.

2025 (May 23): Robert Bakewell born 300 years ago.

2025 (October 24): 200 years of the English Leicester breed in Australia.

1
History of English Leicester Sheep in Australia

'Among the various animals given by the benevolent hand of Providence for the benefit of mankind, there is none, perhaps, of greater utility than the sheep.'[1]

It is difficult to be one hundred percent certain when the first English Leicester sheep arrived in Australia as early records do not necessarily state breeds, but generally just refer to 'sheep', when making reference to early sheep imports. My research indicates that the first documented importation of Leicesters into the colony occurred on 24 October 1825, when the *Mountaineer* arrived in Hobart with *'three fine Leicester rams and two young ewes'* on board belonging to the Bryant Brothers. The first Leicesters arrived in New South Wales in 1827, Western Australia in the early 1830's, Victoria in the mid 1830's, South Australia in 1837 and Queensland in the late 1840's.

From the time of its arrival, the Leicester and Merino have been very closely associated with each other. The early Leicester importers were generally also Merino importers, or perhaps more correctly Merino importers who brought some Leicesters in with the Merinos, using both breeds to compliment each other, a practice which continued until the Border Leicester superseded the English Leicester.

While the English Leicester, along with the Lincoln, was one of the dominant British breeds of sheep in Australia for about the hundred years from the 1830's to the 1930's, it was not the first true longwool British breed to be brought here. The coarse Irish (Leicester improved) longwool had arrived here in 1796, and the more influential Teeswater in 1804.

Until 1824 it was illegal to take live sheep out of the UK. There had for many years been a conflict of interest between the British woollen manufacturers and the British sheep farmers. The woollen manufacturers of Yorkshire wanted their industry protected by prohibiting the export of either wool or sheep whereas the longwool sheep producers of Lincolnshire wanted to be able to sell their wool, particularly the long wool of which Europe only produced a small amount, to the best market. The Charles II government rigorously prohibited the export of either sheep or wool, though the enactment of 1666 restricted the operation of law to seven years. In 1788, the same year that the first fleet arrived in New South Wales, fresh regulations were issued due to pressure from the manufacturers in which all the existing laws respecting the exportation of wool were repealed, to make way for a new act. This act meant that the exportation of sheep (except wethers for sea stock, under special licence) was prohibited, under the penalty of forfeiture of the sheep, and the vessel carrying them, together with three pounds for every sheep, and also three months solitary imprisonment, to be inflicted on every person concerned, or assisting, for the first offence; and heavier fines and punishment for repeated transgressions.

> *'Any person or persons whatsoever who shall from after the passing of this Act bring deliver send receive or take or cause or procure to be brought delivered sent received or taken in any vessel any rams sheep or lambs of any sort or description whatsoever of the breed of the Kingdom of the United Kingdom or the Isles being alive to be carried out of the said Kingdom shall be guilty of an offence.*[2]

England had, for upwards of one hundred and forty years, set a legal wall around her sheep and their wool under the illusion that the English wool was the best, and if foreign cloth manufacture was denied access to this magic staple they would be condemned to permanent inferiority. The only exception to this was conceded to be, in some quarters at least, the fine staple of Spain. But the fine wool sheep of Spain were as rigidly protected as the long-woolled sheep of England. This policy of protection, which tended to lower the price of wool, was favoured by the manufacturers but disliked by the wool growers.

The repeal of the British Wool Act in 1824 finally meant that sheep could be exported from Great Britain. While this happened with the passing of the *Wool Importation and Exportation Bill* on 21 May 1824, the new laws did not come into effect until 10 December 1824.[3] This date of December 10 is important as it means that it would be well into 1825 before stock legally exported from England would reach the colony.

Just how effective these strict regulations were at preventing breeding sheep from leaving Great Britain is perhaps questionable. Some ships captains must have taken the risk in smuggling sheep out of the country and some people of influence did receive special permission to export sheep, but for the average person the penalties would have been a deterrent to even try. We know that Leicester Longwools were smuggled out of England to the United States as George Washington indirectly ended up with possession of some of these smuggled sheep. The first longwool sheep to come to New South Wales, the Teeswaters were sent here as a gift to Major Johnston, Commanding Officer of the New South Wales Corps. by the Duke of Northumberland, *a breed that must prove essentially advantageous to the general interests of the Colony.*[4] The first Southdown's to come to the colony were imported from England in 1800.[5] Even Macarthur had difficulty in transporting to Australia the seven Merino rams and two ewes that he had purchased at the first auction sale held on 15 August 1804, of sheep from King George III's Spanish flock. After Macarthur purchased Lot 1 of this sale, Joseph Banks drew Macarthur's attention to the one powerful and inclusive Act of the Statute Books debarring the export of sheep from England.[6]

On 5 October 1804, Lord Camden applied to the Lord Commissioners of the Treasury on Macarthur's behalf for the easement of the sheep

Macarthur had purchased through the embargo legislation which prevented their export.

> *'... Mr McArthur has procured in this country, and proposes to take out with him, seven rams and two ewes of the Spanish race, and has applied to me for a dispensation for that purpose. I am therefore to desire that your Lordships will receive His Majesty's pleasure for issuing such directions as may be necessary thereupon.'*[7]

The request was obviously approved as Macarthur sailed from Portsmouth aboard the *Argo* on 29 November 1804, with his precious sheep aboard.

In the early references to the longwool breeds of sheep in the colony there was often some confusion as to the actual breed being referenced to. This was particularly so between the Teeswater and the Leicester. The strong influence that the New Leicester had on other breeds in England, particularly the other longwool breeds such as the Teeswater, Lincoln, Irish Longwool and to a lesser extent the Cotswold meant that they were somewhat similar in type and appearance and therefore easily confused. The early Teeswaters which came to the colony were often also referred to as Leicesters. In 1804 it was reported that *Major Johnston has received three capital Tees Water ewes and a fine ram*[8], but in 1820 the progeny of these sheep were also referred to as Leicesters by J. Gordon, a farmer, in giving oral evidence before Commissioner J.T. Bigge in answer to the question 'What is the present breed of sheep in this settlement?'

> *'The best sheep are of the Leicester Breed or improved Teeswater. These last were brought down to Port Dalrymple by Col. Patterson, and were the Produce of some sheep that were sent out by the Duke of Northumberland to Col. Johnstone. Those of the coarsest kind are a cross of the Teeswater and the Bengal, but there is not much of the latter.'*[9]

In a letter to Sir Joseph Banks, written in January 1805, the Reverend Samuel Marsden states his desire to obtain both Leicesters and Lincoln's which he implies are not at that time present in the colony. He also mentions the difficulty in obtaining sheep from England due to the export embargo on sheep.

> *'... it would be a favourable opportunity for me to have two Rams sent out by him* (Mr Robert Campbell) *and Permission be obtained to ship them. We have got the Spanish, South-down & Tees-water Breed already; tho' not general. I think the Leicestershire & Lincolnshire Breed would very much improve our Flocks, could we obtain them. I have written to my Agent to purchase me two Rams; and requested Mr Campbell to bring them out provided Government will give Consent for them to come out to this Country.'*[10]

There is one early reference to Leicester or Lincoln wool, but again this may well be a case of not knowing exactly which breed the wool was from. It was

THE HISTORY OF ENGLISH LEICESTER SHEEP IN AUSTRALIA

most likely Teeswater wool as by 1821, when the comment was written the Teeswaters had a wide influence on the sheep of Van Diemen's Land.

> *'Hitherto the wool sent home from Van Diemen's Land, has been the produce of the Lincoln, or Leicester breed, and did not produce more than 1s.2d. per pound.'*[11]

R.B. Skamp, writing for the 1889 Year Book of Australia, states that:

> *'Small additions to this flock (*Bengal sheep from East India*) were made from time to time, and prior to the year 1791 a few Leicesters, Southdowns, and low class Irish sheep, brought out in some of the Government ships, were established at Port Jackson.'*[12]

I do not think that too much credence can be placed on this report, as he goes on to say that sheep were not introduced into Van Diemen's Land until 1807, which is incorrect.

There is also another somewhat similar reference to the early importation of Leicesters and Southdowns, this time involving Macarthur.

> *'About 1792, some Southdowns and Leicester sheep were imported from England, which produced an important change both in the fleece and the carcass. Captain McArthur, who at that time possessed about thirty ewes of the Bengal breed,*

purchased eight or ten of the latter importation, with which he commenced sheep-farming, and finding the wool of these sheep to improve in quality by this cross, he continued the system; and very shortly effected a very considerable improvement in the length of staple, and peculiar softness ...[13]

This information is in contrast to Macarthur's own correspondence which indicates that in 1794 he did not own any sheep at all, although in the following year he did own 'several hundred'[14]. I have not been able to find any further evidence that this importation occurred. A lot is recorded about Macarthur and his sheep farming but I cannot find any mention of him using Leicesters.

Any sheep Macarthur had acquired during the previous year must have been, in the main Bengal or Indian sheep from the Cape, which carried hair and not wool.

If he received any Irish sheep, it probably was from his friend Captain Hogan, of the *Marquis Cornwallis*, in February 1796[15]. It is known that some of the long-woolled Irish sheep did come to the colony very early on. Culley described the original long woolled Irish sheep as '*I never saw such ugly sheep as these ... they are almost in every respect contrary to what a well formed sheep should be*'[16]. The Dishley Leicester was used to dramatically improve this breed, and it would have been these Leicester improved sheep that came to the colony. Once again it may well have been a case of incorrect breed identification that led these early writers to call them Leicesters.

Sheep were certainly carried on most sailing ships (legally only wethers) as a source of fresh meat, so it is possible that Leicesters and Southdowns could have arrived in the colony this early if they had not been eaten during the voyage. I would suspect that if they had in fact been alive on their arrival, that they would have quickly been consumed by the local residents as the colony was very short of food at this time. If they had been either rams or ewes and spared a trip to the butchers, and been used for breeding, then their genetics would very quickly have been diluted, as the main breeds in use at that time were the Cape and Bengal type.

Henry Parker, writing in 1833, also makes reference to Leicesters being in Van Diemen's Land early on, but as these comments were not written until 1833, some eight years after the first documented import, he may well be correct in his reference to the Leicester.

> *'... Colonel Paterson, the Commandant at Launceston, who, foreseeing it would prove a profitable speculation to import a flock from New South Wales, obtained one of the Tees-water breed, which was subsequently crossed by Leicester and Bengal: the latter not remaining long in favour, soon became extinct, but the Leicester breed was much esteemed from the size of the animals carcase, as well as the excellence of the meat.*'[17]

While there are these early references to Leicesters being here, I cannot find any further substantive evidence to prove these claims. Henry Parker's comments in which he mentions both the Leicester and Teeswater would give some weight to the fact that he was not getting the two breeds confused, so perhaps they were both here in the first decade of the 1800's. Colonel Paterson was the commanding officer at Launceston from

March 1806, when the settlement was moved from Port Dalrymple to Launceston, until December 1808.

We do know that the Teeswater's were here, so what were these Teeswaters that were so valuable to the early development of the sheep industry, particularly in the north of Van Diemen's Land? The Teeswater were an indigenous breed to Teesdale in the County of Durham and derived their name from the river that separates Durham from Yorkshire. Most early writers say that they originated from the same stock as the Old Lincolnshires. The ewes were highly prolific, an unusual trait for sheep of that time.

David Low gives perhaps the best description of the Old Teeswater as:

> *'The most remarkable of the inland breeds* (of long-woolled sheep) *was the Old Teeswater, so named from the valley of the beautiful river which separates the counties of York and Durham. This valley is exceedingly fertile, though of limited extent; but the breed to which it gave a name extended, with some change of characters, northward into Durham, and southward through the greater part of Yorkshire, until it merged in the heavy-woolled sheep of the marshes on the one hand, and those of Leicestershire and the other midland counties on the other. The true Teeswater sheep, as reared in their native valley, were of the larger class, very tall, bearing a long but not very thick fleece, inferior only in toughness and length of filaments to that of the ancient Lincolns. The wool was, however, more hard, less uniform in the staple, and*

very coarse towards the extremities. These sheep were of an exceedingly uncouth form. They had coarse heads, large round haunches, and long stout limbs. They were slow in fattening, and required for their support good pastures, with a supply of hay and corn. They were the most prolific of all our races of sheep, bearing usually two, and not unfrequently three, lambs at a birth; and they were surpassed by no other sheep in the faculty of milk yield. This coarse and heavy breed has now (1845) entirely disappeared in its original form.'[18]

It was at Port Dalrymple (the name Port Dalrymple referred to the entire Tamar River area including Launceston) that the Teeswater really stamped its influence on the early sheep industry of the colony, and indirectly had an influence on the future sheep of the mainland. The Teeswater being a large, prolific breed of sheep with long, lustrous but coarse wool obviously suited the conditions that prevailed at Port Dalrymple. Their numbers rapidly increased by crossing with the imported Bengal sheep and by 1817, just 14 years after the first sheep were introduced into Van Diemen's Land, the colony had more sheep than New South Wales. The Teeswaters influence was recorded by Surveyor-General G.W. Evans when he wrote that while wethers at Port Jackson averaged not more than 40 pounds, at Port Dalrymple it was common for yearlings to weigh from 70 to 80 pounds and three year old's 150 pounds or more.

'But this great disproportion of weight arises in some measure from the greater part of the sheep at this settlement having

become, from constant crossing, nearly of the pure Teeswater breed.[19]

Site of the York Town settlement at Port Dalrymple, Tasmania. Brenton Heazlewood.

The Teeswater influence on the sheep bred at Port Dalrymple led to these sheep being referred to as Port Dalrymple bred ewes.

> *One Thousand Sheep for sale by private contract.*
> *1000 sheep, principally* Port Dalrymple *bred ewes, either in lamb, or with lambs by their side. Much attention has been paid to the improvement of the wool of these animals, they are particularly recommended to the notice of new settlers.*
> *Apply to Lewis & Collicott.*[20]

THE HISTORY OF ENGLISH LEICESTER SHEEP IN AUSTRALIA

Sadly, the Teeswater influence slowly died out as the influence of the Leicester became stronger.

By the mid 1830's the Merino and English Leicester were the two main breeds of sheep in Australia. The success of each breed was in many ways determined by the success of the other. The English Leicester complimented and crossed well with the Merino, giving it a greater length and weight of wool, better constitution and more meat. The Leicester breeders relied on the ever-increasing number of Merino ewes to which they could supply flock rams.

The dominance of the English Leicester and Lincoln as the prime crossing sire to be put over the Merino continued until the early 1950's, when their numbers went into a sharp decline due to the popularity of the Border Leicester, which is a faster maturing breed, developed from the Leicester.

2
Tasmania

'It is quite impossible to find a climate more congenial to the health of sheep than Tasmania.'[21]

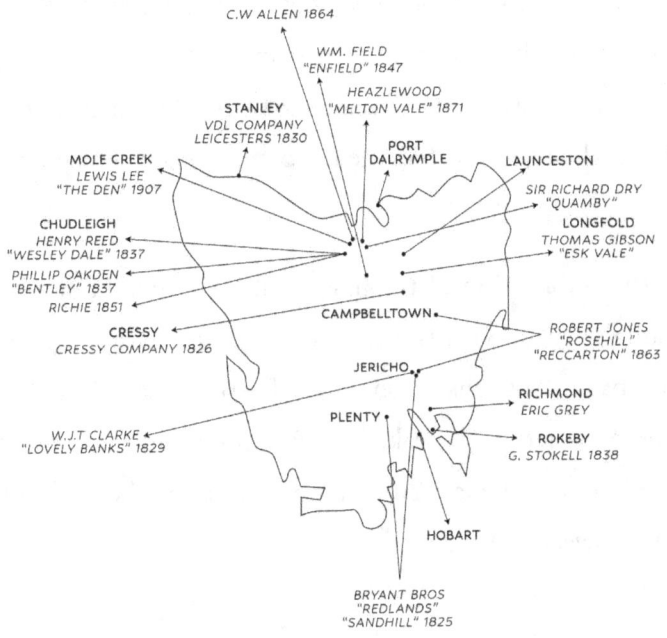

While it is remotely possible that Leicesters were present in Tasmania before October 1825, it is in that year that the first documented importation into Australia occurred when the Bryant brothers imported three Leicester rams and two Leicester ewes. They also had on board Thoroughbred horses, Devon cattle and Merinos.

Tasmania was the main source of English Leicesters for the rest of the country, both for stud sheep and flock rams, for about 80 years after their introduction. Being second settled after New South Wales, Tasmania had the advantage of having good numbers of Merinos and Leicesters to help increase the sheep population of the other states as they were settled. The South Australian Company sourced a lot of their early sheep from Tasmania as Western Australia had done to a lesser extent some 7 years earlier. Very large numbers of sheep were sent across Bass Strait to Victoria as soon as it was settled in 1836. It was not until about the 1890's that large numbers, particularly flock rams, were being sent to New South Wales to put over Merino ewes. Tasmania ceased to be the dominant state from the early 1900's, when Victoria became the powerhouse for English Leicester numbers.

Up until the publication of the first combined flock book in 1935, all of Tasmania's studs, with the exception of one, were registered with the Sydney based Australian Longwool Sheepbreeders' Association. The number of registered studs peaked in 1948 at 43, but after that there was a sharp decline. For the last 20 years there has only been about 4 studs registered each year in Tasmania.

THE HISTORY OF ENGLISH LEICESTER SHEEP IN AUSTRALIA

Bryant Brothers

Mr James Bryant and Mr Edmund Bryant first arrived in Hobart on the *Anguilar,* on Wednesday, 4 February 1824.[22] They brought with them a letter of recommendation from Under-Secretary of State Wilmot to Lieutenant-Governor Sorell. James' statement of property showed he possessed assets of one thousand and twenty-four pounds. Edmund had nine hundred and fifty-five pounds. They received adjoining grants of 700 (Edmund) and 800 (James) acres of land near Jericho soon after their arrival. In August 1826 they were granted a further 700 and 800 acres (giving both 1,500) *in consideration for their importation of valuable livestock.* In September 1828 they again applied for more land, justifying their application by the extensive improvements they had made consisting of a stone barn and granary house valued at six hundred pounds, a considerable amount of fencing, and several huts. They were running 2,200 sheep (including 76 pure Saxon Merinos), 60 head of cattle, and 9 horses. They were employing 4 Free and Ticket-of-Leave farmers and one Free ploughman. The Land Board recommended that in view of this that they should be granted an extra 2,000 acres (being 1,000 acres each).

On 4 April 1824 (exactly 3 months after his arrival), Edmund Bryant left Hobart for London on the *Brig Woodlark.* Accompanying him was Mr G. Gavott, who was to marry Edmund's sister. The reason for the return trip to England was for Edmund to marry (he was married in Somerset on 15 February 1825) and to purchase livestock.

Edmund, his wife, and his brother Francis and his wife and 5 children, as well as Mr and Mrs Gavott, plus their livestock, sailed from Plymouth on the 25 April 1825 per the *Mountaineer*. The *Mountaineer* arrived in Hobart on 24 October 1825, with its significant cargo of livestock.

> *'The Mountaineer struck upon a rock at Symon's Bay, at the Cape, and sustained so much injury as to be obliged to unload and undergo repairs, which detained her nearly two months. ... She brings a valuable importation of stock in the Colony ... a thorough bred two year old horse, and a filly one year old; a thoroughbred Devon bull, two years old, and a heifer of the same age; nineteen merinos, consisting of 17 ewes, and 2 rams, 7 having died on the passage; three fine Leicester rams, and 2 young ewes, the property of Messrs. Bryant and Gavott. All this stock is in most capital condition.*'[23]

These five Leicesters may well have been the first Leicesters brought into the colony. This supposition is backed up some 18 years later in 1843, when a press article relating to the insolvency of both Edmund and Francis Bryant gives the credit to Francis for *'the introduction of our finest breeds of sheep and cattle, the Hereford and Leicester.'*[24]

While the Bryant's may have been early importers of Hereford cattle, the Cressy Company is credited with their first importation into the colony in 1826.

The Bryant's Leicesters are mentioned by Henry Widowson who had come out as an agricultural expert with the Cressy Company.

THE HISTORY OF ENGLISH LEICESTER SHEEP IN AUSTRALIA

'As yet [1829] there are but few long-woolled sheep that have been imported from England; the best I have seen were those brought by the Messers. Bryants; whether these gentlemen reported them to be pure Leicesters, or not, I cannot say, but such a title was given to them, from having been bred in a village long famed for the finest breed of Leicesters. I have, however no hesitation in declaring them to be nearly devoid of that blood; still the cross hit well, and there was an evident improvement in the lambs ... the Leicesters stand pre-eminent, on account of their superior frames and fine symmetry: from either of these breeds make a selection, and never, if you can avoid it, permit any mongrel breed of sheep to approach your flock.'[25]

The Bryant's became prominent land holders owning, or leasing several properties in both the south and north of the state. The eldest brother, James, died in 1833 and the press noted that he had farmed at Jericho for many years and that he had introduced superior horses and Leicester and Teeswater sheep from England. At the sale of his estate in April 1833 the following Leicester sheep were advertised for sale.

800 Leicester sheep of very superior quality (these would have most likely been Leicester cross)
2 pure Leicester rams
3 pure Leicester ram lambs
2 pure rams 1 year old
7 pure ewes, the produce of imported sheep[26]

Francis was leasing 'Redlands' at Plenty, in the mid 1830's and there he was one of the first farmers to use irrigation in the state.[27] Edmund Bryant at one stage owned or leased property in both the north and south. These included, but not necessarily at the same time, 'Bentley' at Chudleigh, 'Trafalgar' at Evandale, 'Weston' near Railton, 'Carrington' near Campbell Town, 'Kingston' near Ben Lomond, where he built an irrigation system that covered some 200 acres, as well as 'Sandhill' at Jericho.

When the rural economy was in recession in the early 1840's the Bryant's were forced into insolvency and had their properties on the market. On the same page of the Van Diemen's Land Chronicle, on Friday, 19 November 1841, both Francis and Edmund were advertising their properties and livestock for sale at 'Redlands' and 'Kingston' respectively.[28]

Among the 4,000 sheep listed to be sold from Francis Bryant's 'Redlands' property on 9 December 1841, were:

> *100 Pure Leicester Ewes, with lambs by their sides, the lambs being the product of sheep imported from England and purchased from that celebrated Leicester breeder, Thomas Ewer, Esq., they will be sold in pens of five each.*
> *100 Maiden ewes, pure Leicesters.*
> *8 Imported Leicester rams, not to be surpassed in the colony, and previously referred to as being bred by Thomas Ewer, Esq.*[29]

The report of the 'Redlands' sale appeared in the press the day after the sale.

THE HISTORY OF ENGLISH LEICESTER SHEEP IN AUSTRALIA

EXTENSIVE SALE OF STOCK – The great sale by auction at Redlands, of stock belonging to F. Bryant, Esq., took place yesterday. The day was propitious, and there was a very numerous assemblage of gentlemen, a large portion of whom however were attracted to the spot more from a desire to view the justly celebrated stock of Mr Bryant, than with a view to become buyers. At eleven o'clock the company sat down to a very handsome cold collation, when they had an admirable opportunity of testing the quality of both the beef and mutton grown on the estate. The sale commenced at one o'clock, by Mr MacMichael, with the pure Leicester rams imported by Mr Bryant. There were eight of these beautiful creatures, which, with the exception of one that was passed over, were knocked down at twenty guineas each. Number nine commenced the pure colonial bred rams, scarcely, if anything, inferior in size of carcass or symmetry of form, to the pure imported. These rams, which extended to number fifty-five, were knocked down at prices ranging from thirteen pounds to four guineas each, the average being at the rate of five pounds ten shillings per head. These were succeeded by the crossed, which varied in price from two pounds to six pounds ten shillings. The pure Leicester ewes, in pens of five, with five lambs, went off at from 15s. to 23s. per head. A small lot of beautiful Leicester wethers sold at 16s. 6d.[30]

The sheep for sale at 'Kingston' included:

'A considerable number of pure Leicester Rams imported from England, a breed daily becoming more sought after in the colony to counteract the ill effects of loss of weight in carcass and wool; also a flock of South Downs.'[31]

The Bryant's not only imported Leicester sheep over a long period of time but also Thoroughbred horses, Hereford cattle and Merino rams, all of which played an important part in improving the early livestock of the colony. They also, through their partnership with John Aitken, played an important part in the introduction of Leicesters to Victoria in 1836.

While the Bryant's would have no doubt imported pure bred Leicesters, it appears that they only used them on a commercial basis, basically to be put over their Merino ewes. Their main aim with the Leicesters would be to breed enough rams for their own use, perhaps selling any surplus. This was the custom with early imports as it was not until perhaps the 1880's that stud breeding as we understand it today really started. I have therefore not been able to trace any present-day Leicester flocks back to the Bryant imports.

Edmund Bryant moved to Victoria in about 1844 and died on 20 April 1849, at his residence on the Leddon, Port Phillip. Francis also moved to the Port Phillip area in 1848 and I think he returned to England in 1858, dying in 1860. Their contribution to the early agriculture of the colony was considerable, not only through the quality of the livestock that they introduced, but also by their modern farming practices such as irrigation.

The New South Wales & Van Diemen's Land Establishment (later known as The Cressy Company)

The New South Wales & Van Diemen's Land Establishment company was formed in London in 1825 in response to the glowing and often exaggerated reports of the suitability of New South Wales and Van Diemen's Land for the growing of wool. The company was formed with a capital of fifty thousand pounds and it was to receive grants of 20,000 acres at Western Port, Victoria (this was later disallowed) and 20,000 acres in Van Diemen's Land.

The expedition to establish the company sailed from England on board the *Albion* on 17 November 1825 (one year after the lifting of the ban on the export of sheep) and arrived in Hobart on 4 May 1826. They had purchased quality livestock from the leading breeders in England to bring with them. These included Thoroughbred, Flemish (Belgian draught) and Cleveland (Carriage) horses, Durham (Shorthorn) and Hereford cattle as well as Merino, English Leicester and Southdown sheep.

When the Establishment's expedition sailed, it not only carried breeds that would prove to have a lasting influence on Australian agriculture, but also individual animals of merit representing those breeds. The thoroughbred stallions Buffalo, Bolivia and Aladdin, while not the first into Tasmania, had a lasting influence in the state.

The Shorthorn cattle we are told were purchased from Charles Champion, a leading breeder who had already exported to the USA. These were perhaps the first Shorthorns into Australia. The Herefords certainly were. Joan Austin Palmer, the breed's Australian historian wrote, '*The Cressy*

Stud is the most important in Australia's history. Not only were they the first Herefords to be imported into this country, but their excellence made them the foundation stone on which almost every other herd is based'.

The Merino sheep were both Saxon from Germany, and Anglo-Merinos which were descendants of the Spanish Merinos which entered England following the defeat of Spain in 1809. The Merino was of course the most important sheep breed to come to Australia as it became the foundation of our wool industry.

The Southdowns, which the Establishment landed in 1826, while not the first of this breed to arrive in Australia, were the most important importation of this breed as they comprised the foundation stud for Australia. It is most likely that they came from the Ellman flock.

The company settled on land in the Cressy-Bracknell area in northern Van Diemen's Land.

The Leicester sheep that were brought in by this company were most likely the second importation of this breed into the colony. It is almost certain that these sheep would have also come from Charles Champion. It would make sense that they would have purchased both the Shorthorns and Leicesters from the same breeder. Toosey did his farming apprenticeship with Charles Champion and assisted in taking some of his Leicesters to Ireland, so this association was perhaps also important in where stock was sourced from.

The descendants of The Cressy Company Leicesters are not well documented, but some are recorded in the early flock books via Donald Campbell, who managed 'Richmond Hill' after Toosey's death.

The three breeds of sheep on board the *Albion* when it arrived in Hobart in May 1826, would come to represent the three-tiered system which early developed, and continues today. The wool breed (Merino), the first cross sires (Longwool, Leicester) and the prime lamb terminal sires (Southdown).

When Captain Bartholemew Thomas, one of the original partners in the company, wrote to Governor Darling (NSW) in June 1825, stating that, '*Our great object is to lay a foundation of the finest and most approved Breeds of every description of animal useful to the Colony, which we mean to increase by yearly importations from England*'.[32] Little did he realise that they would indeed achieve this aim.

Henry Widdowson, in his book 'The Present State of Van Diemen's Land', says of the Cressy livestock, '*The Saxons cannot be equalled in fineness or quality of wool, the Leicesters stand preeminent on account of their superior frames and fine symmetry*'. In summary he states that: '*The gentlemen of the Van Diemen's Land Agricultural Establishment deserve the thanks of the colonists at large for having imported more better bred beasts than any other individuals.*'[33]

We can presume with some accuracy as to who bred these Leicesters imports and some present-day registered sheep can be traced back to these Cressy Company imports.

George Stokell

George Stokell was another early importer of Leicester sheep. He emigrated to Van Diemen's Land in 1822, when 35 years old, arriving in Hobart Town on 25 December aboard the *Venerable*. He had owned several farms in England and was obviously a man of means.

During Stokell's early years in Van Diemen's Land, grants of land were fairly easy to obtain and some of the 16 properties that he ended up owning, covering some 29,000 acres, may have been acquired in this way. 'Rokeby', his first property was purchased. Rokeby is now a suburb of Hobart, on the eastern shore of the Derwent River.

> *George's first years in Van Diemen's Land were very prosperous. The farms were being cleared and fenced, houses built and Durham cattle and Leicester sheep imported (one lot of eight sheep from Newbiggin, Durham, cost 93 pounds).*[34]

The eight Leicester sheep mentioned here were most likely imported into Van Diemen's Land in 1856 or 1857. They are referred to in a letter addressed to John Stokell (son of George), Jerusalem, Hobart Town, dated 16 February 1857 and signed by H.A. Harrison of Newbiggin, Sadberge, Darlington. Another letter from Mr Harrison, dated 27 September 1857 mentions Leicester sheep purchased by Stokell from Mr Hodgson, Mr Carter and Mr Mason, all from Yorkshire.[35]

These were not the first Leicesters that the Stokell's imported. In his journal George Stokell refers to four Leicester rams being moved from Stokell to Sadberge on 3 January 1840. I do not know if these four rams were the first Leicesters that they imported, and we can only presume that these would have also been purchased from the same breeders as the later imports.

Again, we are fortunate in that we have a good record of who bred the Stokell imports and these can be traced to some present day sheep.

W.J.T. Clarke

W.J.T. Clarke (William) who arrived in Hobart on the *Deveron* in December 1829 brought 20 pure Leicester sheep with him as well as other livestock. He would eventually become the greatest user of Leicester sheep and their crosses in Australia, and perhaps the world.

William was born in 1805 and both his parents died when he was young. At 15 he went to work for a drover who moved livestock from the west of England to the great Smithfield Market in London. While in London he would stay with a great-uncle who was a butcher and before long Clarke had become highly competent at assessing the weight and value of sheep and cattle.

When he was 21 the firm he worked for went broke due to a sharp recession so Clarke established his own business of droving, butchering and dealing in stock. Because his London base was near the docks, he began to do business with ship's captains, supplying them with animals for slaughter on their long voyages and even on occasion buying valuable breeding stock for export to the colonies. Through this he heard about the pastoral industry opportunities in New South Wales and Van Diemen's Land and decided to emigrate.

In 1829 he secured the vessel *Deveron* for the voyage and then returned to Somerset to buy livestock and also eloped with the curate's daughter. In London he fitted out the vessel for his livestock which included his favourite saddle mare 'Jessica', the draught stallion 'Champion', two cart-mares recently in foal, a red bull named 'Comet' and some good cows in calf. He also purchased 20 pure Leicester sheep to bring with him.

Clarke arrived in Hobart on 23 December 1829 with his precious cargo of livestock, having not lost any on the voyage out due to his good preparation and care for them during the voyage. He had cash and assets of about two thousand five hundred pounds which would normally entitle him to 2,500 acres. He was granted 2,000 acres which he eventually selected at 'Windfalls' near Campbell Town.

Clarke put his experiences of dealing in stock and his negotiating skills to good use as he had soon leased 'Merton Vale' at Campbell Town (6,000 acres) and 'Lovely Banks' (700 acres) north of Hobart. He had also purchased 6,000 sheep and 800 cattle.

In the early years of the Victorian settlement Clarke first squatted, then leased and bought land until he controlled some 250,000 acres there. He also acquired 220,000 acres in South Australia as well as having 80,000 acres freehold and 50,000 acres of crown lease in Tasmania by 1856. In 1871 he also purchased 50,000 acres in New Zealand.[36]

He used Leicester sheep throughout his career. In 1840 he returned to England to purchase more Leicester sheep. He visited: *'the leading Leicester studs, making a careful selection of sheep to take back with him. He was seeking large-framed animals, with a long growth of wool.'*[37]

In 1860 he returned to England for the second time and again bought Leicester sheep. Clarke was perhaps the first person to introduce the Leicester into Victoria, as *'up till 1852, it is doubtful if there was another* flock of long-wools in Victoria besides the Leicesters kept at Dowling Forest by Mr. W.J.T. Clarke.[38]

There is no evidence that W.J.T. Clarke ever showed his Leicesters. The Flock Book for British Breeds of Sheep in Australia 1913, Vol. 5 records the registration and history of two flocks belonging to Sir R.T.H. Clarke (a grandson). Flock No. 16, Somerset Farm, Mooroopna, has its history listed as:

> 'This flock was established about 1838 by the late Hon. W.J.T. Clarke, who imported some sheep to Tasmania, and in 1851 a number of rams and ewes, being progeny of these sheep, were brought to Victoria, and formed the foundation of this flock.[39]

Sir R.T.H. Clarke also registered a second flock in 1913, Flock No. 15, at Bolinda Vale, Lancefield Junction. This flock was established by the purchase of 104 ewes and 13 rams from the flock of Mr W.T. Williams of New Zealand. This was a smaller stud than Flock No. 16, only mating 58 ewes in 1914 whereas his Flock No. 16 was mating 1,020 ewes.[40]

We may ask why two studs were registered at the same time. I think the answer has to do with the fact that Sir R.T.H. Clarke's grandfather, W.J.T. Clarke purchased the station 'Moa Flat' in Otago, New Zealand in 1871 and his eldest son William (Will), the father of Sir R.T.H. Clarke inherited the property upon his father's death. It is recorded in a Weekly Times article relating to the disposal of the Bolinda Vale stud that the English Leicester flock at Sunbury was transferred to Moa Flat but after several years the nucleus of what was to become the Bolinda Vale stud came back to Victoria, and that flock was added to by the purchase of more New Zealand sheep.[41] In looking through the flock books relating to the Bolinda Vale stud it can be seen that there were many imports of rams and ewes from New Zealand into the stud during its existence. It was kept very much a New Zealand based stud with rams being imported as late as 1924, just two years before the death of Sir R.T.H. Clarke, whereas the Somerset Farm (flock No. 16) stud was basically kept as a closed flock, that is an Australian based flock.

The use of English Leicesters, which was started by W.J.T. Clarke was carried on by his son and grandson until the dispersal of the Bolinda Vale stock on 16 March 1928. The notice of the clearing sale listed the following ewes, *4444 yearling, 454 two years, 420 three years, by English Leicester rams out of Comeback ewes.*' The stud flock is listed as having 150 ewes and 75 rams.[42]

The stud dispersal was a result of the death of Sir Rupert and the fact that his young heir was being brought up in England.

It is a pity that Clarke did not leave a record of who he purchased his sheep from in England, but many present day sheep can be traced back to his purchases.

Van Diemen's Land Company

The Van Diemen's Land Company (VDL Co) was formed in England by an Act of Parliament in 1825. Its formation closely followed the formation of the Australian Agricultural Company (AA Co). Partly as a result of the AA Co receiving an assurance that no rival incorporated or joint stock company with similar objects could be established in New South Wales for twenty years, the VDL Co turned its attention to Van Diemen's Land. Its shareholders, eleven in all, were closely connected with the woollen industry of the West of England and, since 1819, had been associated in agitating for a repeal of the tax on foreign wool. The VDL Co was set up with the objective of producing fine wool for export back to England to help reduce that country's reliance on supplies of fine wool coming from Europe. It was initially granted 250,000 acres, later to become 350,000 acres in the north-west tip of Van Diemen's Land.

The relationship between the VDL Co and the AA Co were not always harmonious. Foiled in their first attempt to prevent the establishment of the VDL Co by raising official doubts as to the existence in the island of a sufficient quantity of suitable unallocated land, the AA Co immediately planned to obtain an injunction to restrain the VDL Co from purchasing sheep in Europe for five or six years on the ground that the competition between the two would materially raise the price of sheep. Eventually after a considerable time in which the two companies could not come to an agreement between themselves, Bathurst was forced to intervene. As a result, the markets of Spain and Portugal were allotted to the VDL Co and those of Germany to the AA Co for a period of three years. This arrangement was to prove impracticable and by 1826 the AA Co had relented and in 1827 the VDL Co was buying Saxon Merinos.

The first sheep that the VDL Company landed at Circular Head in 1826 were not fine wool Merinos but *'50 head of the Cotswold or fine Long Woolled Gloucestershire sheep, a breed which is thought by some very competent judges will be of great benefit to the flocks.'*[43]

There were many shipments of fine-woolled sheep to follow with almost 1,000 Saxon and Merino sheep landed in 1829 alone. The company also purchased many local sheep and by March 1830 had 6129 in total.

The March Annual Report of the company states that: *'the only live stock which has been sent* (out in 1830) *were five rams and twenty ewes of the most improved Leicester breed, as they are likely to answer well on the Surrey and Hampshire Hill district, it being probable that the wool will there retain its length, but the staple will become finer.'*[44]

As well as the Merinos, Cotswolds and Leicesters, the company also introduced Cheviots in about 1834.

From the early 1830's, Edward Curr (agent for the VDL Co, we would now use the term manager) was selling large numbers of rams. In 1831, 51 Saxon ram lambs and 3 Cotswold ram lambs were sold. In the last quarter of 1836, 323 Saxon, Negretti, Cheviot and Leicester rams were sold along with 200 ewes. These high sales figures of 1836 reflect the expanded market that the new colony at Port Phillip had created.

LEICESTER RAMS FOR SALE.

BY

The Van Diemen's Land Company

THE COMPANY *will dispose of a few year old Leicester Rams, of very superior quality, at Ten Guineas each. Wethers* *of the first cross from this breed, out of the ordinary ewes of the country have averaged 90 to 100 lbs. weight, and their fleeces have realised 9s. 8d. each in England. A liberal credit will be given. Apply to Messrs. HENTY & Co., Launceston, or to the undersigned.*

EDWARD CURR[45]

In 1840 the company listed their sheep numbers as:

Merino   5014
Cotswold   101
Leicester   913
Cheviot   326
Improved 2384[46]

James Gibson followed Edward Curr as company agent and he advised the directors to lease much of the land. In November and December 1851, advertisements appeared each week in the *Examiner* newspaper offering to let, Woolnorth 100,000 acres, Emu Bay 300 acres and Circular Head 15,000 acres.

A separate notice offered the pure Durham cattle and Merino sheep. The details of the sheep were:

280 Pure Leicester Breeding Ewes 2 to 5 years
60 Maiden Leicester Ewes
25 Pure Leicester Rams 2, 3 and 4 years
50 Ram Lambs
100 Pure Merino Rams[47]

It is evident that few of the livestock sold on the properties in response to these advertisements for in 1852 they were being offered by auction in Launceston. On 6 March 1852, Messrs Robinson & Alison advertised that they were to receive from Stanley in a few days the following Pure Stock to be sold by auction on 9 March:

One hundred and fourty five Leicester Breeding Ewes
Twenty Pure Leicester Rams 2, 4 & 6 tooth
Fourty three Pure Leicester Ram Lambs[48]

The Leicester sheep of the VDL Co seem to have just disappeared as no present day sheep can be traced back to them, most likely they were dispersed through many of the Leicester breeders of the day. The VDL Co did register Shropshires (foundation date not known), Lincolns in 1924, Southdowns in 1939, and a Border Leicester stud in 1984.

What was the VDL land at 'Woolnorth', is now prime dairying country.

Robert Jones

Benjamin Jones, who arrived in Van Diemen's Land in 1816, was already an experienced Australian farmer. He had owned a farm at Rosehill near Sydney. He obtained a grant of land at Lower Marshes (Jericho) in 1821 and named it 'Rosehill' after his New South Wales farm. In 1829 his sons John, Robert and James were each granted 500 acres in the same area.[49]

It was Robert who became one of the noted breeders and exhibitors of Leicesters in the south of Van Diemen's Land, along with George Stokell and Samuel Blackwell. His original sheep were imported from England in association with W.J.T. Clarke, *'and were the finest Leicesters that could be procured in England. They were selected by Mr W.J.T. Clarke, who was an excellent judge of long-woolled sheep.'*[50]

The 'Reccarton' (now spelt 'Riccarton'), Campbell Town stud of Robert Jones (the second) was established in 1863 with ewes from his fathers Lower Marshes flock. He adhered to the original blood lines until about 1873, when he introduced 31 ewes purchased from John Stokell, of Stockdale, Tasmania.

'The sheep obtained from Mr Stokell were very handsome and symmetrical, but not so dense in the wool as Mr Jones's original sheep. The quality, however, was very fine, and they cut 7 lbs. of washed wool. With these ewes he mated a fine ram, named 'Sandy', who possessed great beauty of form. His wool was 9 inches in length, and he cut a fleece weighing 11 lbs. of washed wool. 'Sandy' was a very well bred ram, being from a

select Stokell ewe by one of three rams which were purchased, regardless of expense, by Mr John Barton, of Barton House, Yorkshire, for Mr JD Wood, of Dennistoun, Tasmania. The sire of this ram was bred by Colonel Pijes, and took third prize at the Royal Agricultural Society's Show in Manchester, being afterwards sold for 65 guineas. The dam of this imported ram was by 'Sandy', the winner at the Royal Agricultural Society's Show at Battersea, who was sold for 140 guineas at Mr Sandy's sale. Another ram used in this flock was purchased from Mr Thomas Gibson, who obtained his original sheep from Mr George Farmer, of Beacon Downs, Exeter, whose sheep had occupied a high position as a stud Leicester flock for over 70 years. This was a very symmetrical sheep, but not so fine in the wool as 'Sandy', and he was only used with those ewes that showed a tendency to thinness in the wool.

Of the lambs got by 'Sandy', one was reserved for a stud ram, which exhibited a remarkably fine fleece, long in the staple, curly, and very lustrous. This sheep, named 'The Earl of Leicester' made a marked improvement in the flock, his stock having great length of staple, combined with quality and lustre. 'The Earl of Leicester' was remarkably even in the fleece, the quality being almost as good on the thigh as on the shoulder. As a hogget, his wool measured 16 inches in length, and he cut a fleece of 19.5 lbs. of washed wool. He was a very handsome sheep, with short legs, splendid breast, and fine, well-shaped head. The rams since used in the stud flock have always been selected from the Earl's descendants.

Mr Jones has, for many years past, bred for lustre wool, and the flock is now well known for its brilliant lustre. When matured, the Reccarton sheep weigh extremely well. A lamb of the season 1877, killed at 7 months old, weighed, when dressed, 103 lbs. The soil at Reccarton is of a light, loamy character. The pasture is very dry in summer, the district being noted for the scanty rainfall. The sheep run in paddocks all the year, never receive any artificial food, and are never housed. The average weight of wool from the stud flock is about 13 lbs. per head of washed wool, which has realized as high as 1s. 10d. per lb. the stud flock now numbers 158 ewes and 40 rams. The rams from this flock are held in good estimation by the sheep breeders of Tasmania and Victoria. They have realised prices ranging from 10 guineas to 50 guineas each, the highest prices being obtained in Tasmania. Young ewes from this flock have been sold from 10 guineas to 25 guineas each.

Mr Jones has been very successful in the show-yards, having taken a great many prizes. Among others – one first prize at the Midland Show, three first prizes at Melton Mowbray, and three first prizes at the Southern Show, in 1875; first prize and champion at the Midland Show, three first prizes at Melton Mowbray (at these two shows there was very strong competition), and two first prizes at Elwick, in 1876; two prizes at Elwick, and champion cup at Launceston in 1877.[51]

There was a big interest in Leicesters at this time, particularly in southern Tasmania. At the Southern Show of 1875, mentioned above, there were 13 exhibitors of Leicester sheep.

In common with several other Leicester breeders of the period, Robert Jones was a noted Merino breeder and his name regularly appears in show press reports both as a judge and exhibitor of Merinos.

Two flocks were established in 1864 from a combination of Robert Jones and Samuel Blackwell sheep. They were Flock No. 8 (Longwool Flock Book) of C.W. Allen and Flock No. 14 (Longwool Flock Book) of B. & E. Sadler of 'Drumreah', Deloraine.

Ritchie

The oldest continuing stud of British sheep in Australia is the English Leicester flock which has been run on the 'Mayfield' property at Chudleigh, Tasmania by the Ritchie family since 1851. It is twenty years older than any flock currently registered in the Flock Register, being formed two years before transportation to Van Diemen's Land ceased.

Ritchie Bros. Ram lambs at "Mayfield" 1904. The Weekly Courier, March 19, 1904.

It is registered as flock No. 5 but when the Australian Longwool Association produced their first Flock Book in 1909, No.'s 1 to 4 were allotted to New South Wales flocks which had been founded between 1903 and 1907. The Ritchie flock may have been allotted No. 5 in the previously published Tasmanian Longwool Flock Book and therefore continued with that number in the Australian Longwool book and subsequently with the Australian Society of Breeders of British Sheep.

Registered in the name of M.H.J. Ritchie, it is at present in charge of Mr Neon Ritchie, a great-great-grandson of James Ritchie who established the flock with sheep purchased from Sir Richard Dry.

Members of the Ritchie family were prominent in the development of the Port Dalrymple settlement, and Captain John Ritchie was Commandant of settlement in 1812.

When Governor Macquarie visited Van Diemen's Land in 1811, he gave instructions for a road to be surveyed between Launceston and Norfolk Plains (Longford). Ritchie was awarded the contract for the forming of the road, for which he was paid with a cow, presumably from the government herd.[52]

His brother, Captain Thomas Ritchie, R.N. arrived in Hobart in 1818 aboard his own brig *Greyhound* in which he had traded for the previous three years between India and Sydney. Captain John Ritchie died in 1820, leaving his Van Diemen's Land property to his brother Thomas.

Thomas prospered and acquired further property in the north, including 'Scone' at Perth where he built a flour mill. It was powered by a water race and weir across the South Esk River.

He was one of the first to push his flocks and herds west of the Meander River and in 1829 he selected 1,300 acres fronting onto the Lobster Rivulet at Chudleigh.

Writing of the property in a very lively account of Deloraine history, Daniel Griffin in 1893 describes it as *'one of the finest for its size in the colony'*, and of the animals he said:

> *'Mr James Ritchie keeps a pure herd of Shorthorns and a pure flock of Leicesters on his 'Mayfield' estate and how much better this herd and flock looks than the scraggy, wretched-looking cattle and sheep on which good grass is wasted in other places.*
>
> *One thing is only too palpable: our reputedly rich men of today*

have not half as much enterprise in them as had the early colonists, who spent so much money in order to improve the breed of stock by frequent importations. [53]

Our oldest registered flock obviously had a good home and the care of very able stockmen.

At the 1875 Campbell Town Show, Mr George Ritchie and Mr Maurice Weston judged the Leicesters.

Mr James Ritchie was a member of the Tasmanian Longwool Sheepbreeders Association and his name is listed with those who shipped sheep to New South Wales in 1893. He received six pounds, sixteen shillings and sixpence for two sheep.

Members of the family have shown sheep frequently during their long history, Mr M. Ritchie being a consistent exhibitor at Launceston in the 1920's and 1930's.[54]

THE HISTORY OF THE MAYFIELD FLOCK
(Australian Longwool Flock Book, Vol. 1, 1909)

FLOCK No. 5

Messrs. RITCHIE BROS., Mayfield, Chudleigh, Tasmania

This flock was founded in 1851 by the purchase of pure sheep, the progeny of pure Leicesters imported from England by the late Sir Richard Dry, of Quamby, Tasmania. The Quamby sheep were carefully selected by Sir Richard from one of the purest English Leicester flocks in England. From 1851 down to 1880 rams were purchased from the flock of the late William Field, of Enfield. These two flocks, to which the Mayfield stud owes its origin, run back in pure descent to Mr Robert Bakewell's famous Dishley flock. In 1880 the late Mr James Ritchie introduced a ram called 'No. 3', imported to Tasmania, and bred by Mr John Barton, of Barton House, Malton, Yorkshire. 'No. 3' was by Fuljamb, winner at all local shows; dam by Sandy, winner at the Royal Show held at Warwick; g. dam by Old Sledmere. Fuljamb was got by Sandy, dam by Owston.

Today Neon Ritchie keeps a flock of approximately 25 ewes. Neon's sheep are of excellent quality and his blood lines are frequently sought by other breeders.

C.W. Allen

Eight of the 21 Leicester flocks recorded in Volume I of the Longwool Flock Book mention the 'Leicesterville' flock of C.W. Allen, Westbury as a source of some of their sheep. He was born in America. His father had served in the British Army at Waterloo and brought his family to Van Diemen's Land in 1848.

Although trained as a carver and gilder, Charles, like many other young Tasmanians, succumbed to the lure of gold and went to the diggings in Victoria in 1853. On his return he took up farming, first at Cressy and then at 'Harvey Dale', Westbury.

He established his English Leicester flock in 1864 from Robert Jones of Jericho and Samuel Blackwell of Melton Mowbray. They had imported from Burgess, Stubbins and Stone. He also used sheep from the John Trewthie flock which had originated from importations from Turner and Tremain.

Mr Allen showed regularly through the 1870's, 1880's and 1890's and also sent sheep to Victoria, New South Wales and New Zealand.

His flock was a source for a number of studs founded between 1870 and 1907. Some of these became premier English Leicester studs of Australia and reflected the high quality which Mr Allen had achieved in his breeding and selection.

On semi-retirement he moved to the 40 acre property, 'Leicesterville', on the outskirts of Westbury. This property is now called 'Culzean'.

Mr Allen is remembered for his refined, gentlemanly disposition and his skill and achievement as a stud master is recorded in several publications.

A councillor and Warden of Westbury, he was instrumental in securing a water system for that town and he was a member of the House of Assembly between 1903 and 1909.

When Mr Allen dispersed his flock in 1907, many breeders took the opportunity to found new studs or augment existing ones with sheep of proven quality.[55]

Sir Richard Dry

While we do not know when, or from whom, the first Leicesters came to Quamby, the estate owned by Sir Richard Dry, we do know that these sheep formed the foundation of the Ritchie flock.

Dry's father supported Irish peasants against the British. He was captured in the Battle of Vinegar Hill in 1798 and was banished for life by the British Government.

In New South Wales he was a political exile, not a convicted prisoner, which was a fortunate distinction for when Colonel Paterson established the settlement at Port Dalrymple (1804) at the mouth of the Tamar River, he was able to take up a government appointment as Collector of Customs and Storekeeper. He was pardoned in 1809 for good behaviour and received a small land grant. He did well and purchased land including the 170 acres on the old site of the Launceston Showground which he called 'Elphin Farm'.

In 1819 Governor Macquarie (NSW) gave him an additional grant of 500 acres. Richard Dry continued to prosper, purchasing a lot of land, most of it on the Meander River, until his holdings eventually amounted to 30,000 acres.

THE HISTORY OF ENGLISH LEICESTER SHEEP IN AUSTRALIA

In 1828 he decided to build a house fitting his wealth and status in the community. It took 10 years to complete and included 7 sitting rooms, 11 bedrooms, halls, ballroom, conservatory and two coach houses. It also had stabling for 26 horses, a chaff house to hold 40 tons and barns to hold 20,000 bushels (500 tonnes) of grain.

Richard senior died in 1843, and it did not take long for Richard Junior to start getting into serious debt. In 1854 there was the first of the forced land sales to help lower the mortgage. I do not think that Richard Jr took a strong interest in the farming side of things, being more interested in politics, horse racing and socialising.

While Leicesters, and perhaps Teeswater's may have been on Quamby for some time, it is stated that it was Richard Dry (later Sir Richard) who provided the Ritchie foundation Leicesters. The flock history of the Ritchie stud states that '*The Quamby sheep were carefully selected by Sir Richard from one of the purest English Leicester flocks in England.*'[56] Richard did not go to England until 1857, so he could not have personally selected sheep to come to Quamby at that time.

Richard Junior entered politics in 1844, aged 29, where he soon became a leading figure in the anti-transportation movement. Van Diemen's Land was still regarded as no more than a large prison by the British Government, and as transportation to other colonies ceased, the flow to Van Diemen's Land increased. It was not until 1853 that transportation to Tasmania ceased.

Richard went to England for the sake of his health in 1857 and was knighted by Queen Victoria, making him the first Australian born Knight.

After some years of retirement he re-entered politics and in 1866 became Premier of Tasmania.

The 'Quamby' estate was a bustling farming enterprise. Eight hundred people lived on the property either as tenants or servants. We do not have any record of when or from whom he imported his Leicesters, but we do know that he imported Southdowns from the Duke of Richmond's flock in 1860.

William Field, Enfield (1814 – 1890)

William was the eldest son of William Field who had been transported to New South Wales and then Port Dalrymple after being convicted of stealing sheep. After completing his sentence in 1814 he quickly began to purchase land and herds of cattle which he used to supply meat for the government stores.

During the 1820's he acquired the properties 'Enfield' at Bishopsbourne, 'Westfield' at Westbury, 'Eastfield' at Cressy and 'Woodfield' at Cressy. As well as these properties he owned others and rented large areas of land to run cattle on. At his peak he had 10,000 cattle grazing on 300,000 acres.

William, his eldest son took over 'Enfield' in 1825, and established his Leicester stud in about 1847, purchasing the foundation sheep from leading English flocks.

In 1864 Willian sold 191 Leicester sheep to New Zealand, making him perhaps the first person to export pure British sheep from Australia. In

1864 he was also showing his sheep at the Port Phillip Farmers Show, at which he won 1st and 2nd prize for a pair of Leicester rams.

In 1867 he sent 90 two-tooth Leicester rams to Victoria for sale, so that *'Flock masters in Victoria will have an excellent opportunity of getting a change of blood as they will be submitted to public competition in about a fortnight.*'[57]

The Enfield sheep played an important role in the establishment of many later leading flocks. These included the flocks of MacGregor (Dalmore), Ritchie (Malton) and Heazlewood (Melton Vale).

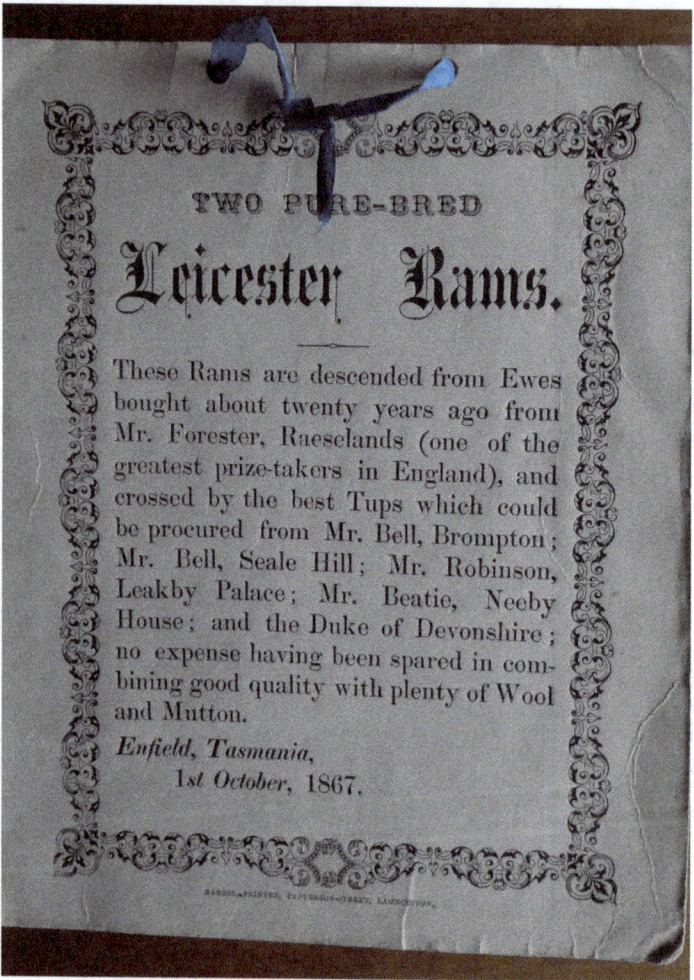

Enfield display card. Brenton Heazlewood.

Lewis Lee (Flock No.63 'The Den')

The sale of 'The Den' English Leicester stud in mid 1988, signalled the close of a long and remarkable chapter of exhibition at the Royal National Launceston Show.

The owner, the late Lewis Lee of Mole Creek, Tasmania, who had recently died, had exhibited at 67 consecutive shows, initially as assistant to his father, and soon after as stud master in his own right.

From a young age I had the pleasure of knowing Lewis quite well. I remember that his sheep were always meticulously prepared and paraded with a masterful touch. Lewis was always quiet when handling his sheep, and this was reflected in his sheep always being very placid. Lewis could always tell you what was not correct with his sheep and why yours were better than his. He made you feel that you had the best sheep at the show, although this was seldom reflected when the ribbons were being handed out. Lewis was known for politely telling the judge of the day why he had 'got it wrong' if he thought that someone else's sheep deserved to be placed above his. When the judge of the day came to open up the wool of Lewis's sheep, he would say, regardless of the judges age, 'this side boy', pointing to the side that he had opened in a perfect straight line.

Lewis' participation was characterised by cheerfulness, modesty and dignity, which endeared him to fellow exhibitors, judges, stewards and patrons alike. He knew his sheep well and he exhibited them not only to win prizes, but to win friends as well. He was outstandingly successful at both.

Missing at the 1988 show were the quality of sheep neatly penned, and the slight, erect, neat figure of the owner moving through the shed.

In 1986 Lewis was made a life member of the Royal National Agricultural Society.

It is indicative of this attitude to life, that he chose to sell his stud prior to his death to the Heazlewood family, who had been competitors and associates for three generations, so that he could see them safely in good hands for continued stud management.

Lewis Lee had become widely known as the owner and custodian of 'Lees Paddocks', an isolated cattle property run 70 km south of "The Den". This land was taken up by his father George Lee in 1888. The Lee cattle had been driven there annually for 100 years for summer grazing.

The battle with the conservation movement to maintain this tradition is widely publicised in the wonderful history of the area, and the Lee family, 'Snarers and Cattlemen of the Mersey High Country' written by Lewis's great-nephew, Simon Cubit.

In the book Lewis says, 'If there is a God, I'm closer to Him here, (The Paddocks) than I would be in any church. This is His valley, and I am its keeper'.

For Lewis Lee, the love of wilderness and nature, together with the stewardship of his beloved 'Paddocks' blended with a lifetime of devotion and dedication to his equally loved English Leicesters.[58]

Norm Badcock "Connaughtville" (left), Lewis Lee "The Den" (right). Brenton Heazlewood.

Eric Gray (Flock No. 347 'Marengo')

The death of E.B. Gray in October 1997, marked the passing of a major contributor to the revival and rejuvenation of the English Leicester breed in Tasmania. Eric Gray spent much of his working life at the helm of a major farm machinery and distribution business which was based in Hobart.

In the post-war period he pioneered the use of heavy machinery for land development on his east coast wool growing properties.

His English Leicester stud which was based at Belmont Lodge near the historic town of Richmond was founded on Tasmanian ewes. It was the rams which he subsequently imported from New Zealand that so much improved the scale and early maturity of the breed.

He was a regular and successful exhibitor at Hobart, Launceston and Melbourne and promoted the breed with enthusiasm. In 1990 he supplied some English Leicester ewes as part of the Tasmanian consignment to the Colonial Williamsburg Foundation of Virginia, USA, thereby helping to re-establish the breed into the USA.

Eric Gray was a highly respected, energetic and vital figure in both rural and business circles in Tasmania.

Other early breeders

Some of the other early Tasmanian breeders who did not register flocks, or who had dispersed their flocks before breed societies were formed are:

- Basil Archer, Woodside, Cressy. Founded 1870.
- Miles Bennett, Esk Farm, Longford. Founded 1868.
- George Hogarth, Raeburn, Breadalbane. Founded 1876.
- JW Brumby, Ashton, Cressy.
- Thomas Gibson, Esk Vale, Epping.

Flock Rams to the Mainland

The frozen meat trade to Britain, which started in 1880, had a flow on effect to the Tasmanian long-wool sheep producers. This took about a decade to be felt by the Tasmanian stud long-wool producers, but the influence came in the form of an increase in demand for long-woolled sheep in New South Wales in particular. This in turn resulted from the markets the NSW graziers had for lamb to go into the frozen meat trade to Britain.

> *The demand for the Tasmanian longwool sheep at present is very strong, and continually large consignments are being forwarded to the mainland states, principally New South Wales. Messrs Robert Gardner Ltd., shipped by s.s. Sydney from Devonport on Saturday 70 Lincoln flock rams on account of Mr P.C. Best, Deloraine, and 20 Leicester flock rams on account of Mr R. K. Heazlewood,* [my grandfather] *Glenore, while the s.s. Rotomahana took yesterday an exceptionally fine draft of 85 Leicester flock rams on account of Mr R.G. Heazlewood,* [my great grandfather] *Glenore. This consignment comprised very well grown and beautifully woolled sheep, and were admired by those who saw them, and must prove satisfactory to the purchaser. All the above are for private orders from New South wales. Few realise the extent or value of the longwool industry to Tasmania. Messrs Robert Gardner Ltd., who handle practically all these sheep, state that up to the present date* [August 19, 1913] *2161 longwool sheep have been shipped since the beginning of the year.*[59]

This report is typical of the reports of the exports of flock rams from Tasmania to the mainland in the early 1900's. In November of the same year it was reported that 2,666 longwool sheep had been sent out of the state so far that year.[60] Rams were not only sent to supply private orders, but also to the Sydney and Melbourne sales. Some stud breeders also took sheep to the major shows with the aim of selling some, or all of the sheep after the show.

In 1916, my grandfather R.K. Heazlewood sold 99 rams to Tasmanian buyers as well as 252 rams through auction sales and private orders in NSW and Victoria. Of the Tasmanian sales 92 were flock rams. They all went to wool growing areas in the Midlands and the south of the state, at prices ranging from 2 ¾ guineas to 3 ¼ guineas, 40 to one property and 35 to another. Stud rams ranged from 9 to 25 guineas. Interestingly all those stud rams were purchased by Merino properties, indicating that either they were prepared to pay extra for top quality, or most likely they were used for breeding Leicesters for their own use.

'GMB' wrote in the Weekly Times, August 1913:

> *For years past Tasmania has been noted as the home of stud sheep ... The breeding of longwools promises to become a very important industry. The demand for these sheep is rapidly increasing, and breeders are supplying that demand with a sheep that is eagerly sought...Breeders are very particular about the purity of their strains, and some of the flocks are descended from importations in the early days of the state. From time to time the very best sheep procurable are imported from Great Britain and elsewhere. The industry, which is growing day to day, promises to become a very substantial addition to our national wealth ... Ram lambs from nine to ten months old bring from two and a half guineas to four guineas, f.o.b., at the*

> *shipping port ... This season one order alone accounted for 210 Leicesters, and last year a private order was received for 450 Lincolns. At present about 55 breeders are supplying sheep, and nearly all of them belong to the Australian Longwool Breeders' Association, for which Mr A.J. Stewart, of Robert Gardner Limited, acts as agent, and receives copies of the minutes of all meetings. In order to keep in touch with what is being done, the local breeders hold meetings to discuss them, and also offer any suggestions that may benefit the Association. The number of sheep sent to the Sydney sales this year was from 700 to 800 in excess of last. So far, the figures for the first six months of this year are less than 300 behind the total of the whole of last. When those sent to the Melbourne sales are taken into account, and private orders filled, the total for 1913 will considerably exceed that of any previous year.*[61]

While these reports did occur during the golden decade (1910 – 1920) for the longwool breeds of sheep, and in particular the Leicester, they do reinforce the close association that the Leicester had with the Merino up until it was slowly superseded by the Border Leicester.

The sale of Tasmanian bred rams to the mainland had started much earlier and was in full swing by the 1890's. On the third day of the 1893, Sydney Annual Stud Sheep Sales, held at Kirribilli Point, North Sydney, 20 Leicester breeders from Tasmania were selling sheep. In total they sold 310 rams and 90 ewes. In studying the list of vendors it is interesting to note that, of the 20 breeders selling sheep only one is from the south of the state, the rest all from the central north.[62] This is in contrast to the early concentration of the Leicester breeders being in the south of the state where they were primarily Merino breeders. The shift had been made from those importing the Leicester along with the Merino, their main focus being fine wool production, to the stud master concentrating on flock ram production, in an area of the state that more suited the breeds involved.

In 1892, 2,565 stud sheep of all breeds were shipped from Tasmania to Sydney for the annual sheep sale. These were shipped on three vessels, the *Corrina*, 516 sheep, the *Burrumbert*, 1,249 sheep, and the *Lindus*, 800 sheep. The Tasmanian Long-Woolled Sheep Breeders' Association had sent 130 long-woolled sheep on the *Corrina*. This association had been formed in 1892.

> *A meeting of gentlemen interested in the breeding and exportation of long-woolled sheep was held at the offices of Messrs. J.H. Geddes and Co., St. John Street, on Saturday afternoon, there being ten breeders present. The object of the meeting was to secure a reduction in the charges previously made by agents in Sydney and Melbourne on the sale of long-woolled sheep, and to arrange for a floor separate from the Merinos, on which to sell long-woolled sheep in Melbourne. It was decided to form an association to be called the Tasmanian Long-Wool Sheep Breeders' Association. The following were elected a committee of management:- Messrs. RC Gibson, R Hogarth, JW Brumby, C Walker, W French, and Miles Bennett; Mr GL Meredith being appointed hon. secretary. Mr Meredith informed the company that he had the offer of the*

show room in one of the leading wool warehouses in Melbourne, and submitted a scale of charges supplied by well known Melbourne agents, which were considered highly satisfactory. Concerning the future of this undertaking it may be mentioned that a large number of sheep have already been promised to the association, for export both to Sydney and Melbourne.[63]

At the 1893 Melbourne sheep sales the averages for pure Leicester stud sheep sold by Powers, Rutherford and Co. were:

- W.G. Hogarth: 10 rams – 2 Pounds 12 shillings 6 pence ($450 Approx. 2020 value)

- W.G. Hogarth: 4 ewes – 2 Pounds 2 shillings ($360 Approx. 2020 value)

- Falkiner Bros.: 3 rams – 2 Pounds 12 shillings 6 pence ($450 Approx. 2020 value)

- C.W. Allen: 10 rams – 11 Pounds 15 shillings 3 pence ($2,000 Approx. 2020 value)

- C.W. Allen: 10 ewes – 3 Pounds 16 shillings 8 pence ($550 Approx. 2020 value)

- Ben Gibson: 13 rams – 3 Pounds 4 shillings 6 pence ($550 Approx. 2020 value)

- R.G. Heazlewood: 10 rams – 3 Pounds 5 shillings ($560 Approx. 2020 value)

- R.G. Heazlewood: 10 ewes – 2 Pounds 17 shillings 9 pence ($490 Approx. 2020 value)

- R.C. Gibson: 3 ewes – 3 Pounds 3 shillings ($540 Approx. 2020 value)

- R. Hall: 2 ewes – 2 Pounds 17 shillings 9 pence[64] ($490 Approx. 2020 value)

The number of long-woolled sheep sent to the Sydney sale increased to 501 in 1894, English Leicester's being from:

- Mr Miles Bennett, Longford, 56 rams, 30 ewes.

- Mr J. McFarlane, Kentishbury, 3 rams.

- Mr E. Walker, Clairville, Westbury, 20 rams.

- Mr W.R. Jones, Brookside, Sheffield, 23 rams.

- Mr George Hogarth, Raeburn, Breadalbane, 3 rams, 6 ewes.

- Mr R.G. Heazlewood, Melton Vale, Glenore, 30 rams, 5 ewes.

- Mr E.H. Heazlewood, Glenore, 24 rams.

- Mr Robert Hogarth, Newstead, 3 rams, 4 ewes.

- Mr Ben Gibson, 16 rams, 10 ewes.

- Mr A. Oliver, Brookhill, Chudleigh, 24 rams, 16 ewes.

- Mr James Ritchie, Mayfield, Chudleigh, 8 rams.

- Mr W.V. Field, Enfield, Bishopsbourne, 5 rams, 5 ewes.

- Mr H.R. Falkner, Wickford, Longford, 5 rams.

- Mr W.C. Grubb, Tolarno, St.Leonards, 15 rams.

The association also sent Lincolns, Shropshires, Cotswolds and 20 Border Leicester rams from Mr Vincent Newton, Clover Hill, Hagley.[65] Mr Newton was perhaps the first person to have Border Leicesters in Tasmania, and one of the early ones for Australia.

A summary of the 1894 sales showed that, from Tasmania, 363 English Leicester rams averaged two pounds, six shillings and two and a half pennies (approx. $362 in 2020), 101 ewes averaged one pound, eighteen shillings and three pennies (approx. $300 in 2020). From New Zealand 150 rams averaged two pounds, eight shillings and one and a half pennies, and 172 ewes averaged three pounds.[66] It is interesting to note the similar price for rams and ewes, and that the ewes from New Zealand sold at a higher price than the rams. Obviously, there was a very strong demand for ewes.

Tasmania is now home to only five studs. The most recently registered stud belongs to Fiona Hume who for many years has run a large commercial flock of pure English Leicesters. These have been inspected and passed to enter the flock book.

Shipping stud sheep from Launceston 1908. The Weekly Courier, 1908.

Campbell Town Show 1921. Weighing an English Leicester ram in the Heavy Weight Competition. Ram held by R.K. Heazlewood. The winner, an English Leicester at 126kg. The Weekly Courier, June 16, 1921.

B.M. Badcock sheep Longford Show 1933. Weekly Courier, October 19, 1933.

Judging Rams Launceston Show 1918.

Judging young Leicester ewes Launceston Show 1918. The Weekly Courier, October, 1918.

THE HISTORY OF ENGLISH LEICESTER SHEEP IN AUSTRALIA

Leslie Heazlewood with ewe hoggets in the sheep yards at Valmont, Whitemore, Tasmania 1916. Archives Office Tasmania.

B.M. Badcock (Tasmania) ewe. Australian Farm & Home, February 1950.

Henry Badcock & Son (Tasmania) ewe hoggets. Australian Farm & Home, February 1950.

Boyes Bros. Champion Ram Hobart Show 1912. The Weekly Courier, October 17, 1912.

Judging Yearling Leicester Rams, Launceston Show 1914. The Weekly Courier, October 8, 1914.

Boyes Bros., Longford, Tasmania. Champion ewe at Royal Melbourne, Longford and Hobart shows 1907. Flock Book for The Longwool Breeds of Sheep in Australia, Vol. III, 1911.

John Badcock Champion Ram Hobart 1909. The Weekly Courier, *October 21, 1909.*

3
New South Wales

'To be sold by auction by Mr James Simmons

A Family of Pure Leicester Sheep, consisting of the parents, two rams and three ewes, their offspring. The heads of the family were imported into the Colony by the ship Portland, in 1828, and are, with their produce, considered the best specimen of pure Leicesters ever seen in New South Wales. This breed feeds to a great weight, and are, without exception, the best stock for the Shambles (slaughterhouse). The family would form a great acquisition to the lawn of any Gentleman's Estate, being objects of great beauty and interest, when the fleece attains its full length.'[67]

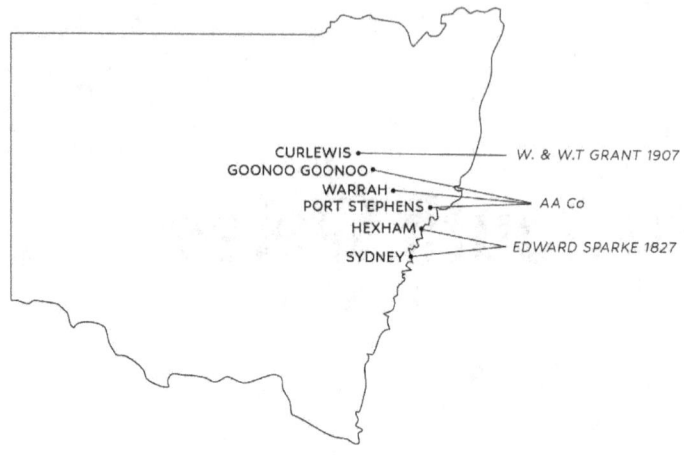

Edward Sparke

My research indicates that a Mr Edward Sparke Jr was the first person to import Leicester sheep into New South Wales, in April 1827, some 18 months after Edmund Bryant had landed the first ones in Van Diemen's Land.

The story of Edward Sparke Jr, and Edmund Bryant, the first person to import Leicesters into Van Diemen's Land, have many similarities. Both sailed from Falmouth on the *Anguilar*, arriving in Hobart on 4 February 1824. Bryant remained in Van Diemen's Land while Sparke sailed on to Sydney. Both later returned to England, married and then returned to the Colony with their wives and Leicester sheep, and both became insolvent in the 1840's. One can presume that on the long sea voyage to the Colony that the Sparke's and Bryant's would have had long discussions about farming and the best types of livestock, so perhaps it is not too surprising that they were both involved in the very early importing of Leicesters.

Edward Sparke Sr (1769-1844) married Mary Hosking (1772-1850) in 1795 in South Brent, Devon. Probably encouraged by his brother-in-law, who had previously been in NSW, Sparke wrote to Robert Wilmot Horton, Under Secretary at the Colonial Office, in February 1823 seeking information about land grants and other encouragements in NSW, stating:

'I have been for many years an extensive farmer in Devonshire and am desirous of Emigrating to the Colony of New South Wales. I am possessed of property to the amount of Fifteen Hundred pounds and have a grown-up family.' [CO201/149, f 270]

He received the standard advice and a letter of introduction to the Governor – and sailed from Falmouth, Devon on the *Aguilar* with his wife and five sons.

- Edward Sparke Jr (1797-1871)
- John Sparke (1799-1852)
- Andrew Sparke (1802-1830)
- William Sparke (1804-1878)
- George Sparke (1806-1875)

Also on this sailing of the *Aguilar* were brothers Edmund and James Bryant. The Bryant's stayed in Van Diemen's Land with Edmund later returning to England, marrying, purchasing Leicester sheep and arriving back in Van Diemen's Land in October 1825.

In Hobart Town, the Sparke family transferred to the *Belinda*, arriving in Sydney on 23 February 1824. The family established itself at 98 Pitt St, William Sparke Sr opening a butchery business in King St with the assistance of his sons Andrew, William and George.

Edward Jr started dealing in cattle, which were grazed on leased land at Liverpool. He also travelled about the Colony looking for land, and was, evidently, impressed by the Lower Hunter.

They all applied for grants:

- Edward Sr, 2,000 acres in the parish of Alnwick, county Northumberland, granted 29 November 1825 (called Woodlands).

- Edward Jr, 1,920 acres in the parish of Willis, county of Brisbane, granted 14 June 1827 (called Dunwell).

- John, 1,920 acres in the parish of Stanford, county of Northumberland 18 July 1828, near Mulbring.

- Andrew, 960 acres near Polkolbin (which lapsed on his early death).

- William, 1,920 acres in the parish of Hexham county of Northumberland, 20 May 1929 (called Webland Park).

The names Woodlands, Dunwell, Woodbury and Webland all come from family farm names in Devon.

Edward Sparke Jr returned to England in the brig *Colombia*, sailing on 7 March 1826. On 29 August in that year, he married Susan Abbott, daughter of Robert and Elizabeth Abbott at St Petroc, Brent, South Devon.

In November 1826, they sailed from Plymouth on the *Elizabeth*, with 60 Leicester and Merino sheep, arriving in Sydney on 7 April 1827. I do not know how many of these 60 sheep were Leicesters.

A fortnight later, the following notice appeared in the *Sydney Gazette* (23 April 1827):

> '*MR. EDWARD SPARKE, JUN. begs to announce to the Public, that he has returned to the Colony, per Ship Elizabeth, and has brought with him some of the finest EWES and RAMS that he could possibly select from several Flocks of the greatest note of the Leicestershire Breed; and as he has brought more Rams than what he will require for his Flocks, should have no objection to let any Gentleman have one, two, or three for the Season, on moderate Terms.*
>
> *No 98, Pitt-street, April 17, 1827.*'

A couple of months (June 1827) later Edward Jr's grant was gazetted. Dunwell is on Sparke's Creek, a tributary of Dartbrook in the Upper Hunter.

The 1828 Census shows them as Edward (30) and Susan (23) Sparks with their daughter Elizabeth (5 months), at Dunwell, with 200 cattle, 300 sheep and 7 assigned convicts.

Despite developing Dunwell and leasing the 6,000 acres adjoining, in 1831, Edward Sparke Jr put in a manager, and removed his family to Ravensfield, 2,000 acres near Maitland, which he leased until 1841.

From 1836, he held several runs on the Namoi River. He was also involved in a partnership with Phillip Wentworth Wright, trading as Phillip Wright & Co, in various businesses in and around the town of Aberdeen.

In January 1839 Susan Sparke died at Ravensfield, aged 34. In November 1840 Edward Sparke Jr and Philip Wright put to auction – land in Aberdeen, two properties of Edward Sparke's, allotments in West Maitland, sheep, cattle and horses.

A month later, December 1840, the partnership of Wright and Sparke was dissolved. In 1842 Phillip Wright was declared insolvent; in 1843 Phillip Wright & Co and Edward Sparke were similarly declared.

The lease on Ravensfield expired at the end of 1840, Edward Sparke selling off his stock and furniture in January 1841. Sometime before this, he sent his daughter, Elizabeth Mary, then aged eleven, home to her maternal grandparents.

Edward Spark's next venture was to establish an extensive butchering and salting down works on the family property Woodlands, within three quarters of a mile of the Hexham wharf on the Hunter River, extending it with a boiling works – and arranging to accept wool and other produce for shipment. He lived at his parents' home, Barrahineham.

> *'At Hexham, too, a situation every way suitable from its proximity to the river, an extensive slaughtering and salting establishment has been commenced by Mr E Sparke, and arrangements have been made for carrying on business to a very great extent.'*[68]

In 1849 he announced that he was about to depart for New Zealand, and offered:

> *'... his services to take charge of any SHEEP, CATTLE, or HORSES that may be entrusted to him during the passage, and make sale of the time on their arrival, if so required; and he flatters himself that he is competent, having imported stock of his own from England to this colony, under his own management on board of ship, satisfactorily; and also having had 25 years' experience in the colony, which gives him full knowledge of the treatment and value of stock.'*
>
> Maitland Mercury, 10 March 1849

What became of this venture is not known.

He was still in NSW in the late 1850's. A note appears in the *Maitland Mercury*, 15 January 1859, to the effect that his nephew William Andrew Sparke of Newcastle would act as his agent.

Nothing more is known, except that he died in Blackawton Devon on 13 March 1871, and is buried in St Michael's Church, near his daughter Elizabeth Mary (1828-1888) who had married William Cutmore Baker (1821-1887) in 1846.[69]

We do not know what became of the Leicester sheep that he imported. I can only presume that the majority of the rams were mated to the local ewes, so that *such graziers as may desire to preserve a little mutton along with the wool in their breeds of sheep*[70] could do so, with perhaps some put over the ewes that he imported to keep some pure sheep.

The Australian Agricultural Company

The Australian Agricultural Company (AA Co) was formed in London in 1824 by an Act of Parliament and Royal Charter. The purpose of the Company was to run fine-woolled sheep on a one-million-acre grant of land in New South Wales. Its directors and major shareholders included directors of the Bank of England, the East India Company and Members of Parliament.

The company based its plans on the Bigge's Report on New South Wales (1822). In his report, Bigge's noted that in relation to the growth of fine wool that he saw it as:

THE HISTORY OF ENGLISH LEICESTER SHEEP IN AUSTRALIA

'The principal if not the only source of productive industry within the colony from which the settlers could derive the means of repaying the advances made to them from the mother country, or supplying their own demands for articles of foreign manufacture.'

It intended that most of their shepherds and labourers would be assigned convicts, supervised by 'free and experienced persons', many of them sent out from Europe on seven-year contracts. Robert Dawson was appointed the Company's Agent in New South Wales, and in March 1825 he, along with the London Clerk, H.T. Ebsworth, and a consultant N.A. Nilson went to France to purchase sheep. Ebsworth and Nilson then continued to Saxony to make further purchases. The group sailed from Plymouth on two chartered ships, the *York* and *Brothers* in June 1825. They had on board 720 French and Anglo merino sheep, eight horned cattle and seven horses. The ships arrived in Sydney on the 13th and 15th of November 1825. On arrival most of the servants and stock went to 'The Retreat' (now 'Kelvin Grove') at Bringelly near Camden, rented for their reception by the Company's Colonial Committee (James Macarthur, Hannibal Hawkins Macarthur and James Bowman).

Dawson, the Company's agent, was under pressure to choose the site of the Company's Grant and get things under way. In January 1826 he went to inspect land in the Port Stephens area, one of the most promising locations suggested to him.

Influenced by the access to water, Dawson chose to establish the Company's main settlement at Carrabean (later Carrington) on the northern shore of Port Stephens and to take up the whole Grant in one block on the land stretching north from Port Stephens to the Manning River.

In the following years the area was explored, a village grew up at Carrabean, farms were established at Booral and Stroud, more indentured servants and stock arrived, and by 1828 the number of assigned convicts had reached 180. Between June 1825 and July 1827, AA Co imported 2,170 Merinos from Europe.

The unsuitability of some of the Port Stephens land soon began to surface, and by 1833, after long and complicated negotiations between the Governor and the Colonial Office, an agreement was reached where some of the Port Stephens land was exchanged for more suitable land inland. As a result, the Company's pastoral grants were to be: the western part of the original Port Stephens Estate (464,640 acres), the Liverpool Plains/Warrah (249,000 acres) and Peel River/Goonoo Goonoo (213,000 acres).

The Company also obtained a 2,000-acre land grant with the right to mine coal at Newcastle. Under the 1828 Agreement with the Colonial Office, the Governor in NSW was not to grant any further coal rights or assign convicts for the working of any other coal mine for the next 31 years. In return the Company was to supply the Government's coal needs at prime cost, to one quarter of the year's output, and those of the Public at reasonable prices.

The line of sheep stations stretched north to the outstation and dairy at Gloucester and west into the foothills of the Barrington Tops. Each station had two flocks, each of approximately 500 sheep, in the care of 6 shepherds and a hut keeper. In 1832 there were 32,000 sheep at Port Stephens in the care of a Chief Overseer, 13 free overseers, 2 free shepherds and 102 convict shepherds.

During the 1830's the Company used bullocks and drays to deliver stores and rations to their inland stations. By the late 1830's the losses in bullocks and the expenses in repairs to the drays resulting from negligence of the drivers and the deterioration of the roads were considerable.

Phillip King (then Commissioner) thought that mules working along the Company Line would soon recover the expenses of bringing them and their drivers from South America. He arranged for 20 mules, 2 asses (a male and a female) and 3 'huasos' (or 'country people') to be contracted in Valparaiso, Chile. They arrived in Sydney in April 1840. The mules coped well with the Company Line and their upkeep was considerably less than that of the bullocks.

The early 1840's were a difficult time for the Company and the Colony in general, with drought, especially severe on the Liverpool Plains, and an economic depression. Many sheep were boiled down for tallow, the prices for fat cattle collapsed and the demand for coal fell away.

In February 1856 the Directors decided that the sheep should be removed from the Port Stephens Estate and that the Company's Establishment at Stroud should be disbanded. By the 1920's all but a few acres of land had been sold at Port Stephens.

In the late 1850's the decision was made to develop Warrah for sheep, but it was not until 1864 that a sheep breeding program began, financed by a special call of thirty thousand pounds on the shareholders. The Stud Merino Flock was established in 1866 with the purchase of Mudgee rams.

The other great area of expenditure from 1869 onward was fencing. Merewether (General Superintendent) visited properties in Victoria and returned convinced of the advantages of raising sheep in paddocks rather than by shepherding. Huge quantities of wire were sent from England. Labour to put up the fences was always a problem but the Warrah Estate was divided into two sections, East Warrah (122,600 acres) and West Warrah/Windy (127,000 acres). Their boundary fences were later rabbit proofed. The first paddocks were 6-7,000 acres for sheep and 3-7,000 acres for cattle, but they were later subdivided. The paddock fed sheep were 'larger in carcass and in better condition', the wool 'better grown, more mellow to touch, cleaner and therefore much more easily and quickly washed'. The first paddock lambing occurred in 1872.

The sheep numbers grew steadily: 35,000 in 1865, 68,000 in 1870, 104,000 in 1875, 155,000 in 1890 and 181,000 in 1900. In 1901 the Windy woolshed was built, described at the time as one of the largest and most elaborate in the state.

From the turn of the century there was a movement for Closer Settlement and the breakup of the large estates, with Warrah and Goonoo Goonoo being prime targets.[71]

While the AA Co was formed to grow fine wool to be sold on the London market, their use of English Leicesters, though limited, to improve the carcass and wool of the Merino is interesting. The correspondence between New South Wales and London gives us a detailed insight into the thinking behind the use of Leicesters, and in this case, the implementation of these plans and the results. AA Co were not the only fine wool breeders in the colony thinking along these lines. The Van Diemen's Land Company also imported and used English Leicesters for the same reason, as did many individual graziers running Merinos.

In October 1835, James Stephens at the Colonial Office, forwarded to the AA Co Secretary (in London) a letter and prospectus from Mr Frederich Barthels about a project to improve wool in Australia. In the prospectus he mentions crossing Leicesters with Merinos to increase the quantity of wool. The matter was referred to a Director, Richard Hart Davis, who remarked: *'This appears to be a useful hint.'*[72]

The Directors decided to contact Barthels about his plan.[73] In December 1835, Barthels was told, should he decide to go to NSW, the Company would provide an introduction to their Commissioner, Colonel Henry Dumaresq.[74]

In February 1836, the London Secretary, forwarded Barthels' plan to Colonel Dumaresq, with the comment:

> *'I beg leave to enclose you a copy of a letter of introduction which I have given to Mr Barthels – with a copy of the correspondence and of a prospectus relative to the breeding of Merinos in NSW.*
>
> *Since that correspondence took place, circumstances of a family nature have induced Mr Barthels to relinquish for the present his intention of emigrating to NSW. I have, however, thought*

it advisable to forward the enclosed papers, in the hope that they may contain some hints which might be of service in the future management of the Company flocks.'[75]

Dumaresq received this letter in June 1836 on his return from a trip from Port Stephens to the Peel (Goonoo Goonoo) and the Liverpool Plains (Warrah). He promptly requested, with two Durham bulls:

> *'... (as suggested by Mr Barthels) 3 close woolled boney rams of Leicester blood and 6 to 12 ewes. Barthells' plans not generally practicable – flock of the best ewes already kept apart for breeding.'*[76]

This dispatch was received in London in December 1836. In January, Richard Hart Davis, laid three letters before the Committee of Management, concerning the purchase of 3 rams and 12 ewes of the Leicester breed:

> Letter from Philip Oakden[77] of Bentley Hall with the names of breeders – W.R. Smith, Dishley Grange (this is William Smith who had taken over Dishley Grange from Robert Honeybourne, Robert having taken over Dishley Grange from Robert Bakewell); W. Smith, Swarkestone, Derbys; Wilkinson, Lenton, Notts; W. Burges, nr Home Pier Point, Notts; Harrow, Loughborough, Leics; Parr, Wanslip nr Leics; Hokes, Kingston, Notts.

Letter from Bennet dated 20 Dec recommending the flock of Earl Spencer at Althorp Park.

Letter from Earl Spencer 20 December saying he believed he could supply rams and ewes but waiting for return from bailiff.[78]

Later that month, the Committee of Management, received a note from their agent, Wilkinson, that in a letter from Mr Wilkinson of Lenton, 27 January 1837 – had selected 3 rams and 12 ewes, former at 8 guineas and at 4 guineas per head, and that they had been put on dry food preparatory to trip to NSW. Had not succeeded in procuring a man to travel with them.[79]

Arrangements were made for a vessel to take the sheep (and two Durham bulls) to NSW.

In March, Mr Wilkinson reported that 'the sheep would be moved per Pickford's boat, to arrive in London on the 11th' and the money was paid to him via Smith, Payne and Smith's Bank in Nottingham [The AA Co Governor/Chairman, John Smith, was a senior partner is this bank].[80]

The barque, *Achilles*, sailed from Gravesend on 23 March 1837, arriving in Port Jackson, 21 July 1837. On board were 3 Leicester rams and 12 Leicester ewes in the charge of a shepherd, A. Nathaniel Barraclough. Dumaresq reported they were in excellent condition and arrangements were being made to move them from Sydney to Stroud.[81]

In August 1837, Dumaresq reported that that stock had arrived at Stroud, 'the bulls and sheep are precisely what were required'.[82]

The importation of these Leicesters was a genetic experiment by the Company in an attempt to get a more robust sheep with heavier-cutting wool. From a mixture of the initial three Merino strains, in 1837 Dumaresq established an inland breeding stud at the Peel. Though this stud was formed by taking some of the best flocks of French Merinos up to the Peel, by the end of the third year the inland stations were producing a high quality but coarse-woolled sheep with greater bone, constitution and heavier fleeces than the stations on the coast. This would have been partly a product of the environment and the Leicester influence.[83]

> *The experiment of crossing the Leicester rams imported in 1837 with the Merino flocks was being watched with some care in London; and Mr J.T. Simes [an early agent and broker influential in Australia] ... reported that at the last sales the crossbred wool had sold at better prices than their Merino wool, but that nevertheless considered the Leicester blood too robust a type and the cross too violent ... [Simes'] advice was followed, and the cross was henceforward restricted to one-third of the number of ewes stationed at the Peel.*[84]

To effect such a widespread change in the Peel flocks as was reported, Leicesters must have been used in the stud flock, also in an F2, F3 or more generational crossing, and when one sees in later decades the successful dilute infusion of Leicester and Lincoln genes into flocks across Australia to create a well-adapted combing-wool Merino, then, the possibility and success of such an AA Co development is likely. It must rate as one of the first influential long-wool infusions involved in the creation of an Australian combing Merino.[85]

In February 1838, Mr Wilkinson [of Lenton, UK] wrote to the Hon. JTL Lesley Melville:

> *'... Both the bulls and the sheep were very good animals and were also well bred, so that they themselves can well be depended on for breeding. The wool of the sheep, also was before they were put to their dry food, of a very nice quality. But owing to their food, and the fatigue of the long journey, the first clip would of course be spoilt. After that, the animals being in health I have no doubt, but that the wool will be very good, and that they will be of great use for improving the Company's flocks.*
>
> *They will give length of staple to the wool of the South Downs or would improve the quality of the mutton of the Merinos, and would give early maturity and swiftness of growth to both.*
>
> *I never recommend any great number of animals myself, but my advice always is "to let them be very good". In a new Colony, in particular, this must be of great advantage, because there are many individuals of course, who will not import, and will therefore go to select from the best flocks and herds at home.*

I am in hopes the <u>No. of ewes</u> you have will be sufficient to go with the breed; and should the Company at any time wish for a <u>few more rams</u> to put to the <u>descendants</u> of those they now have, or to cross <u>their flocks still further</u>. If they will only let me know in time, I will select <u>the best</u> I possibly can for them: I have some <u>very good young Rams this year</u>.'

This letter was duly forwarded to Colonel Dumaresq, arriving in June 1838, three months after he had died at Port Stephens on 5 March 1838.[86]

That month, the Directors wrote:

London Despatch 33, 16 March 1838, London Minutes

We shall look forward with anxiety to result of the projected cross of the Leicester breed of sheep with the Company's flocks, and we feel confident, that you will proceed with every necessary caution, in this experiment about to be made.[87]

In May 1838, the London Secretary forwarded J.T. Simes & Co's Report, of 9 April. In it, Messrs Simes recommended that:

'... the longest and healthiest woolled sheep should be separated from the rest, and crossed with long woolled sheep from the neighbourhood of Launceston, it may perhaps, be advisable to wait the result of the intended experiment with the Leicester *sheep shipped from England last year; when we shall be better enabled to estimate the advantages or otherwise, to arise from such a breed.*'[88]

The 'long woolled sheep from the neighbourhood of Launceston', would be referring to the Teeswater cross sheep, also known as 'Port Dalrymple type'. This type of sheep gained a good reputation for doing well under the local conditions with many being shipped across Bass Strait when the Port Phillip district was settled.

In October 1838, J.E. Ebsworth, the Acting Commissioner, reported:

'As recommended in [the Secretary's letter] I will delay purchasing rams from VDL, till the results of the Leicester cross known. The success of the pure bred Leicesters has not hitherto been great, only three rams having been reared, others have been dropped but weakly constitutions. The cross between 100 French and Saxon ewes and the Leicester rams has been very satisfactory, about ninety lambs. We need to see their fleeces.'[89]

The next year, Phillip Parker King, the next Commissioner, reported:

'Every care is taken with experimental cross between Saxon and Leicester sheep not to let them mingle with the others. Cross promises well – size, character and figure of the Leicester and great improvement in the fineness of the wool.'[90]

And in January 1840, King summarised the wool classer's report:[91]

> Report from Mr Koelz on the Leicester fleeces, also the cross, now yearlings – wool of latter shorter but finer and closer than Leicester, and coarser and longer than the merino, but equally close in pile very successful – 2 lbs. more wool and 15 to 10 lbs. to carcass.
>
> I have been much surprised to find the Co's wethers so small and light average 44 lbs. and we have been obliged to kill some under 30 lbs. – the general average in the colony is 50 – 60 lbs.
>
> Consequence of high breeding but does give some difficulty with sales, for butchers prefer larger carcases – those yours may be better in flavour
>
> Thinks it desirable to throw dash of Leicester blood into the Peel flock – heavier fleece and carcase but not loss in staple
>
> Ten first cross rams saved to be put to another flock of fine-woolled sheep – Mr Koelz of opinion we should not breed from first cross rams, but only the imported rams
>
> Very unfortunate in the increase of the sheep sent out first year (1838) we only saved
>
> 2 ewe lambs

THE HISTORY OF ENGLISH LEICESTER SHEEP IN AUSTRALIA

1 ram lamb
2 deaths
4 missed

The increase last season was

2 ram lambs
2 ewe lambs
3 ewes slipped
3 ewes missed

So that the pure Leicester flock is now

3 rams
2 ram lambs
11 ewe
2 ewe lambs

Question: the propriety of crossing your, already justly approved of breed, with Leicester blood, to give a larger size of carcass and an increased weight of fleece. By doing so a small number of sheep, and consequently less expense of shepherding will produce an equal return of income. By crossing only the Peel River ewes in the way I propose, the purity of the flocks at Gloucester and Telligherry may be preserved.

Later in 1840, King wrote:

> The Leicester cross seem to have delicate constitutions or else the heat of the climate has weakened them. I am sorry to say that the pure ram lamb dropt in the Colony is dead, and the one of last year is likely to go the same [margin: *he is since dead*]. The pure ewe lambs are also sickly. Every care has been taken of them. They have been housed every night that has been wet.
>
> It is believed that the rams and ewes are from the same flock; and if so, the cause may be attributed to their breeding 'in and in'. The parents do not breed so surely as could be wished.
>
> It would be very desirable to send out another ram or two from another flock and of a superior character as to wool.
>
> The cross, however, are healthy – I shall send them to the Peel next season.
>
> I have put the cross rams to a flock of Saxon ewes to see what will be the result. It is necessary that something should be done to increase the size of your sheep. The wedders are so very small. One of 45 lbs is considered a good sheep. We are now by necessity killing them at not more than 36 lbs and some are lighter.[92]

And in November 1840:

> Your advice respecting the cross of the Leicester Breed with the Peel River ewes will be attended to. I had some weeks since decided upon sending 30 of the cross Leicester rams to the Peel for the next season (April) but before this I shall have the benefit of your opinion of the wool sent home by the *Meanwell*.
>
> Ten of the same rams will be sold at the next Maitland sale to enable me to ascertain the price they ought to fetch and twelve I shall retain for private sale at Port Stephens.
>
> In my Despatch No. 31 paragraph 97, I alluded to the delicacy of the Leicester sheep and the deaths of the two ram lambs. The ewe lambs then reported to be sickly recovered and are now healthy but I cannot report more favourable of the produce. We have this year as yet only 4 lambs (2 being very male and two females). 2 more of the ewes appear to be in lamb.
>
> Mr Koelz assures me that the greatest care is taken of them particularly in wet weather when they are housed. I think they would do better at the Peel and when the increase sufficiently I shall send some.[93]

Report from London wool brokers Simes & Co 22 Dec 1840 [AA Co 1/19 DK66c&d]:

We beg herewith to transmit account sales of the wools sold for you at the public sales for the general result of which we cannot do better than refer you to our price current already sent.

Our attention has naturally been drawn to the two bales of coarse wool, the produce of your Leicester crossing, and we still incline to the opinion which we ventured to give you in our letter of 24th June last, that the Leicester breed is too robust for crossing with the delicate constitution of your Flocks. However desirable it may be on the score of profit to increase the weight of the fleece, still, with a flock of such superior quality as your, we feel bound to repeat the recommendation given in our above mentioned letter, to proceed with the greatest caution and in the first instance, to have recourse to some breed more congenial than that of Leicester. Indeed so far from endangering the deterioration of your established quality some consumers and buyers of your wools have hinted the propriety of culling from your flocks, some of the old sheep and of throwing in a fresh supply of pure blood which notice we think it right to lay before you.

THE HISTORY OF ENGLISH LEICESTER SHEEP IN AUSTRALIA

In March 1841, King wrote:

> I am sorry to report that one of the imported Leicester rams has lately died from 'red water'. One only is left. We have a yearling ram, but he is not fit to tup until next year.
>
> I have heard of some very good Leicester rams which have been lately imported and are for sale – the price is 25 guineas for each. I am about to visit Sydney to see them and if they are good sheep to purchase to.
>
> The increase of the pure Leicester ewes last season is
> rams lambs
> ewe lambs
> others were dropped but were weakly and soon died.
>
> Mr Koelz thinks that the climate of Port Stephens is too warm and relaxing and recommends the cross bred sheep being sent to the Peel. As I agree with him in this opinion I have decided on sending the ewes to the Peel, where the feed is much more nutritious and the climate equally adapted to their character.
>
> Mr Koelz has examined the sample of half-bred Leicester wool forwarded from London, he does not think it 'much superior in fine-ness to the Company's and certainly not at all in softness'. 'No doubt (he thinks) the second cross will

far surpass it in either and still keep up the weight of the fleece much the same'.[94]

In May 1841, King added:

With reference to paragraph 21 Despatch 46 I have purchased two Leicester rams bred by Mr Pratt [of *Caldwell*] 'got by a famous ram of Mr Spencer's of Snarestone in Leicestershire'. The price was £20 each. They have been undergoing a dressing to prevent any chance of scab.[95]

In June 1841:

Two of the Leicester rams alluded to in my letter No. 46 paragraph 21 were purchased for £20 each. They are fine sheep and of a good quality of wool. The rams originally sent out have died: I think they must have suffered from a constitutional ailment like consumption – I trust their progeny will not be affected in a similar way. The young cross-bred sheep have been sent to the Peel; the climate of which will certainly be more favourable as well as the pasture more nutritious for them. They have not done well at Telligherry [Port Stephens].[96]

London Despatch No. 33, 26 March 1841, AA Co 78/3/1:

THE HISTORY OF ENGLISH LEICESTER SHEEP IN AUSTRALIA

... [Referring to] your Despatch No. 31

[#94-102] These paragraphs relate to the state of our stock generally, and refer to the cross of the Leicester bred sheep which we had enabled you to introduce. We shall attend to your suggestions relative to sending out more rams and also a succession of carefully bred bulls. There appears to us some little contradiction in this part of your Despatch which we cannot reconcile, probably from the omission of some work: in [#97] you say "the Leicester <u>cross</u> seem to have delicate constitutions" while in [#100] you say "the cross (alluding we presume to the same) however are healthy". [Margin: *cross should have been 'sheep' meaning those imported*].

London Minutes, 26 March 1841, AA Co 160/94:

[Received] Letter from John Wilkinson to H.T. Ebsworth (London Secretary), 22 March

I am sorry that the rams & ewes which you had from me sometime since have not succeeded so well as could have been wished. The description of those required to be sent, they were from my flock (which Mr Melville saw part of last summer) which is as healthy and with as good a constitution as any Leicester flock. Two of the rams were by a different sheep from the ewes but if they had all been by one they would have bred as regularly and produced as many lambs as if the rams had been from the one Leicester flock and the ewes from another. I never had better sheep (which is now about a year old) than at this time for size, symmetry, constitution and long wool of a fine quality. With respect to me purchasing a few Leicester rams for the Company, if it is their wish, I will endeavour to do so, but I should like in that case to have instructions from you what they would like and what price I might go to, if I could purchase what they want at a less price than you mention I should.

Ordered: that a description of the rams required be sent to Mr Wilkinson, and that he be requested to state at what price he can procure them.

London minutes, 23 Apr 1841, AA Co 160/94:

> Letters from Wilkinson, Lenton near Nottingham, dated 8 & 10 April, had seen Mr Burgess, a celebrated breeder of Leicester rams who would dispose of 2 or 3 of his finest and closest woolled rams at 20 guineas.
>
> Resolved: Wilkinson to purchase three of Mr Burgess's rams at 20 guineas, and that a few Leicestershire lambs would be required hereafter.

London minutes, 7 May 1841, AA Co 160/94:

Deputy Governor [John Studholme Brownrigg] authorised to by a few Leicester and Cotswold ewes and rams for the Company.

London despatch No, 36, 21 May 1841, AA Co 78/3/1:

We regret to find that so little success has attended the experiment with the Leicester sheep.

We have not, however, deemed it desirable, at present, to adopt the suggestions of Messrs Simes & Co detailed in their reports of the 24th June (now in your possession) and of the 22nd December, a copy of which is herewith forwarded (**A**).

In accordance with your recommendation, we have purchased from a very superior Leicester flock, three of the finest rams, which will be shipped per first opportunity. The experience of another season and the quality of the wool of the 2nd year's clip from the Leicester and cross Leicesters, will enable us to judge more accurately as to the extent to which it may be desirable to extend those breeds.

Three Leicester rams were purchased from John Wilkinson for £60 [London minutes, 21 May 1841, AA Co 160/94]. These were shipped on the *Hero of Malown* (under the care of Robert Williamson) and arrived in Port Jackson on 23rd December 1841.

In September 1841, King wrote to the Company Secretary:

> Be so good as to correct an error in my Despatch No. 31 [July 1840] at paragraph 91 – for 'Leicester cross' write 'Leicester sheep' – meaning the imported pure sheep.
>
> From the Court's Despatch No. 32 paragraphs 7 onwards it would seem that a report has been made to me of the wool of 1838-40 – none however has been received nor have any other information relative to the sale except that it realised an average of 1s/10d per lb. It would be satisfactory to know the opinion of the brokers & manufacturers upon the Leicester cross.[97]

At the same time that King wrote this requesting more information regarding the sale of wool, London sent the following dispatch.

London Despatch No. 41, 17 Sep 1841, AA Co 78/3/1:

> We have now to inform you that the wool per *Louisa Campbell* the clip of the Company's flocks in 1840 was sold by public auction on Wednesday the 25th ultimo at the following prices

341 bales PS washed
Average 1/11.25d

82 bales PR washed
Average 1/8.5d

18 bales unwashed
Average 1/0.5d

60 bales locks & fribs
Average 1/5

14 bales damaged
Average 1/6.5d

2 bales Leicester
1/7.5 and 8d
517 bales

A considerable improvement has taken place in the Leicester fleeces the Clip of 1840 having produced 4d to 5d per lb. more than the clip of the previous year.

In January 1842, King commented on a Report by Messrs Charles Hall (Superintendent of Sheep then at Goonoo Goonoo), C.F. Koelz (Woolsorter) and Edward Robins, Deputy Superintendent of Flocks at Gloucester):

You will observe that much difference of opinion exists between Mr Hall and Mr Koelz with respect to the necessity of importing fresh blood: Mr Koelz thing it unnecessary since the rams from the Peel will be a sufficient change. So far, he agrees with me but then we differ in the consequence of bringing the Peel rams to Port Stephens; he does not think that for the short time they will be required for tupping they will materially fall off in condition. To remedy this Mr Robins who agrees with me, proposes that the ewes should meet the rams half-way on the Dividing Range. But the objection to this arrangement in my opinion is great – because the ewes would so much dislike their old pastures after tasting the more nutritious grass of the higher grounds that they would fall off at the time when their strength is of the most importance. Mr Hall does not question the improvement that would ensue by crossing pure ewes at Port Stephens with his Peel rams but thinks that it would require a constant repetition.

Upon mature consideration of the reports of these gentlemen I am of the following opinion
That a change of blood is necessary for the Port Stephens' pure sheep.

That it will be desirable that the Peel rams should be sent down to tup them at the approaching season.

That the crossbred Leicester rams should at present only tup the Colonial ewes at the Peel.

That an importation of pure blood should be sent to the Colony from the most approved Anglo-Merino flocks – in the selection of which great care should be taken.

I do not agree with Mr Robins as to the crossing with South Downs, at any rate until the experiment of the Leicester cross shall be fully and fairly tried.

In the 21st paragraph of my Despatch 48 I informed you of the necessity for purchasing two Leicester rams which were then for sale in Sydney. I effected this a £20 a piece and I am happy to say that have turned out very good sheep. They were bred by a ram of Mr Spencer's at Snareston in Leicestershire and said to be descended from the celebrated breeder of stock Mr Green, Lesmarton Leicestershire. They were purchased by Messrs Hellpina & Co from Mr Thomas Pratt of Caldwell in August 1840.

The arrival of the 3 rams (one of which however died soon after his arrival although landed in apparent health) has provided us with an ample stock of rams, so that I shall be able to send some pure males and females to the Peel, the climate of which is certainly more adapted to that breed.

The increase of pure Leicester for 1841 is 4 rams and 2 ewes – got by the rams purchased by me in Sydney.

It is very much to be regretted that the season at the Peel should have been so unfavourable for the sheep of the cross-Leicester that were not from Port Stephens – but it has been a very trying season everywhere to the north of Sydney. On the whole the sheep here are generally in fair condition but I trust rain will fall before the season for putting the rams to the ewes or they will be in bad order – and what is worse there will be no grass for the winter.[98]

To this, the Directors replied in part:

... we are disposed to to adopt the conclusions at which you have arrived, namely 3rdly. That the crossbred Leicester Rams should at present only tup the Colonial ewes at the Peel.

The remarks of Mr Hall relative to the wool of the Leicester breed seem deserving of serious consideration for if his premises are correct, the deterioration in the quality of our wool would be most disadvantageous, & not to be compensation on the increased weight of the fleece and size of carcass. We doubt not you will have been careful in trying the experiment of the Leicester cross to confine it within the limits indicated in former dispatches until the results are sufficiently ascertained.

You will of course have acted on your own conclusion relative to sending the Peel Rams to tup at Port Stephen, & confining the cross-bred Leicester rams to the Colonial ewes.[99]

In April 1842, King noted that he had received the letter from Simes & Co, the London Wool Brokers, commenting on the Report:

'The crosses of the Leicester we think decided improving.'

and further comments:

It is satisfactory to observe that the wool from the cross-bred Leicester sheep has been sold at an improved price. The improvement also of the carcase as to age and weight is not less satisfactory. I have yet hopes that the cross may be persevered in with advantage.[100]

In June 1842, King wrote:

I am glad to be able to make a favourable report of the cross Leicester sheep as regards the increased weight of carcass. The wedders of the first cross now 3½ years old weigh 70 to 80 lbs which is double the weight of the pure merino wedders.[101]

In December 1842:

The report of the sale of the half Leicester sheep, as contained in Mr Henry Ebsworth's letter of the 23rd April is very satisfactory. The carcass of the first cross has very much increased. Some have weighed upwards of 70 lbs which is at least 30 lbs more than those usually weigh from which they have been bred. In good condition I think they will weight 75 to 85 lbs.[102]

Almost two years later, in October 1844:

I am afraid that the expectations that have been entertained for an advantage to be derived from crossing your flocks with Leicester rams will end in disappointed. When confined to the first cross a very great difference is manifest – the carcase being larger and the fleece more weighty but more open and coarse. The second cross produces a larger stapled fleece than the French merino, but being more open is not heavier – and the carcase not larger. The third cross approaches still nearer to the merino in quality but they are invariably coarse in the breech – uneven in size and the fleece lighter than the thoroughbred sheep. Mr Hall remarks, and I coincide with him, that the breed has been given a fair trial and no advantage appears to have been gained by it. After crossing so as to get the wool to a fair quality, a sheep is produced neither larger in carcase nor heavier in fleece – but with the disadvantage of an uneven fleece and invariably a coarse breech – an evil to be avoided as much as possible.

Mr Hall strongly recommends that the breed should not be persevered in – to sell the thorough bred Leicester sheep – and fatten and slaughter or boil down the 2nd and 3rd crosses. As I think that this recommendation is for the interests of the Company I have no hesitation in submitting it to the Court, but until I receive its decision I have desired that a trial be made by putting the pure merino ram to the thoroughbred Leicester ewe, as well as to a flock of the 1st cross; and see what will be the result. Hitherto the cross had been carried on the opposite way – the Leicester rams has been put into a selected flock of the best merino ewes.

In consequence of the climate at Port Stephens being too warm for these sheep, they have, with the exception of the pure Leicesters, been sent to the Peel.[103]

In July 1844, the Directors wrote of the need, at a time of drought and depression, to reduce the size of the Company's flocks by sales or boiling down for tallow – and to reduce expenses. They added:

We shall be glad to know whether the tendency of the Leicester sheep to fatten, tells in NSW as it does in this country, for now that tallow has become an article of export (and undoubtedly it will continue to be so) we conceive the Leicester breed will rise in value... and it may be worth consideration, how far it is desirable to promote the extension of that breed.[104]

King replied in December 1844:

(Court's Despatch 69 para 12). As to the Leicester sheep and their fitness for breeding for tallow I beg to refer you to my Despatch 107 (paras 42 to 44). I shall cause a trial to be made but I believe it will be found that the Merino is a more profitable breed for that purpose. The climate and more nourishing artificial food which they get in England are more favourable than those of this Country for this class of sheep.

As regards the wool the experiment has had a fair trial, and it is found that after the first cross the carcass is no longer than the Merino's. We shall see whether the richer pastures of the Peel will cause the animal to lay-on more tallow than the merino. If it does it may be worth consideration whether the Peel should not be kept for that particular breed. It will not do to have both breeds at one place, for it is not possible, on account of the negligence of the shepherds to prevent their becoming mixed.[105]

By the same mail, he wrote the Company Secretary:

> I yesterday had the pleasure of receiving your letter of 30th July with intelligence of the sale of our Port Stephens clip which I am delighted to find seems to have given satisfaction to the Court – at least so far as it is compared with the sale of the former year.
>
> The improvement seems to have confined itself principally to the low wools – and how the Leicester came to be in such demand and risen in value above others is unaccountable – unless the manufactories have been working up more coarse than fine wool. I trust their speculation will not meet with disappointment or we shall suffer – I am always afraid of these <u>rises</u> above the common market price.[106]

In September 1845, King wrote, commenting on the 1845 Clip being sent to London:

> The lamb's wool is an experiment to ascertain whether the lambs will not produce as much, or nearly as much at the end of the year as if they had not been shorn; and also to ascertain whether it will not be profitable always to shear

them at six months old. At Port Stephens 2,075 Saxon and French lambs were shorn and produced 2,277 & 1,941 lbs which afterwards hand-washed was reduced to 1,230 lbs. In the grease the average was 1.097 lbs. 140 first cross Leicester and Merino lambs produced (in grease 190 lbs) 140 lbs of hand washed wools.[107]

In August 1846, King reported that, of the 3,542 sheep boiled down for tallow, 570 had been Leicester rams, and 1,251 Leicester ewes – before they lost condition.[108]

No further reference until 1864, when the General Superintendent, Edward Christopher Merewether, wrote that he was:

> '... well aware of experiment to cross Saxon merinos with Leicesters and other coarse sheep. more than once tried here but not successful. Given type of wool he aims for, had no plan of resorting to such a violent cross—just meant to indicate did not intend to cultivate fineness and expense of quantity.'

By this time, all the sheep had been removed from Port Stephens, Goonoo Goonoo was now the property of the Peel River Land and Mineral Company, and Merewether was engaged in re-stocking Warrah.[109]

While the AA Co only made two importations of Leicesters, the use they made of them was to have an important influence on the Australian Merino. Today the AA Co is purely a cattle operation.

The importing of both sheep and cattle into New South Wales was prohibited for many years up until 1888, due to the fear of introducing 'scab' and other diseases. When the ban was lifted it resulted in the importation of large numbers of sheep. In 1891, 354 American Merino, 60 German Merino and 18 English stud sheep as well as 14,026 stud sheep from other parts of Australasia entered New South Wales. In 1891 it was estimated that there were 220,700 Leicester sheep in New South Wales, of which nearly 80,000 were ewes. This compares with there being 150,000 Lincoln ewes and 293,800 long-woolled/Merino cross-bred ewes.[110]

W. & W.T. Grant, Booloocooroo, Curlewis. Flock No. 25

Mr Grant, who came from New Zealand, ran three large studs on the 8,800-acre Booloocooroo property which were all based solely on New Zealand blood lines. They were English Leicesters, Border Leicesters and Romney Marsh.

The English Leicester flock (ALSA Flock No.25) was purchased from the estate of the late Mr J. Fleming Douglas in 1909. It was founded in 1907 based solely on New Zealand blood lines from Grant, Nixon, Threlkeld and Reid. Mr Grant only ever used New Zealand rams, other than some that he bred himself. Up until the stud was dispersed in 1946, Mr Grant imported some 75 stud rams from New Zealand, as well as one consignment of 113 ewes. It would have been the only Australian stud solely based on New Zealand rams and ewes, in essence a New Zealand stud based in Australia as no Australian ewes or rams were introduced.

Mr Grant also purchased the Douglas Border Leicester stud which was also founded in 1907 (Flock No.10). This stud was also solely based on New Zealand blood lines with at least 102 stud rams and 51 ewes being imported from New Zealand during its existence.

Likewise, the Romney Marsh stud (Flock No.85) was also solely based on New Zealand blood lines with at least 39 rams and 176 ewes being purchased from New Zealand.

Mr W.T. Grant's cousin was Donald Grant of Timaru, New Zealand, from whom a lot of the sheep were purchased. Over the 28-year period that Mr Grant ran the Booloocooroo studs, he imported at least a total of some 556 stud sheep. This must surely be a record of importing stud livestock of any kind that will never be beaten.

Mr Grant said that he preferred the New Zealand type better than the English or Scottish types as they were closer to the ground and more compact.

Due to the Grant's decision to subdivide the Booloocooroo estate into smaller properties in 1937, the three studs were dispersed by auction in March of that year. A total of 3,500 ewes were put up for sale which consisted of 950 stud English Leicester ewes, 1,450 stud Border Leicester ewes and 1,100 stud Romney Marsh ewes. Also 103 stud rams were offered for sale with 2,700 young sheep being sold later that month. The Border Leicesters sold very well but the demand for the English Leicesters and Romney Marsh was very restricted with the bulk being passed in for private sale.

New South Wales has never seen the large numbers of registered studs or stud ewes that Victoria has had. In the early days, the majority of studs were registered with the Australian Longwool Sheepbreeders' Association, which is to be expected as this was a Sydney base association. Stud ewes peaked in 1916 at 1,856 ewes. Numbers remained at about this level until about 1949, when during the next three years there saw a sharp decline to approximately 300 ewes. Since then, there has only been very low numbers in NSW, with some periods of no registered studs.

H.K. Nock Champion ram. Flock Book for The Longwool Breeds of Sheep in Australia, Vol. XVII, 1928.

A.L. Bennett English Leicester Ram, "Oaklands", The Oaks, NSW. Flock No.1. 1912. Flock Book for The Longwool Breeds of Sheep in Australia, Vol. IV, 1912.

Scott Bros., Wellington, NSW. Stud ram 1915. Flock Book for Longwool Breeds of Sheep in Australia, Vol.VII.

English Leicesters at Wagga Wagga Experimental Farm 1910. State Library of New South Wales.

English Leicesters at Wagga Wagga Experimental Farm 1910. State Library of New South Wales.

F.W. Moffat NSW 1917

Sydney Royal Easter Show 1998 line up of Champions. L-R Judge Mr Geoffrey Halliday, Ethel Stephenson, Jan Brown, Peter Stephenson and Judy.

4
Western Australia

'Sheep without horns are counted the best sort, because so much of the nourishment does not go into the horns.'

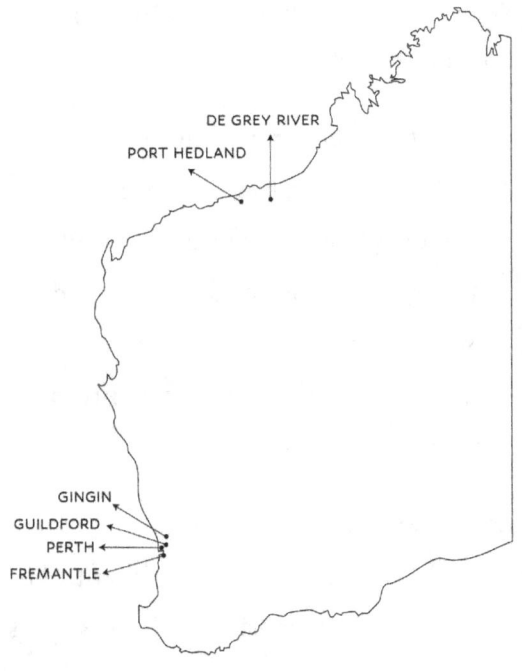

On 2 May 1829, at Arthur's Head on the southern side of the Swan River mouth, Captain Charles Fremantle took possession of the whole continent for Great Britain. The whole continent had not been included in the boundaries of New South Wales and therefore the western region was not prior to this time a British possession. The British were concerned that the French might form a settlement on this part of the continent and therefore control the sea lanes to Asia.

It was Captain James Stirling who persuaded the Colonial Office, with the backing of some capitalists, to form a colony at the Swan River. Stirling had visited the area in 1827. While Stirling was en route from England with the first settlers to form the colony, an act of British Parliament was passed which authorised him to make laws in His Majesty's Settlements in Western Australia. The use of the name 'Western Australia' for this new colony on the west coast of New Holland was the British Parliament's first official use of the name Australia.

The recorded livestock that were brought into the settlement in the first two years included 42 horses, 85 cattle, 471 sheep, 69 goats, 120 pigs, 44 rabbits and assorted poultry, dogs and a hive of bees. While it was common for the origins or breeds of animals not to be recorded, in this case sheep breeds mentioned were Merino, Saxon, Southdown and Leicester as well as Cape sheep. So this means that Leicesters were introduced directly from England sometime during 1829 and 1830.[111]

Van Diemen's Land also supplied early sheep to the colony with some being shipped there in its first year of settlement. These sheep were genetically based on the Bengals transferred from Norfolk Island to Van Diemen's Land before 1810, but they had been improved dramatically in the previous few years by the influence of British and Merino sheep. It is very doubtful if any of these early sheep were pure Leicesters, but there is the possibility that there may have been a Leicester influence in some of them. It was not until 1936 that the first documented pure Leicester's from Tasmania were sent to Western Australia.

Leicester's continued to be imported directly from England.

> 'Early in 1830, when there were only two wooden houses and no stone ones in Fremantle...From England they imported horses...cattle...Leicester rams and ewes, and superior pigs.'[112]

By 1832 over one million acres of land had been granted to settlers, but clearing the land proved extremely difficult and there were no convicts to do the work. Many farmers left the colony before they were ruined. In 1837 there were only 8,500 sheep in the colony.

The first stud to be registered in Western Australia was started in 1913 by William Padbury, Guilford. He purchased 30 ewes from H. Morphett, South Australia and 2 rams from Sir R.T. Clarke, Bolinda Vale, Victoria.[113]

A.W. Edgar & Co established a stud in 1916, also based on Bolinda Vale bloodlines. In the early 1870's Mr Edgar came to WA from Victoria and was one of the pioneers of the North-West. He was instrumental in opening up the De Gray River country, north of Port Hedland. In about 1899 he moved south and purchased Strathalbyn at Gingin. There he established his English Leicester stud as well as a noted Shorthorn stud. Mr Edgar also served on the council of the Royal Agricultural Society.

W.G. Spencer established his flock in 1921. All these early studs were registered with The Australian Longwool Sheepbreeders' Association.

In 1966 Mr Mervyn Reynolds, Wimbledon farm, Meckering, established an English Leicester stud, one of six British Breed studs, plus a Poll Dorset stud that he ran on his property. This is perhaps a record for the number of British Breed studs run by one person. Mr Reynolds began breeding with a Southdown stud in 1953 and then bought a Dorset Horn stud in 1962 followed by a Poll Dorset stud in 1963. He later established Shropshire, Suffolk, English Leicester and Romney Marsh studs running a total of 1,900 stud sheep. In 1966 Mr Reynolds exhibited 83 sheep at the Perth Royal Show winning awards with six of the seven studs.[114]

The last English Leicester stud ram to be imported into Australia before live imports were banned from England came to Western Australia. It was imported by F.N. Everitt (F140), Duranillin, in 1951.

Strong numbers, both in the number of studs and ewes, were maintained right up to 1970. This is in contrast to the other states that saw sharply reducing numbers twenty years earlier. From 1991 until 2014, when Peter and Jo Gelmi founded their stud, there were not any registered studs in Western Australia.

THE HISTORY OF ENGLISH LEICESTER SHEEP IN AUSTRALIA

Western Australians, Mr. and Mrs. John Scone with one of the English Leicesters they took to Melbourne by road, for exhibition.

Mr & Mrs John Scone from Western Australia with their sheep at the 1962 Royal Melbourne Show. Australian Stud & Farm, October 1962. Photo credit: Stock Journal.

5
South Australia

'The quality of a ram can usually be determined from his conformation and from his get.'[115]

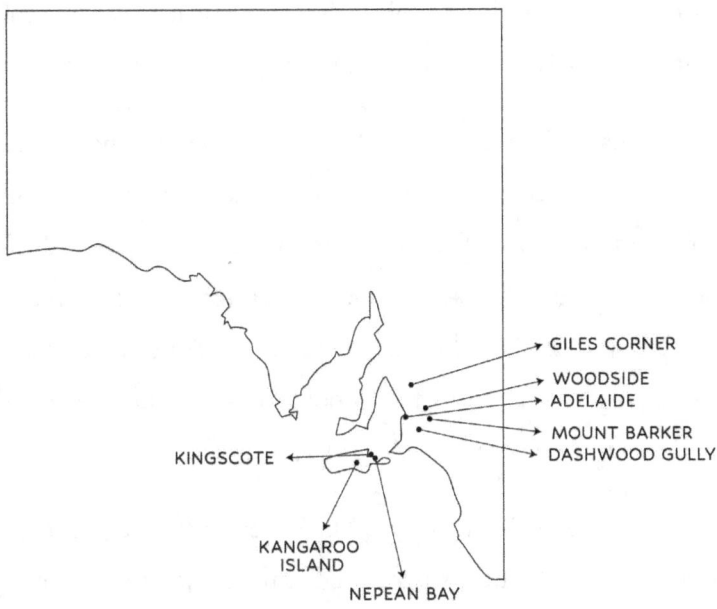

South Australia came into being in 1834 when the British Parliament passed the South Australian Act, 1834, but the colony did not really begin until 1836 when the first settlement by The South Australian Company was established on Kangaroo Island. This site soon proved to be unsuitable and it was moved to the mainland.

The idea of a settlement being wholly for free and orderly people and therefore free from being a dumping ground for British convicts was being discussed as early as about 1830. The first meeting of the friends of this new movement was held at Mr Tooke's office, 39 Bedford Row, London, on 3 August 1831. Mr Torrens, afterwards the Chief Commissioner of the colony, presided at that meeting. The proponents did not want convicts and wanted to avoid the vast squatting system that had developed in New South Wales. They envisaged a settlement based on agriculture rather than the pastoral system. South Australia was thus instituted as a free colony of agriculturalists largely based on the English system of stock keeping.[116]

The idea was that all free settlers would purchase the land they would farm. This was good in theory but by the time the settlement had come into operation the squatting districts of Victoria had spread west. Who would therefore want to purchase land in South Australia when it could be freely taken in nearby Victoria? Thus, by necessity the South Australians took up the squatting mania, resulting in wool production superseding the less profitable agricultural base.

'The South Australian Company' (9 October 1835 – 17 March 1949), also referred to as the 'South Australian Company,' was formed in London on 9 October 1835, after lobbying by the South Australian Association. The founding board, headed by George Fife Angas, consisted of wealthy

British merchants in order to develop a new settlement in South Australia; its immediate purpose was to encourage the purchase, in advance, of land in the planned colony of South Australia. The South Australian Company ended business in its own right on 17 March 1949 when it was liquidated by Elders Trustee & Executor Company Ltd, which had been managing its Australian affairs since the death of the last Colonial Manager, Arthur Muller in 1936.

In the first report of the Directors of The South Australian Company, it is stated that the fourth object contemplated by the Company was the Growth of Wool.

> *'As a beginning, they have purchased a very fine lot of rams and ewes of the finest and purest breed of Merinos, which were selected with great care and at much expense in Saxony, by the son of a great sheep-holder of Van Diemen's Land for his own stock.*
>
> *They have also purchased and sent out, in their different vessels, a supply of pure Leicesters and South Downs; and for the further supply of this kind of stock, they look to the neighbouring colonies of Van Diemen's Land, New South Wales, and elsewhere, as they can be best and cheapest supplied.'* [117]

The diary of John Brown, who sailed out on the *John Pirie* gives an interesting insight into the work involved in looking after the sheep and other livestock once they had arrived in the new colony. This diary starts eight days after the *John Pirie* arrived at Nepean Bay.

August 24th 1836

> Four of the Company's labourers and myself were sent up to the Salt Lagoon, for the purpose of erecting Sheds, and making a Fence round two small plots of Ground, for the live Stock which have all come up here, there being plenty of both Water and Grass at present, and where it is intended to establish a permanent Station.

Salt Lagoon, Kangaroo Island, South Australia. Brenton Heazlewood.

August 29th

On Saturday Evening all the Men went down to Kingscote for their Wages and fresh supplies of Provision, leaving me quite alone until this Afternoon (Monday) when two of them came back named Bates and Powell, with a few Sheep, Pigs, & grey Peas, but the other two Men called Jones, who are Brothers (and were brought out, in the 'John Pirie') refused to come with the 'Stock', because Mr Stephens would not allow them to bring any Spiritous Liquours. The Man Bates mentioned above, has been 13 Years on this Island, and is a very active, civil sort of Fellow.

August 30th

This Afternoon all the Merino Sheep came from Kingscote but have been most dreadfully ill used by the Persons who had charge of them across the Bay to this Station, indeet two Rams, 1 Ewe, and a Lamb, are nearly lifeless, having entirely losss'd the use of their Limbs, by being roughly drag'd a very considerable way in the Salt Water from the Boat to the Beach, after which not being able to stand, they were carried up here in the most careless manner, as though of no value, nor had any feeling There was also 2 Boars, & a Sow, in the Boat, but they only brought 1 Boar, up to the Station, having loss'd the other two, in the Woods.

August 31st

I had the three unfortunate Sheep & Lamb, laying by the Fire all Night, and find the Ewe & Lamb, considerably better this Morng being able to stand upon their Legs, and eat a little choice Grass, but One of the Rams is dead, & the other not much better – I therefore must attribute the loss of this fine Animal to be caused by the greatest negligence of the party, who brought them from Kings–cote Yestdy, as even the poor Things that could walk, were completely saturated with Salt Water and unable to eat a bite of any kind of Fodder on their arrival here.

September 10th

During the Night, the other poor Merino Ram that came up from Kingscote in such a miserable condition died, His Loins were very much bruised, and Kidneys swell'd, which I have no doubt was caused by the rough usage he experienced on his Journey to this Station the 30th Ult.

September 20th

... The Merino Sheep having strayed away Yestdy Eveng, G, Bates and his two Women have been employ'd seeking them all this Day, without success.

September 21st

The Merinos were found this Forenoon in the Woods, at a considerable from here, and on their arrival were secured by Cords to Tethering Irons ...

September 24th

Powell and Chandler have been engaged all Yestdy & this Day, strengthening the Sheep Sheds, and commenced putting a good strong Fence round the two Paddocks, as the temporary One that was made at our first coming here, have become of no use whatever tokeep in the Sheep – G, Bates, has been employ'd since Thursday assisting me with the live Stock, and occasionally at the Fence.

September 25th

Chandler was employ'd this Forenoon among the live Stock and Powell at the new Fence – G, Bates arrd from Kingscote at Noon, and bringing with him a few Rope Yarns.

October 1st

This Morng I got G, Bates to Saw off part of a Horn, from One of the Merino Rams, in order to get at a Wound that was underneath it, and which the Animal recd on the first Day of being Tethered, by plunging about to get his liberty, when the Cord slip'd under his Horn and cut his Neck severely, since which time it has contd getting worse, on acct of not being able to apply any remedy , for the above mentioned piece of Horn being right upon the part most Wounded – Powell and Chandler have been engaged at the new Fence, Yestdy and to Day, while G, Bates has been employ'd among the Stock and sometimes at the Fence, during all the Week, but this Eveng left us to accompany Mr Stephens, on a visit to the Main-Land, with which he is well acquainted.

October 15th

The Men, have been engaged all this Week at repairing our Sheep Sheds, and helping me with the live Stock.

October 22nd

The two Men have been employ'd all this Week at making a new substantial Fence round the Paddocks and attending to the Stock – on Tuesday last I had all the So Down Sheep shear'd, and gave them a complete dressing with Ointment, as they were very bad indeed, of the Scab, but think they are now cured.

October 29th

Chandler has been assisting me all this Week with Stock, and occasionally at the new Fence – Powell did not go to Work untill Thursday – On Tuesday last, two of the So Down Ewes, stray'd away and were lost untill this Day, when we found them in the Wood near to the Beach.

November 5th

On Sunday last One of our larger Sows was found dead, She was going about as usual the previous Day, but in miserably lean condition – Chandler did not go to Work untill Wednesday, during which time Powell assisted me with the live Stock, and commenced building a Goose House that was finished Yestdy.

November 12th

On Wednesday last, 16 Wedder Sheep, were sent from Kingscote to this Station, but on Acct of the high blustering foul Wind, they were unable to reach this place, and were obliged to be landed at North Cape, during the Afternoon in the most terrible plight, being half drown'd, and not One of them able to travel from the Spot on which they were landed, The following Morng found 7 of them dead, 2 missing, and 7 alive, which latter, with great difficulty were brought to this place, some of them so very Weak, had to be carried by the Men, a considerable part of the way, which is a distance of 5 Miles, along the Sea Beach – We have searched amongst the Black-wood ever since to find the Missing Ones, but without the least success, so that I have no doubt but that they are dead – One of our little Sow's was found dead this Morng with its throat uncommonly swell'd.

November 19th

On Monday last we commenced shearing the Merino, and Leicester Sheep, putting the Wool of each sort, into a separate Sack, and on Thursday 3 of our Merino Rams had their Horns partly sawn off, as they were beginning to grow into their Heads, in a very dangerous manner indeed – During the Week our Men have been engaged building two more Pig-houses, which are finish'd.

November 26th

One of our wretched Wether Sheep, was found dead, on Monday Morng last, and the other six, were stray'd into the Bush, since which time Chandler has been engaged looking for them and a little Boar, that has also gone astray – since Thursday he has been assisted by Powell (who only commenced work on that Day) – They succeeded in finding al the Sheep, but 2 out of the 6 were dead, and the remaining 4 are the picture of misery – While searching the Beach on Thursday, the also found our large white Sow being dead, with 2 small Pigs, that She had litter'd, – This Sow has been in the habit of going between here and North Cape, for a Month past, and it appears had brought forth young in the Bush, about 20 yds from the Beach, leading to that place – I am very sorry we had not the means of confining here, for want of Hog troughs, untill it was over late.

December 3rd

The Men were employ'd the beginning of the Week, in digging 3 Wells of 6 or 7 Feet each in depth, but got nothing except salt Water in all of them – On Wednesday I recd orders from Sml Stephens Esqr C,M, to get the Stock together in readiness for departing to the Main-Land, by the Brig Emma, Capt Nelson, who would take them on board the following Day or Friday at latest, We therefore on Thursday housed all the Ewes And a Ram lamb of the So Down breed, but which was exceedingly ill and died within an Hour after being brought Home, the causes of his Death in my opinion, is from being for a length of time obliged to live upon very unwholesome Food, and brack'd Water as several of the full grown Sheep, have likewise been very unwell during the last Week, and all of them are greatly falling off in condition for the Grass is so dry and burnt by the Son, that they will nor eat it, but prefer green Leaves & Twigs of the various kinds of Trees & Shrubs, which are growing in the Woods about the place, and have no doubt, that many of them, are of a poison – ous Nature – The Merino Ram lamb was not found with the Ewes as usual, nor have we been able to get sight of him since, although the Bush has been carefully search'd, both Yestdy and to Day, where ever the Ewes were known to feed, or frequent for shelter in the Woods, during hot Weather – We have all the Pigs, except a little Boar which has been missing for the last Fortnight, and a large black Sow that stop almost continuely at North Cape – There has been very little Fodder at this Station for a Week past, and we are now without any whatever, so that the poor Sheep have nothing to subsist upon while confined (waiting, for the Boats coming from Kingscote, to take them away) except the poor dried Grass, that can be collected about the place, which is miserable fare indeed.

December 5th

Mr Beare accompd by a lot of People, arrd this Morng to take away the Stock &c, but on Acct of the very high Wind that was then blowing, they only took away a few Stores, thus leaving the Sheep to hunger another Day – Tuesday Dec 6th The Boats came up this Morng and succeeded in taking away every thing that could be muster'd here, our Live Stock, consisted of 4 Merino Rams, 2 do Ewes, 2 Leicester Rams, 6 do Ewes, 2 So Down Rams, 1 do Ewe, and 3 Wedder Sheep, besides 3 Boars, 7 Sows, 2 small Pigs, and 10 Geese, – Unfortunately 2 of our So Down Ewes and 1 Leicester Ewe died, this Morng, which I can attribute to no other causes but the want of substantial Food, for the Grass they have been living upon since Thursday last, is so dry & burnt by the Sun, that very little nourishment indeed, can be obtain'd from it, and several of our Sheep have been uncommonly Weak in consequence of being so long confined, upon such miserable stuff, as we hourly expected the Boats arrival ever since that Day, to take them from this place of starvation.

December 7th

We got all the Stock and Stores, from the Salt Lagoon Station, safely on board of Ship last Eveng except 1 Wether Sheep, & 1 black Sow, both of which got away from the Sailors that were taking them to the Beach also a little Boar that has been missing for some time past and the Merino Ram lamb, which we have not been able to find, but Chs Powell is to remain at the Station, and therefore will probably secure them again – During this Day we have rec'd on board, an additional supply of Stock from Kingscote, to take with us to the Main Land, they consist of 23 Wedder Sheep, 2 Working Bullocks 2 Saddle Horses, 1 Sow, & 4 small Pigs, as likewise a quantity of Horses, and will proceed to Sea in the Morng should the Wind & Weather permit.

December 9th

We left Kingscote early Yestdy Morng and arrd in Rapid Bay this Eveng at which place 9 Wedder Sheep were landed for Colnal Light, – One of the Leicester Ewes died Yestdy and a Merino Ewe is uncommonly Weak – Sunday Dec 11th Having remained at Anchor all Friday Night in Rapid Bay, we weighed again early Yestdy Morng, and arrd at our destination Hold fast Bay this Eveng thus making 4 Days Works of a 4 Hours Passage and have every prospect of a heavy Gale of Wind, to keep us on board a Day or two longer.

December 16th

On Monday last, the Wind blew so very strong that not an Article could be landed, but during Tuesday it abated considerably, when part of our Stores were landed, and have got every thing on Shore this day, except 2 Flour Barrels, that are either lost, or have not been put on board at Kingscote as we can only get two instead of four, which are mentioned on the Invoices – While bringing the Sheep ashore on Wednesday last one of the Wedders unfortunately got his leg broken & was therefore killed and sold by Capt Martin – The grey Mare was deliver'd up to Mr Gilbert, who will take charge of her untill Mr Morfit arrives.

December 19th

Early this Morng and during Palmers Watch, all our Sheep got out from the Park netting, but were found during the Day scatterd about in every direction, and 3 of then were dead, which no doubt have been killed by the wild Dogs, they consist of the weak Merino Ewe, a Leicester Ewe, and a Wedder, the So Down Ewe is likewise bit in several places, but think not dangerously.

December 21st

Both Yestdy and to Day, have been employ'd Shearing the Wedder Sheep and afterwards rubbing them as likewise all the others, with boil'd Tobacco, they being much affected with the Scab.

December 25th

During last Night, our Mare broke her tethering Rope, and stray'd away unperceived by those on Watch (Chandler & Palmer) and has not been seen by and Person all this Day.

December 27th

This Morng out large black Sow, was found dead near the Tents, – She has been shot in the left shoulder apparently with a Ball, by some Person unknown, this Sow was heavy in young, there being 8 fine [?] Pigs nearly full grown, found inside of this poor Animal, – Likewise One of our white Sows has recd a severe wound behind the left shoulder during this Forenoon, which has been done by a Spear or other Weapon of that Sort – All the Swine were housed last Night, except the black Sow which we could not find, – The Mare has not yet been found although She has been searched after in every direc – tion for 6 or 8 Miles round.

December 28th

I have been Out all this Day, looking for the Mare, accompd by 3 of our own Men, and 2 other Persons who volunteer'd, we went in two different directions towards the foot of the Mountains but have not been success – ful – During last Night one of our white Sows litterd 8 Pigs, and the So

Down Ewe, brought forth a fine strong Ewe lamb, all of which are doing well – About Noon of this Day, the Ship Buffalo anchor'd in Hold fast Bay, she has on board "Capt Hindmarsh" our Governor &c &c &c.

December 30th

The Mare was searched for Yestdy again in vain, but this Forenoon I accompd by 2 of our Men, found her in an excellent pasture about 5 Miles from this place, by the bottom of the Mountains, and improved in her condition.

January 11th 1837

On Monday Morng last, a Lamb, that must have been Dead for some time, being partly rotten was taken from one of Leicester Ewes – she is now quite well again – And this Morng the Mare again broke her tethering Rope, & has not been seen since – The 2 Bullocks & her, are under the especial care of Mr Stuart.

January 14th

We have lost 2 of the white Sows little Pigs, during this Week, by the Mother being down upon them at Night, and we have heard no Acct of the Mare yet.

January 28th

Our Bullocks have been employ'd all the Week in assisting to get Mr E, Stephens Property from the Beach to his Tent, as also 2 or 3 of our Men occasionally – On Wednesday 4 small Pigs belonging to Mr E, Stephens came up to this Station – but one of them was quite exhausted from the "ill" usage it had recd on the Journey – & died very shortly after arriving On Thursday we recd 1 Boar, 3 Sows, 7 small Pigs, 1 Milk Cow, and 4 Shepherds Dogs, from the Brig "Wm Hall", Lately arrd from England, all of which are in good health & condition – The Mare has never yet been found – One of our Sheep (a Wedder) has also been missing since Wednesday, and was supposed to have got amongst the Commissioners Flock, but as we cannot find it there, I have little hopes of seeing him alive again – Several of their Sheep, which lately came from "Hobart Town" bt the Barque "Africaine" have died since being landed.[118]

To establish its base flock of ewes the company turned to Van Diemen's Land and continued to buy and ship sheep for a number of years, with it being the main early source between 1836 and 1839.[119] These early base ewes would have most likely had some degree of Leicester infusion in them as most of the Merino breeders in Van Diemen's Land at that time were using Leicester rams.

W.H. Dutton

Another early influence of the Leicester in the development of the South Australian Merino came through W.H. Dutton. Dutton arrived in Australia in 1826 after an apprenticeship under one of Europe's greatest Merino studmasters and agricultural scientists, Albrecht Thaer. In 1839 Dutton visited South Australia, where at Mount Barker he bought a part share in a 4,000-acre Special Survey. In 1840 Dutton and Finnis arranged for two large drafts of nearly 16,000 sheep to be brought overland for speculation in Adelaide. Due to the depressed market at the time these sheep were not sold. In the second mob brought overland were a large draft of half-bred Leicesters.

J.B. Hughes purchased some of these half-bred Leicesters from Dutton.

> *'I saw that Mr Dutton's were larger, both in fleece and carcass ... I therefore paid him a high price for forty rams...and I was subjected to much ridicule because I selected, firstly, some half-bred Leicesters and then the largest sheep with largest fleeces.'*[120]

Charles Massy states that the conscious use of English long-wool breeds, Cotswolds and especially Leicesters for carcass qualities, and the added bonus of increased staple length and heavier wool cuts, was a significant factor in the evolution of the South Australian Merino.[121]

Sheep numbers rose fairly quickly in South Australia, from 28,000 in 1838 to 6.1 million in 1877.

Fat long wooled ewes, shorn, first prize SA 1870-1875. State Library of South Australia.

Fat long wooled wethers, second prize SA 1870-1875. State Library of South Australia.

Leicester & Merino ewes and lambs, SA 1870-1875. State Library of South Australia.

A.S. Fotheringham

The first English Leicester flock to be registered in South Australia was that of Mr A.S. Fotheringham, 'Hillyfields', Dashwood Gully. He registered the flock with The Australian Longwool Sheepbreeders' Association and is recorded in volume 1 with Flock Number 2. It states that the flock was founded in 1903 with the importation of 6 ewes and 2 rams from New Zealand. Importations from New Zealand were nearly an annual occurrence for this stud as in 1904, 40 more ewes were imported; 1905, 3 rams and 50 ewes; 1906, 4 rams and 200 ewes; 1907, 3 rams and 28 ewes imported.[122]

In 1909 Mr Fotheringham exhibited, at the Adelaide Show, an English Leicester/Merino crossbred ewe with a two month old lamb at foot, the lamb being by a Leicester ram, in other words a three-quarter-bred Leicester.[123] This was the popular New Zealand freezer lamb, the Canterbury lamb. Mr Fotheringham was displaying the suitability of the Leicester to cross with the South Australian Merino ewe to produce an export lamb.

In February 1911, due to the death of his father, Mr Fotheringham had to sell the stud. At the sale 620 ewes and 212 rams were offered for sale. The rams sold from 30 guineas and the ewes from 12 guineas each. Among the buyers were Messrs. W.B. Hill (Terowie), J.D. Johnston (Oakbank), J.G. Habel (Hamilton, Vic), A.J. Walkley (Dashwood's Gully), W.S. Kelly (Giles Corner), D.M. Aitken (Cranbourne, Vic), W. Padbury (Guilford, WA), D. McArthur (Vic), W. Richardson (Woodchester), and S. Hutchinson (Vic).

Mr A. Francis purchased the estate in May, 1912, and carried on with English Leicesters. Although he did not register a stud until 1915 (Flock No. 26), he had by 1913 sent 40 flock rams to Western Australia and 25 to Victoria. In 1913 he also imported 8 stud rams and 4 ewes from New Zealand.

Francis first Leicester ram, SA 1870-1875. State Library of South Australia.

Five Leicester rams "Francis" SA 1880. State Library of South Australia.

W.S. Kelly

Another early registered South Australian flock was that of W.S. Kelly, Merrindie, Giles Corner. He established the stud in 1907 by purchasing 10 ewes from P.C. Threlkeld, New Zealand, and a ram from Lincoln College. Further sheep were purchased from R.G. Heazlewood, A. Oliver & Sons and A.S. Fotheringham. The stud was first registered with the Australian Longwool Sheepbreeders' Association in 1911 as Flock No. 29. Mr Kelly also ran Merino and Shropshire sheep and exhibited these along with the Leicesters. He also imported 2 Dorset Horn rams from England in 1923. His English Leicester stud was dispersed in 1927. He was involved in community affairs, being a Methodist lay preacher and also Chairman of the South Australian Advisory Board of Agriculture. In 1920 Mr Kelly published the book, 'Beef, Mutton and Wool', a practical handbook on meat and wool production for the Australian farmer.

THE HISTORY OF ENGLISH LEICESTER SHEEP IN AUSTRALIA

W.S. Kelly South Australia Imported Ram 1924

J.H.W. Mules

It may not be widely known that there is a connection between the developer of the mulesing operation, Mr J.H.W. Mules (1876-1946) and English Leicesters.

In 1931 Mr Mules, while living at Woodside (SA) set out to minimise the blowfly strike in sheep by removing the folds and breech on Merino sheep. On August 10, 1931 he demonstrated his method to Sir Charles Martin (Chief of the Animal Nutrition Division of the CSIRO) and Dr L.B. Bull who were both impressed by the possibilities this operation had in reducing the costly loss to fly strike the sheep industry suffered.

Koonamore and Melton stations in the north-east of South Australia were the first to try out the operation.

Mr Mules was born in 1876 at Semaphore and in the 1890's worked on the Oraparinna and Coongy stations. He later farmed at Strathalbyn, Grenfell and Ideraway (NSW), Emerald and Keppel Island (Qld). In 1916 he returned to South Australia and managed Bon Bon Station and later Arcoona. He then bought Three Creeks in the Flinders Ranges and then property at Woodside and Teal Flat on the River Murray.

It was while he was farming at Ideraway in 1906 that he imported 85 Leicester sheep from New Zealand from which he hoped to breed a good type of ram for producing crossbred lambs from Merino ewes. He exhibited the English Leicesters that he had imported at the Toowoomba and Brisbane shows in July and August of 1906 so those who were interested 'would have an excellent opportunity of examining the new importations'.

In 1936 Mr Mules is again using the English Leicester when he set out to produce a new breed of sheep. To do this he crossed a pure Peppin Merino ram with pure bred English Leicester ewes. The English Leicesters were chosen because of their good mothering and he hoped they would increase the milk yield of the female sheep of the new breed. The breed would be hornless and the wool was expected to be of bright colour and full of lustre. I have not been able to find any reports on the success or otherwise of this venture, but I would assume that nothing came of it.

In 1940 Mr Mules established an English Leicester Stud (F165) by purchasing a ram and ewes from J.B. Stephenson and H.H. Shillabeer.[124] This stud was listed as dispersed in 1943, so was only in existence for a couple of years.

While Mr Mules would have been a true Merino man, his interest in and use of English Leicesters continues to demonstrate the strong connection that these two breeds have had in developing Australia's sheep industry.

Not only was Mr Mules an innovator in the sheep industry but he was also interested in introducing new pasture species to help improve pasture production.

In 1935 while at Woodside he announced in an advertisement in The Mount Barker Courier that he had been appointed the sole South Australian representative of the New England Pure Seeds Association, of Glen Innes, NSW, and as such has a small quantity of Phalaris seed for sale.

There is no doubt that Mr Mules was an innovator in the agricultural scene of his time. Whilst the mulesing operation which he developed has now come in for criticism from animal welfare groups, it cannot be denied that it has saved the lives of many thousands of sheep over the past 100 years.

Like many innovative sheep breeders over the years since Bakewell bred the Leicester, Mr Mules saw the potential that the Leicester had to help him achieve his sheep breeding aims.

The number of registered studs in South Australia has fluctuated over time. The strongest period being in the 1930's and 1940's when up to 11 studs were registered running up to 560 ewes. From the beginning of the 1950's there was a steady decline in numbers, with no registered studs between 1967 and 1980, when there was a small resurgence. There have not been any registered studs in South Australia since 2012.

W.J. Jenkin (SA), Adelaide Royal Show Champion 1949.
Australian Farm & Home, July 1949.

W.J. Jenkin (SA), ewes. Australian Farm & Home, July 1949.

Royal Adelaide Show 1997. Champion English Ram exhibited by David Parker, being sashed by Beverley Heazlewood, wife of the judge. Brenton Heazlewood.

Paringa, South Australia advertisement. Australian Stud & Farm Monthly, 1958.

*A.A. Hill, Yahl (SA), Adelaide Royal Show Champion 1951.
Australian Farm & Home, October 1951.*

V. Smart, Kingston, South Australia. Champion English Leicester Ram, Adelaide Royal Show 1961. Australian Stud & Farm, September 1961. Photo credit: Stock Journal.

6
Victoria

'Of all the animals that have been domesticated by man, none has rendered him more essential service than the sheep. A large part of the food and clothing of the civilised world is supplied by this useful animal. The culture, improvement, and manufacture of its fleece, have constantly accompanied and marked the progress of civilization both in ancient and modern times.'[125]

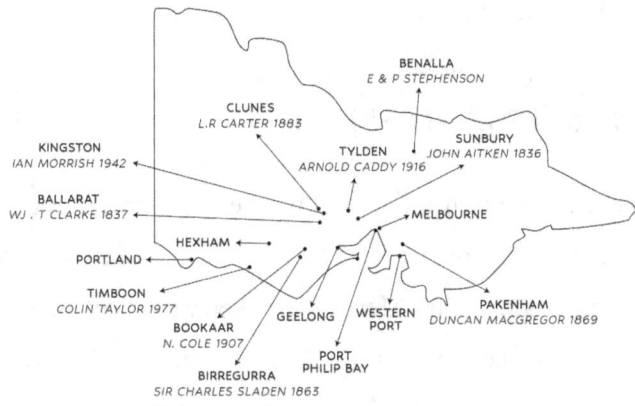

Since about the 1920's, Victoria has been the powerhouse for English Leicester sheep numbers in Australia. The first two registered flocks in Victoria are recorded in Vol. 1, of the Flock Book for British Breeds of Sheep in Victoria published in 1898. These belonged to Mr L.R. Carter, Scale Park, Clunes, Flock No.1 with 50 ewes put to the ram in 1897, and Mr D. McGregor, Dalmore, Pakenham, Flock No.2 with 550 ewes put to the ram in 1897. Both these flocks were based on Tasmanian blood lines which was, at that time, the main source for sheep. It is interesting to note that Duncan McGregor imported 6 rams from New Zealand in 1887, a country from which, up until recent import bans were imposed, present day stud breeders in Victoria were still sourcing outside genetics.

It is very difficult to say with certainty when the first Leicester's would have arrived in Victoria. An article in *The Queenslander*, September 1910, announcing the arrival in Brisbane of two rams and several ewes purchased by Mr C.H. Grove from R.G. Heazlewood, Tasmania, states that Mr Grove's father was the first to introduce English Leicester sheep into Victoria. I have not been able to locate any further evidence to substantiate this claim. I believe that it was either John Aitken or W.J.T. Clarke who would have been the first to take Leicesters there.

John Aitken (1792?-1858)

Not a lot is written about John Aitken, but he was a very influential pioneer Merino breeder in the Port Phillip District. He was also a user of English Leicester sheep to improve his Merino's. Aitken was a Scottish farmer's son who arrived in Van Diemen's Land in 1825. He was farming near Oatlands in 1833-34 and was at this time proposing to take stock to the Port Phillip area, but nothing came of this proposal. It was on 20 July 1835 that Aitken first left Launceston in the sloop *Endeavour* to inspect the Port Phillip country. This was less than six weeks after John Batman returned from his first trip across Bass Strait, and the day before the schooner *Enterprise* began her first trip for John Pascoe Fawkner.

Aitken returned to Launceston on 29 August and made plans to take sheep back to the Port Phillip District. Edmund and Francis Bryant selected the Merino sheep for Aitken to take, with the three of them in partnership with the sheep. At this time Edmund and Francis Bryant were running Merino's and English Leicester's on their properties, as were most Merino breeders.

On 22 March 1836 some 1,600 sheep were loaded on the brig *Chili* at George Town. It was a slow disastrous crossing to Port Phillip, not arriving until the 1 April. About half the sheep were lost during the voyage due to overcrowding and lack of water in the hot calm weather. The *Chili* ran aground on a sandbank under Arthur's Seat, and it is said that Aitken and his men had to carry the remaining sheep ashore one at a time.

After the sheep recovered Aitken drove them to the Gisborne – Sunbury district, becoming the first settler there. Governor Bourke named it 'Mount Aitken' when he visited there in 1837. The sheep that Aitken had taken over with him were Saxon Merinos and cost him five hundred pounds. By 1839 Aitken had bought out Edmund and Francis Bryant's share in the sheep and established Victoria's first commercial Merino stud, and by 1840 was selling rams for five pounds. At the second Melbourne show in 1842 his sheep gained half the awards. Aitken was not afraid to pay high prices for rams. In 1845 he paid 200 pounds for the pick of a new lot of Saxon rams from the VDL Company and in 1850 paid 250 pounds for one Lichnowsky ram imported from Silesia. By the mid 1840's he was selling 600 rams at auction, evidence of a large and commercially oriented stud-breeding enterprise.

> *'The improvement in the Merino sheep of Victoria previous to the origin of the Ercildoune and other stud flocks, was more owing to Mr Aitken than to any other sheep-breeder in the country. His aim was to increase the weight of fleece, and to keep up the fineness and density at the same time. This he accomplished by adding to the length of the staple of the wool. He spared no expense or trouble in introducing fresh blood into his flocks, possessing the qualities he desired.'* [126]

This fresh blood that Aitken used was English Leicester as revealed in a letter from George Russell to John Aitken dated 5 October 1840.

> *'I have sent the bearer Mr Craig and one of my men for the Twenty Rams I purchased from you in June last. I can't get away just now myself to draw them off, but you can do this for me; and if you don't give me good ones I will not have any more from you. I will take the finest Woolled Rams in preference to the Cross Leicesters: give me them as long in the staple as you can; and I hope they are free from Scab. I enclose my acceptance at three months for the amount. One Hundred pounds which I*

hope will be satisfactory. I believe I am entitled to one more for taking twenty.'[127]

John Aitken's close association with Edmund and Francis Bryant, and the fact that he was using Leicester's to help create his Merinos is fairly strong evidence that he would have taken Leicester's with him to Port Phillip early on. Whether he took them as part of his first shipment in 1836 we will never know.

Flocks of sheep were being shipped from Van Diemen's Land to the Port Phillip area on a regular basis. In March/April 1836, flocks were there belonging to John Batman, Messrs. Solomon & Fergusson, Messrs. Wedge, Messrs. Cowie, Stead and Stiglitz, John Drake, and Dr Alexander Thomson.[128] There is no evidence that any of these people did or did not take Leicester's with them. In the two years to the middle of 1837, 100,000 sheep had been sent from Van Diemen's Land to Port Phillip.

W.J.T. Clarke (1805 – 1874)

The other person most likely to have been an early importer of Leicester's into Victoria is W.J.T. Clarke. Clarke had used Leicester's extensively since his arrival with pure-breed Leicester sheep in Van Diemen's Land in 1829. I think it would be safe to say that he would have taken Leicester's, or at least Leicester cross sheep, with him when he first went to Victoria with sheep in 1837.

Sheep had been shipped from Van Diemen's Land to Port Phillip prior to Clark taking some there in 1837, but there had been some heavy losses in transit due to rough weather and unsuitable ships. Clarke made a careful selection from the vessels available and charted the *Hetty* for three voyages to convey his sheep from George Town to Point Henry. He also ordered flour, blankets, slop clothing, tobacco for sheep dressing and several sheep nets, each fifty yards in length. The nets were to be used to restrain the sheep from rushing into the salty sea-water after they were carried ashore. A lot of sheep were lost during the unloading process as they either drowned after being put overboard to swim ashore, drank the sea water or rushed back into the sea after making it safely ashore. Clarke, through careful planning, never lost any sheep.

Clarke chose William Pettett to manage his Victorian operation. Pettett had worked for Edmund Bryant (who had imported the first Leicester sheep into Australia) before being employed by Clarke as overseer at Merton Vale and The Windfalls.

Clarke and Pettett were on board the *Hetty* when she sailed from George Town with the first 600 ewes on 8 April 1837. It was only one month earlier on 7 March that Governor Sir Richard Bourke had named the township 'Melbourne', after the then British Prime Minister Lord Melbourne. Further crossings were done on 25 April and 15 May, a total of 1,958 sheep being transported to Peak Station, which was between Melbourne and Geelong.

In 1838 Clarke abandoned Peak Station and moved his whole operation to Dowling Forest (near Ballarat), which was named after the maiden name of his wife, Eliza Dowling.[129]

THE HISTORY OF ENGLISH LEICESTER SHEEP IN AUSTRALIA

Early newspaper articles also report that Clarke was the first person to take Leicester's to Victoria, but they suggest that he did not do this until 1851 even though they are talking about Clarke's Leicester's being the best sheep for boiling down in 1848. An interesting article was published in The Australian Pastoralists' Review, 15 October 1898, in which it gives an account of Clarke taking sheep to Melbourne. It states that these were the first Leicesters to go to Victoria. They perhaps may have been the first purebred Leicesters into Victoria, but perhaps that may even be disputed as I would suggest that he would have taken purebred rams into the state before then to mate with the large number of ewes that he had there.

'Mr Robert Clarke is a native of Tasmania, having been born in Hobart on 29th April, 1841. On the 20th July, 1851, when only ten years and three months old, he was apprenticed to the late Hon.W.J.T. Clarke for six years to learn the management and breeding of stock. A few days after being apprenticed he sailed with Mr Clarke in a small vessel called The Elizabeth, in which Mr Clarke took some English Leicester sheep bound for Victoria. They were twelve and a half days before reaching 'Cole's Wharf'. Then the work of his life began. He shepherded the Leicesters on Batman's Hill, the site where the Spencer street railway station now stands, when, after resting the sheep about two weeks, he drove them to Jackson's Creek – now – 'Rupertswood' – remaining there about six months, when Mr T. Clarke and he drove them to Dowling Forest, near Ballarat, another of his employer's properties. Arriving there his troubles began. His work was shepherding the Leicester stud flock, and

there being no fences the sheep would double on him and make back to Rupertswood. These being the only Longwool sheep in the colony at the time, the other sheep owners, disliking the breed, would kill them, and for every one lost the manager would give him a good thrashing, the latter trouble occurring to him pretty often during his time on Dowling Forest. About twelve months after he was sent with part of the Leicester flock to another of Mr Clarke's stations, named 'Pyrenees', now Hon. Wm. McCulloch's 'Woodlands' Estate where he remained about six months, returning thence to Dowling Forest. Soon afterwards the latter estate with all stock was sold. While he was at Dowling Forest they had to grind their own flour, and only beef was allowed. Mr Robert Clarke was then sent to Belinda Vale Estate, which was managed by Mr Lewis Clarke, brother of Mr W.J.T. Clarke. After remaining there for about eighteen months the latter estate was let to Mr L. Clarke and Captain Gardiner in conjunction. Mr Robert Clarke was then transferred to Mr W.J.T. Clarke's Rockbank Estate, where his apprenticeship expired ...'[130]

Clarke certainly had pure bred imported Leicester's at Dowling Forest in early 1852 as the above article indicates, and he was having trouble keeping the sheep on his property as is also mentioned, because he was advertising extensively in the local papers for the return of his strayed Leicesters. Most likely these strayed sheep were those referred to in the above article.

THE HISTORY OF ENGLISH LEICESTER SHEEP IN AUSTRALIA

STRAYED

From the station of Mr W.J.T. Clarke, Dowling Forest, 40 pure Leicester imported sheep.[131]

These early Leicester's brought to Victoria were some 62 years later the foundation of the flock registered by Sir R.T.H. Clarke, Somerset Farm, Mooroopna, in 1913 (Flock Book for British Breeds of Sheep in Australia, Vol.5).

Sir R.T.H. Clarke sale catalogue 1917. Brenton Heazlewood.

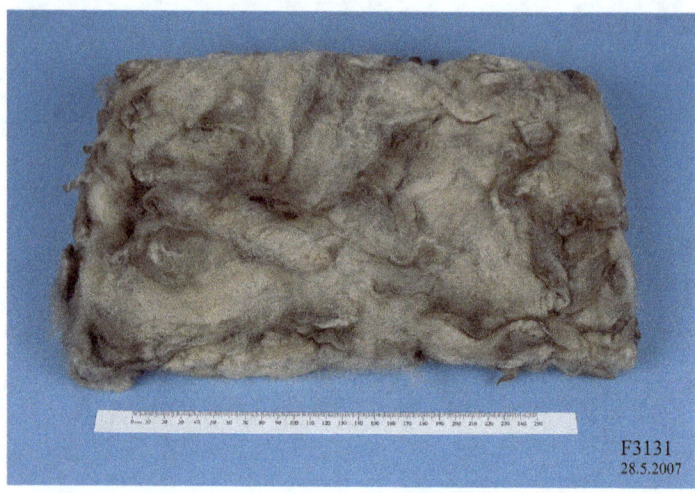

F3131 Wool specimen, Leicester ewe, bred by Sir William John Clarke, Bolimba Vale, Victoria, Australia, 1892. The sample is from a 2 year old ewe and tested at 27.8 microns with a staple length of 135mm. Collection: Museum of Applied Arts and Sciences. Purchased 2003 (originally gift of Sir William John Clarke, 1892). Photo: Chris Brothers.

Brown suggests that prior to 1852, Clarke was the only person running long-woolled sheep in Victoria.[132] It is most likely that Clarke was the first person to run a 'stud' of pure bred Leicesters, but there were certainly purebred Leicesters in the state before this date.

After this date there was a gradual increase in numbers of long-woolled sheep, not only Leicesters, but also Cotswolds and perhaps Lincolns. In 1856, 17 long wool sheep were exhibited at the Port Phillip Farmers' Society Show in Melbourne, with the Cotswolds winning the gold and silver medals.[133]

Thomas Millear (? – 1895)

Thomas Millear was another large landholder who used Leicesters extensively. Thomas Millear and Albert Austin (with whom in partnership, in 1878, he purchased the property Wanganella) were brothers-in-law and business partners. They spent their childhoods in Somerset, came to Australia to pursue farming interests and married sisters from a prominent Western District squatting family. In 1862 Austin married his cousin Catherine Mack and Millear married Nancy Mack.[134]

Millear, along with Albert's brother Josiah Austin and Thomas Maidment bought Green Vale. It was 21,000 hectares and ran more than 70,000 sheep. They also leased two other properties taking their flock to 100,000. Before Millear had purchased Wanganella with Austin he had earned a fine reputation as a sheep breeder but had won more acclaim for his pure Leicester sheep than with Merinos. His pure Leicesters were described as being:

> *'... bred from Hutchison's and Turner's flocks, and have always been first prize takers when shown in this colony, as well as in England. The sheep are well known as yielding heavy fleeces, and bringing amongst the highest prices per lb. in the English market, the wool from these flocks having realised the highest prices ever paid for crossbred wool in the London market...'*[135]

I cannot find any record of Millear importing Leicesters, but that does not mean that he did not import sheep himself. The fact that he had sheep from George Turner, who lived in Devon, and he came from Somerset, may indicate that he may have imported these sheep himself. Perhaps again an example of people purchasing sheep from the area they had come from in England.

At the 1861 Port Phillip Farmers Society Annual Exhibition, Millear, Austin and Maidment won second and third prize for a Pair of Leicester ewes, Four-toothed or upwards.[136] Showing again in 1863 they won first and second prize for a Leicester ram, four-toothed or upwards, first for a pair of Leicester ewes, four-toothed or upwards and first prize for a pair of Leicester ewes, two-toothed.[137] They did not appear to be exhibiting Merinos at these shows.

The partnership was certainly running a lot of Leicester sheep. At the expiration of the Wickliffe lease in December 1876, they offered for sale about 16,150 Leicester sheep.

These consisted of:

- 3,800 mixed weaners
- 1,000 two-tooth ewes
- 900 four-tooth ewes
- 2,000 six-tooth ewes
- 3,050 four years old
- 2,100 five years old

- 600 two-tooth wethers
- 650 four-tooth wethers
- 1,600 six-tooth wethers
- 150 eight-tooth wethers
- 300 cull ewes[138]

Two months later in January 1877, due to the end of their partnership they were selling a further 23,000 Leicester sheep.

These consisted of:

- 75 pure Leicester ewes
- 55 pure Leicester rams
- 100 superior crossbred rams
- 175 nearly pure Leicester ewes
- 7,350 two, four, six and eight tooth wethers
- 9,250 two and four tooth ewes
- 700 two-tooth fat ewes
- 4,200 mixed weaners
- 1,095 mixed ages and sexes[139]

It is interesting to note that Albert Austin who had also competed in Merino competitions, had actually won more prizes for his Lincoln sheep. In 1877 at the Hamilton Show, he won the prestigious trophy, the Hundred Guineas Gold Cup, with his Lincoln ram. He considered the winning ram, '*a magnificent animal, perfect on all points and which he said was the best two-toothed Lincoln ram he ever owned.*'[140]

As neither Austin or Millear lived on Wanganella, but remained on their Western District properties, we can presume that they continued to breed and run long-woolled sheep there, while at the same time breeding superior Merinos on Wanganella.

Tasmania was the main source for the early Leicesters that came into Victoria. An example of the number of Leicesters crossing Bass Strait is given in June 1865 when some 560 Leicester ewes, rams and lambs that had been imported from Tasmania were auctioned at Flemington.[141] Tasmanian stud breeders were selling considerable numbers of stud sheep into Victoria, and to enhance their business some of the leading Tasmanian breeders exhibited their sheep at the important Victorian shows. Mr W Field from Tasmania exhibited sheep at the 1865 Geelong Agricultural Show, winning first prize for both the Leicester ram and ewe from five other exhibitors.[142]

Stud sheep were also imported directly from England. In July and August 1873, 905 stud long wool sheep were imported from England to Melbourne with an estimated market value of thirty-six thousand pounds.[143]

THE HISTORY OF ENGLISH LEICESTER SHEEP IN AUSTRALIA

In September 1871 the first annual exhibition of long-woolled sheep was held at Hexham. Sales of long-woolled sheep had been held there since about 1863 so the staging of a show was a natural progression in Hexham becoming the centre for long-woolled sheep in Victoria. At this first show one hundred and five pure Lincoln and Leicester rams and ewes, and one hundred cross-bred rams, ewes and fat wethers were exhibited. Also 500 nearly pure-bred sheep were yarded for sale.

After the show a meeting was held at which it was decided to form a society to be called the 'Long-woolled Sheep Association of Victoria'. Mr Robert Hood, a prominent Lincoln breeder was elected President. The society's purpose was the development and encouragement of the long-woolled breeds of sheep. At this time Lincolns were more popular in Victoria than Leicesters. After this initial show it appears that the Long-woolled Sheep Association of Victoria took over the running of the Hexham Show as from then on the show was held under the auspices of the Association. At the second show held on Thursday 3 October 1872 there were 130 sheep exhibited.

The Leicester classes were:

- Leicester rams any age; 1st Sanderson, 2nd Charles Gray.

- Leicester ewes; 1st Sanderson, 2nd Anthony Mackenzie.

- Two-toothed Leicester rams; 1st Wm. Armstrong, 2nd Scott, 3rd Bath.

- Two-toothed Leicester ewes; 1st Anthony Mackenzie, 2nd Sanderson.

As well as the classes for the Lincolns and Leicesters, there were also classes for crossbred rams and ewes as well as 'five fat wethers, not Merino'.[144]

By 1873 some of the monthly sales were yarding in excess of 30,000 sheep of all types.

The 1873 show and sale was a great success with 357 pens being erected for the competing sheep and those brought for sale. As two thousand sheep were catalogued for sale, five auctioneers sold the sheep with the priority of sale being determined by lot and only one minute being allowed for the sale of each lot. The first lot of pure sheep sold was 20 Leicester rams from Wm. Murray, of Brie-Brie. These realised from 11 to 15 pounds per head. It is interesting to note that one Lincoln ram sold for 125 guineas while Mr Hood's selected ewes, with lambs at foot, four removed from Merino by pure Lincoln rams brought up to 20 pounds each.

The number of sheep sold and the prices obtained started to fall off in 1874 and 1875. The 1875 Leicester exhibitors included Joseph Ware, Sir Charles Sladen, J. Armstrong, W. Armstrong and Robertson Bros.

The Brie-Brie Leicesters mentioned above are Border Leicesters, although English Leicesters were later introduced into the stud. The Brie-Brie flock was founded by John Sanderson in 1865. The original stock consisted of 8 rams and 36 ewes, which were selected from the stud flocks of the principal breeders of Leicester sheep in the south of Scotland. G.A. Brown describes them as, *'the breed known as 'Border Leicesters'.'*[145]

Some of these imported sheep were exhibited at the Skipton Show, being taken there directly from the ship on their way to Brie-Brie. They took Champion prize for ram and ewe as well as first for best two rams and first for best two ewes. In 1871 a ram bred by Lord Polwarth was purchased at the Kelso ram fair for 115 guineas, this again consolidating the fact that they were Border Leicesters. In 1875, an imported ram was bought at the sale of Sir Charles Sladen's sheep. This ram was bred by Mr Hutchinson of Yorkshire, in other words an English Leicester ram that had been imported by Sir Charles Sladen. The result of the introduction of this ram into the stud flock was described as highly satisfactory. After this introduction 6 ewes were purchased from Thomas Gibson, 'Esk Vale', Tasmania, again being English Leicesters.

The exhibiting of the Border Leicester and the English Leicester's in the same classes at agricultural shows was common at the time of the Skipton show referred to. In England, perhaps the earliest division between the two breeds at a show was in 1858, but this division happened slowly with the two breeds sometimes competing in the same class up until about 1881. With not many Border Leicesters in Australia at this time it is not surprising that they would have been shown in the same classes with the English Leicesters. The fact that the two breeds were mixed in the stud makes one wonder how often this was done to introduce new blood lines, particularly, as in this case of the English Leicester into the Border Leicester stud.

Leicesters were also well represented at the first Ballarat Show held in 1876. Exhibitors were W.R. Scott 'Gledefield' Kiora; W. Armstrong 'Hexham Park'; Francis Edwards, Ascot; William Jeffrey 'Glencairn'; Dr. Plummer 'Wyabun' Gisborne; J. & W. Adams 'St. Enoch's' Stockyard Hill; Thomas Ellwood, Beremboke; William Scott, Bald Hills; Richard Ellwood, Kingston; George G. Morton, Learmonth; Donald Gunn,

Burrumbeet; James Mitchell, Windermere; S. Holgate, Miners Rest; William Thomas, Grenville; Alexander Armstrong 'Warrambeen'; Charles Edmonston, Mount Bolton; John Richardson, Newlyn; C. & A. Finlay 'Glenormiston'; Adam Broomfield, Newlyn; Joseph Hetherington, Cambrian Hill and James F. Strachan, Birregurra.

As previously mentioned, Victoria has been the state with the biggest number of breeders and therefore ewes since the early 1900's. As early as 1917 there were a total of 35 registered studs recorded in both the ALSBA and The Royal Agricultural Society of Victoria's flock books for Victoria, with 4,346 ewes. This compares with Tasmania with only 18 registered studs and 1,408 ewes. Victoria's peak was in 1,949 when it had 63 registered flocks running 5,773 ewes. This was approximately half the total number of flocks and ewes for the entire country.

Sir Charles Sladen (1816-1884)

When Sir Charles Sladen dispersed the whole of his stock in November 1875, due to him retiring to Geelong, it was described as the largest sale of Leicester sheep that had ever been held in Victoria. The highest price for a yearling ram was 130 guineas and the top price for a ewe was 40 guineas. In total 8,925 sheep were sold for over twelve thousand pounds.

Sir Charles had started his flock in 1863, and in 1872 he purchased two rams from T.H. Hutchinson (UK) for 100 guineas each. Sir Charles also bred and showed Lincolns, Cotswolds and horses. He was a regular exhibitor of Leicesters at the various local shows. At the 1874 Geelong and

Western District Agricultural Show he won first prize for both the ram and ewe in wool any age, as well as Champion Leicester (Society's Gold Medal).

Sir Charles was born at Ripple Court, Kent, England in 1816. He studied law before arriving in Geelong with his wife in 1842. He practised law for 12 years, and in 1854 he retired from business and took up Ripple Vale near Birregurra. There he established what was described as Victoria's leading Leicester stud.

In November 1855 as a member of the Legislative Council he took office as treasurer in Victoria's first responsible ministry. He re-entered politics in 1864 and in 1868 was Victoria's Premier for a short period.

Sir Charles was a devout and prominent Anglican layman. A large stained glass window was installed in the Moorabool & McKillop Streets, Anglican Church, Geelong, in 1885 by his friends, in his memory.[146]

L.R. Carter (? - 1906) Flock No1

Flock number 1 in the Flock Book for British Breeds of Sheep in Victoria, Vol.1, 1998, belongs to L.R. (Leonard Robinson) Carter, Scale Park, Clunes. This flock was founded in 1883 by the purchase of 2 ewes and 1 ram from Mr Lyon, Ellerslie. Additional sheep were later added from W. Jeffrey, Glencairn and M. Bennett and W.G. Hogarth, Tasmania.

Mr Carter also bred and exhibited Lincolns (F15 in Vol.1), Southdowns, Merinos and also pigs. At the 1897 Melbourne Show L.R. Carter and Sir Rupert Clarke were the only English Leicester exhibitors. Mr Carter was also involved in the mining industry. He died in December 1906.

The blood lines of this flock were carried on through the formation of two new studs in 1907, both based on sheep obtained from the Exors. of the late L.R. Carter flock. Fred H. Carter registered Flock No.5, and J.R. & A.W. Carter registered Flock No.7. This flock, No.7, was dispersed in 1924, and No.5 in 1930.

Duncan Scott MacGregor (1916-1989) Flock No2

Most English Leicester breeders would have heard of Duncan MacGregor because of the Duncan MacGregor trophy. The MacGregor connection with English Leicesters goes back to Duncan's grandfather, Duncan MacGregor (1835-1916).

A short sketch of this first Duncan is interesting and lays the foundation for 'our' Duncan.

This Duncan left Scotland for Australia in 1857, arriving when gold fever was still gripping Victoria and the pastoral industry was in its ascendancy. Duncan soon gained experience in the pastoral industry of Victoria, New South Wales and Queensland due to his skills as a shepherd and manager of stock.[147] Soon after arriving in Australia he was managing Mount Murchison and Donald MacRae's Culpauline station on the Darling River. He also explored much of south-west Queensland.

In 1869 he returned to Victoria where he married and lived at Glengyle, Coburg, and held the property Clunie. It is here that he formed his studs of Booth Shorthorn cattle and both English Leicester and Border Leicester sheep.

Duncan MacGregor has both an English Leicester and Border Leicester stud registered in Volume 1, 1898 of the Flock Book for British Breeds of Sheep in Victoria. The English Leicester flock is registered as Flock 2, being at Dalmore, Pakenham with 550 ewes put to the ram in 1897. The flock history is recorded as:

'The Dalmore flock was founded on the 2nd of June, 1869, by the purchase of 25 ewes bred by Mr. W. Field, Tasmania. In 1887, 59 ewes and lambs were purchased from Mr. R.Mailer (who had them only a few months), bred by Mr. Field.

The first ram used was bred by Mr.Matthew Wait, and was the progeny of a ewe imported to Tasmania by Mrs.Steele. This ram was used until 1871, when 2 rams were bought from Mr. Field, which were used till 1874, when 4 more rams were bought from Mr. Field, and used till 1877, when 1 ram, bought from Mr. Mailer, was used, with 2 others bred in the flock. On 23rd February, 1887, 6 rams were purchased from Mr. R. Grieve, Branxholme Park, Southland, New Zealand; and 1 of these, bred from Mr. Grieve's original stock, was used, with others bred in the flock, and no rams other than those bred in the flock have since been used.'[148]

Duncan also had Border Leicester Flock No. 7 registered in this first flock book. This flock was actually in the name of Mrs. D. MacGregor of Clunie, Chintin, with 52 ewes being put to the ram in 1897.[149]

By 1874 Duncan was able to move into south-west Queensland. With his widowed mother-in-law (Christina MacRae) he increased the MacRae holdings in the Gregory South and Warrego Districts, and with other partners and through agents took up many of the runs comprising Durham Downs on Cooper's Creek. Most of these leases were applied for and granted between 1874 and 1879, and others not until 1884. By 1893 Durham Downs carried 96,000 sheep, 26,000 cattle and 4,000 horses, all mortgaged in 1894. Duncan also acquired the leases of the twenty-one runs of Glengyle in Gregory North in the late 1870's and early 1880's, and also the leases of Melba Downs, Miranda, Yanko and Mimosa. Glengyle carried 14,113 cattle and 180 horses and the other stations 42,500 sheep, 364 horses and 11,257 cattle. All were mortgaged in 1895.

Duncan's Victorian concerns fared better. In 1875 he bought 4,063 acres of the Koo-wee-rup swamp and by 1880 his drainage scheme had drained 3,871 acres. Enterprising though his drainage schemes were, they flooded his neighbour's properties, which led to two protracted legal cases which he won. By 1889 he had acquired a second Pakenham property, Gowanlea, near Tooradin. In 1891 he turned over the management of his estate and Dalmore to his sons Donald and John (the father of 'our' Duncan), and at Chintin founded the Clunie Border Leicester stud.[150]

Duncan MacGregor died at Clunie on 28 January 1916 and was buried at the Melbourne General Cemetery. He was 81.

He is remembered as one of the first to open up the pastoral areas of western Queensland and was a significant contributor to the European exploration of that region. At his peak he (with his mother-in-law Christina McRae, in the name of the estate of the late Donald McRae) held property totalling more than 10,000 square miles. This was one-third the size of his native Scotland. For the son of a Scottish tenant farmer, this was an extraordinary achievement.

The Federation Drought, the Great Depression of the 1890's and the Queensland land laws combined to destroy Duncan MacGregor's fortune.[151]

Harry Peck described Duncan MacGregor as, *'a man who set a milestone in the pastoral development of Australia. If ever a man earned the name of a pioneer, it was Duncan MacGregor.'*[152]

We know that Duncan MacGregor had a good eye for stock, in particular the type that would do well in the hard country and with cattle in particular, the type that could be walked from Queensland to the markets in Melbourne and still arrive in good condition. Duncan used his Victorian properties as stud properties to breed his Booth strain of Shorthorn cattle, Clydesdale horses and English Leicester and Border Leicester sheep.

We do not know to what extent he used his English Leicester, and later the Border Leicester blood lines in his commercial sheep operation, but it would seem logical that he would be breeding rams to go north for use on his stations because we do know that the Shorthorn cattle and Clydesdale horses were bred for this purpose. He was at his peak about the same time that the other great user of English Leicesters was also at his peak, that being W.J.T. Clarke. Clarke also had properties near Melbourne and close to MacGregor's and there were often articles in the newspapers regarding which type of sheep were the most suitable to be breeding. We do know that the manager of Clarke's properties, Robert Clarke and MacGregor did not get along but that would not have stopped them from noting what each other was doing.

J. MacGregor sheep 1903. The Livestock Annual of Australia 1903, p159.

As already mentioned, Duncan had both an English Leicester and Border Leicester stud registered and entered in the first flock book of British Breeds of Sheep in Victoria. The English Leicester stud was only registered for the years from 1898 to 1906. In volumes I and II of the Flock Book for British Sheep in Victoria the registration is in the name of Mr D. MacGregor, Dalmore, Pakenham, but in the last entry for the stud in Volume III it is in the name of Messrs. J. McGregor and Co., Dalmore. This last entry covers the years 1902 to 1906. His Border Leicester entry in Vol. III is in the name of D. MacGregor, Gowan Lea, Tooradin, Vic. This stud was only registered up until 1909. Why the flocks only remained registered for such a short time I do not know. Perhaps as Duncan got older, he lost interest in them or perhaps as he lost control of the stations, he had no need to be producing the seed stock he once needed. I would guess the latter is the more likely reason.

The Live Stock Annual of Australia, 1903, contains an article on The Dalmore Leicesters stating that they are the property of Mr J. MacGregor.

> *The Leicester flock of Mr J. MacGregor, of Dalmore, Pakenham, is one of the most valuable studs of pedigreed Leicesters in Victoria. ... From the illustration given herewith, the uniform type of the Dalmore sheep will be observed. The proprietor has used considerable judgement in the breeding of his Leicesters, which are particularly remarkable for their lightness of bone and offal.*[153]

Regardless of the reason, a generation is nearly passed by before 'our' Duncan Scott MacGregor takes up the halter and registers a flock of English Leicesters as Flock 115 under the 'Dalmore' prefix in 1936.

We now turn to 'our' Duncan, Duncan Scott MacGregor (1916 – 1989).

There is a gap of 30 years between the last flock book entry of the first Duncan MacGregor's Dalmore flock in 1906 and the re-emergence of the Dalmore name again in 1936 (Vol. 28) in the name of D. S. MacGregor, Baringhup, Victoria. Flock No. 115. The flock book entry states:

> *This flock was started in January, 1936, by the purchase of the ram, 'Viladle Monk No. E.20', (521, Vol 25) and 30 ewes from*

Mr. R. J. Clement (F 60), and 45 ewes (with 39 lambs at foot) from Mrs. O. M. Glover (F 88).[154]

In the next year, volume 29, 1937, the entry is in the name of J. MacGregor & Son with the note that *'This flock was previously registered in the name of D. S. MacGregor* but in June, 1936, it was taken over by the present owners.' This change of ownership occurred only 5 months after it was first registered. I would presume the J. MacGregor is Duncan's father (John) and Duncan the son. Why the change? At the time of registration in 1936 Duncan would have been 20 years old, so old enough to run the stud. In 1937, 78 ewes were put to the ram.

The stud remained registered as J. MacGregor & Son until 1943 (John, Duncan's father died in 1942) when the ownership was changed to D. & M. MacGregor.[155] In 1987 it was again changed to D. S. MacGregor[156] and it remained under this name until its dispersal in 1989 following Duncan's death.

Dalmore Estate water tank. Margaret Kingman.

The Duncan S. MacGregor Memorial Trophy

This trophy is presented annually by the English Leicester Association of Australia to honour the memory of the late Duncan MacGregor.

Duncan MacGregor. Australian Stud & Farm, August 1962. Photo credit: Stock Journal.

Duncan died in July 1989, immediately prior to the ASBA Sheep Show. As if by some divine intervention during the show Mrs D.M. McDonell stood looking at the English Leicester exhibits at the ASBA Show at the Melbourne Showgrounds. There was a good entry that year from 4 breeders. In conversation it was learned that Mrs McDonell had a registered Shropshire flock and was looking to procure some longwools, either Lincoln or English Leicesters. The opportunity to secure the Dalmore flock from the estate of the late Duncan MacGregor was a unique opportunity and, most importantly, it would preserve the Dalmore genetics.

Mr Ian Morrish, 'Bardia' English Leicester stud was appointed to inspect the flock for transfer.

Subsequently 'Old Dalmore', Flock No. 373 was established in 1989 with the purchase of 4 rams, 25 ewes and 1 ram lamb from the estate.

At the next Committee Meeting of the English Leicester Association, Ian Morrish stated his wish to donate his $75 inspection payment to the Association to be used for a trophy in Duncan MacGregor's memory. Colin Taylor suggested it be, 'for best under 1 ½ year old exhibit bred by the exhibitor'. It was agreed that it be held at the Royal Melbourne Show.

The $75 was augmented by a donation from the RASV, sale of handmade sheep halters made by Mrs McDonell and the sale of other promotional material until the amount was $970.60 – sufficient to invest to fund an Annual Trophy.

The first Trophy was competed for in 1990 with visiting judge, Mr Harold Nobes, OBE, UK presiding.

The winners of The Duncan S. MacGregor Memorial Trophy for the best under 1 ½ year old exhibit bred by the exhibitor have been:

- 1990: Eric Gray – Marengo Stud, Richmond, Tas
- 1991: Colin Taylor – Koenarl Stud, Timboon, Vic

- 1992: P. & E. Stephenson – Ostlers Hill, Goorambat, Vic

- 1993: Colin Taylor – Koenarl Stud, Timboon, Vic

- 1994: I.C. Heazlewood & Co – Melton Vale, Whitemore, Tas

- 1995: P. & E. Stephenson – Ostlers Hill, Goorambat, Vic

- 1996: P. & E. Stephenson – Ostlers Hill, Goorambat, Vic

- 1997: Colin Taylor – Koenarl Stud, Timboon, Vic

- 1998: I.N. & E.M. Hay – Karrama, Grassmere, Vic

- 1999: No English Leicesters exhibited at the Royal Show

- 2000: P. & E. Stephenson – Ostlers Hill, Goorambat, Vic

- 2001: P. & E. Stephenson – Ostlers Hill, Goorambat, Vic

- 2002: R. & J. Brown – Jarob, Amphitheatre, Vic

- 2003: Colin Taylor – Koenarl Stud, Timboon, Vic

- 2004: Colin Taylor – Koenarl Stud, Timboon, Vic

- 2005: P. & E. Stephenson – Ostlers Hill, Goorambat, Vic

- 2006: Colin Taylor – Koenarl Stud, Timboon, Vic

- 2007: Colin Taylor – Koenarl Stud, Timboon, Vic

- 2008: P. & E. Stephenson – Ostlers Hill, Goorambat, Vic

- 2009: Colin Taylor – Koenarl Stud, Timboon, Vic

- 2010: Colin Taylor – Koenarl Stud, Timboon, Vic

- 2011: P. & E. Stephenson – Ostlers Hill, Goorambat, Vic

- 2012: Colin Taylor – Koenarl Stud, Timboon, Vic

- 2013: No entries

- 2014: P. & E. Stephenson – Ostlers Hill, Goorambat, Vic

- 2015: No entries

- 2016: Colin Taylor – Koenarl Stud, Timboon, Vic

- 2017: Colin Taylor – Koenarl Stud, Timboon, Vic

- 2018: Colin Taylor – Koenarl Stud, Timboon, Vic

- 2019: Colin Taylor – Koenarl Stud, Timboon, Vic

- 2020: No shows held due to Covid 19

- 2021: No shows held due to Covid 19

- 2022: George Willows – Nant Stud, Evandale, Tas

- 2023: Brenton Heazlewood – Melton Park, Whitemore, Tas

- 2024: Brenton Heazlewood – Melton Park, Whitemore, Tas

Colin Taylor (wearing the trophy scarf) with Duncan MacGregor trophy winning ram, Royal Melbourne 2006. With Colin is the judge, Ivor Robinson, NZ, and his wife Margaret. Brenton Heazlewood.

Duncan S. MacGregor Trophy line up at the Royal Melbourne Show 2001. Ethel Stephenson.

Eric Gray winning inaugural Duncan MacGregor Trophy, 1990. The Muster, December/January 1990/91.

N. Cole, West Cloven Hills, Flock No. 12

Flock number 12 was established in April 1907 by Mr Stansmore, who purchased 59 ewes and two rams from Mr D. MacGregor (Flock No. 2) and subsequently disposed of them to Messrs Cole Bros. in July 1912.

At the turn-of-the-century, 1900, West Cloven Hills was 48,000 acres, running about 48,000 sheep, 6,000 cattle and 500 horses.

When the English Leicester flock was purchased, they already had a super fine flock of Merinos, a Lincoln flock and Shropshires, and for a time a Corriedale flock, which were used over portions of the Merino flock.

During 1945 and 1946 this flock changed ownership three times. In March 1945 it was taken over by Mr A.S. Bradshaw from Nicholas Cole (Snr). Then in January 1946, the flock, consisting of 1 ram, 49 ewes and 24 ewe lambs, was purchased by Mr I.D. Macdonald. In November, 1946, Nicholas Cole Jnr. purchased the stud and transferred it to 'Warra Yadin'.[157]

It was run there for many years, using them to cross over the Corriedale ewes for a prime lamb, and selling the ewe portion as first cross dams, especially to the wetter environments where they could handle the tougher conditions.

The stud eventually came back to West Cloven Hills where it presently is. With this move back to West Cloven Hills, the stud prefix was changed to *West Cloven Hills* in 1990. It is still used as a cross over the Corriedale ewes, and the ewe portion is grown out and sold as first cross ewes.

The stud currently runs in its own 40 acre paddock on West Cloven Hills, so the size of the stud suits the paddock where the sheep are kept, not only because of sentimental reasons, but also because they fulfill a role in producing a great first cross lamb.

Nicholas Cole 4th used to relate that back in the 1930's and 1940's when West Cloven Hills received orders for train truck loads of rams to go to the Riverina and Western Australia – the rams were sorted and walked to Camperdown, 13 miles, and loaded onto rail trucks to go to their destination.

The West Cloven Hills shearing shed is a blue stone shearing shed built in 1851 after Nicholas Cole's first one was burnt down the week before the Black Thursday fires. He lost all his out buildings and most of the grass on his run, only the house being saved. He commissioned a Mr Swan of Swan Marsh to construct a new shearing shed. Construction was going well, when in August/September of 1851 gold was discovered at Ballarat and districts, the workers downed tools and 'bolted to the diggings' leaving the shed uncompleted and the walls at about 5 feet high. A decision was made to pitch a shingle roof on the unfinished walls.[158]

Nicholas Cole in their historic wool shed. Brenton Heazlewood.

Arnold Caddy (1866 – 1948)

Arnold Caddy. *Australian Stud & Farm Monthly*, August 1962.

Dr Arnold Caddy is a person that very few of today's English Leicester breeders would have heard of, yet during the 1920's and 1930's he was a prominent breeder and promoter of the breed.

Arnold Caddy was born in March 1866 in the UK where his father was the deputy inspector-general of hospitals and fleets, Royal Navy. He was educated at King's College School and St George's Hospital. He served as house physician and house surgeon at the Royal Northern Hospital, and senior house surgeon at the Leeds General Infirmary. In 1892 he emigrated to Calcutta where he went into partnership with William Coulter, surgeon to the Marwari Hindu Hospital. He also served as an officer in the Calcutta Light Horse.

In November 1902 he married a daughter of Archibald Currie of Melbourne and in 1912 they moved to Chandpara, Tylden, Victoria to farm. Caddy gave up the practice of surgery, and during the war of 1914-1918 he was in command of a reinforcement camp with the rank of Lieutenant – Colonel, Victorian Light Horse.[159]

Caddy bred Red Poll cattle and English Leicester sheep, becoming president of the Red Poll Cattle Association.

Volume 9, 1917, of the Flock Book records Major Arnold Caddy, Chandpara, Tylden, as establishing English Leicester Flock No. 33a in August 1916 with sheep from Mr J. G. Habel, Messrs. McArthur Bros. and Mr John Langham.[160] He must have quite quickly built up his ewe numbers as in 1925 he was mating 204 ewes, in 1930, 383 ewes and in 1931 456 ewes. Before the formation of the Australian Society of Breeders of British Sheep (ASBBS) in 1925 Caddy was a member of the committee responsible for publishing the flock book for the Royal Agricultural Society of Victoria.

When ASBBS was formed in 1925, Caddy was a committee member of the Victorian branch and also a Victorian representative on the Federal Council of ASBBS. In 1925 Caddy wrote a small booklet entitled 'Robert Bakewell and His Times', which was issued by The Australian Society of Breeders of British Sheep.

Certificates awarded to Caddy for sheep exhibited at The Australian Sheep Breeders' Association Forty-Ninth Show, 1927, have survived, one for English Leicester Ram over two and a half years, third place, and for ewe under one and a half years, again third place.

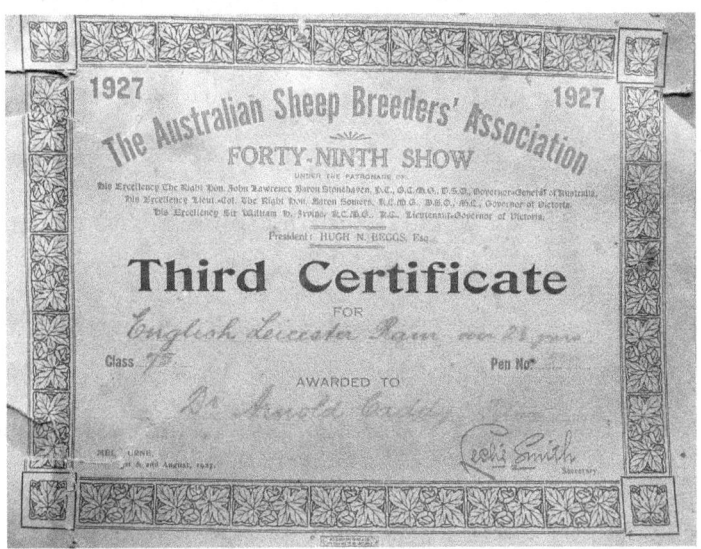

Arnold Caddy Third Prize Certificate. Brenton Heazlewood.

Caddy also imported rams from New Zealand over several years, the first of these being imported in 1924.

Caddy's flock was dispersed by sale in February 1945 at which time he was mating 160 ewes.

Arnold Caddy is perhaps best remembered in ASBBS circles for his writing of a 16 page booklet published by ASBBS in 1925 entitled *Robert Bakewell and His Times*.

Caddy died in April 1948, aged 81, and was remembered as a scientific farmer who used trace elements and superphosphate. Caddy Place in Florey, ACT is named after him.[161]

Ethel and Peter Stephenson, 'Ostlers Hill', F352

When Peter retired in 1976, their week-end farm at Flinders on the Mornington Peninsula became their home, a working farm which offered a complete change of lifestyle. Ethel's dream to have longwool sheep and work with wool had come true.

This was the time when wool crafts, hand spinning and weaving were booming, and the breeding of coloured sheep was popular. It followed that the first flock at Flinders was coloured, very mixed and collected from a variety of sources, often the sale yards. An Association was formed – enthusiastic members set up field days, workshops and classes in shows to educate the black sheep breeders. It was at the 1st National Congress of Black and Coloured Sheep Breeders in Adelaide in 1979 that Ethel and Peter met with one of the speakers, Lawrence Alderson from the UK. His book 'A Chance to Survive – Rare Breeds in a Changing World' had just been published. He had also been involved with The Rare Breeds Survival Trust, which had become a significant body in the UK to conserve the domestic rare breeds. He talked about the minority breeds of sheep in England and Robert Bakewell's work on the native sheep of Leicestershire – breeding up to become the 'proved Leicester', the sheep that went all over the world improving the existing breeds – The English Leicester.

Here they found the direction for the Ostlers Hill Flock.

They became regulars on Judging Day at the Melbourne Royal Show and the Sheep and Wool Show, meeting up with English Leicester breeders and their sheep. One exhibitor at that time was Beverley Scott, 'Aracoola' stud, Dunkeld in the western district of Victoria.

When the Scott's moved to Queensland the sheep were put on the market and at their dispersal sale on 4 December 1984, they purchased 14 ewes including 2 show sheep and 5 ewe lambs. That was the foundation for the Ostlers Hill Stud.

A ram was purchased from Norm Badcock's 'Connaughtville' stud, Westbury, Tasmania for the first season – followed by a further ewe purchase from the Badcock studs and also from Eric Gray's 'Marengo' stud.

When the first ram purchased died, Ethel made the decision to bring their next sire over from New Zealand, and in due course Riverside 'Bishop' from Harold Bennett's flock joined the stud. He was later followed by Riverside 'Deacon'.

In 1985, Ethel and Peter moved from Flinders to NE Victoria, Goomalibee, near Benalla. Warmer climate, more acres and irrigated pastures agreed with their Leicesters and they thrived. As well as 100, 1st

cross Border ewes they ran 500 Merinos crossed with the English Leicester, run especially for wool for hand spinners.

In 1988 a ram and 6 ewes were purchased from Colin Taylors 'Koenarl' stud to carry out an embryo transplant program. Additional Koenarl bloodline was added with the purchase of 5 ewes at the 1996 'Koenarl' reduction sale.

In 1995 they retired to a smaller property of 100 hectares 'The Homestead' on Broken Creek at Goorambat, also near Benalla.[162] After Peter's death Ethel continued breeding and exhibiting her Leicesters.

Ethel was not born into agriculture as her career was as an architect but from the time that she purchased her first English Leicester sheep she worked tirelessly in the promotion of the breed and the use of its wool. As an example, she helped set up a market for the wool overseas where it was used for dolls' wigs as well as always being on the lookout for the best market that the Association members could get from the local wool buying companies.

Ethel was a great believer in putting the sheep in front of the public and exhibited her sheep at numerous shows. She always prepared and presented her sheep to perfection and was always willing to assist and mentor other exhibitors for the betterment of the breed. Ethel was also a very good show judge of not only English Leicesters but also other Heritage breeds of sheep.

Ethel always took a very keen interest in mentoring and assisting younger members. She was very generous with her time and sheep, often giving young members sheep so they could exhibit and breed from them.

Ethel also put considerable energy into the administration of the English Leicester Association of Australia (ELLA). She was an early member of the reconstituted ELLA, later becoming Secretary and or Treasurer for 26 consecutive years. She was also a financial benefactor to the Association, quietly paying for goods and services without recompense or awareness by the other members. She donated money to the Association for the purpose of providing a perpetual trophy to be competed for at the Australian Sheep and Wool Show.

Her interest was not confined to the ELAA as she was a founding member of the Black and Coloured Sheep Breeders Association as well as Heritage Sheep Australia. Ethel was also involved in the Handweavers & Spinners Guild of Victoria, Collingwood Children's Farm, Benalla Rose Garden and the Benalla Art Gallery.

Ethel Stephenson. Brenton Heazlewood.

Ian Morrish F181, (1927-2009)

Ian will be remembered for his dedication to his breed, the English Leicester. His 'Bardia' stud flock No. 181, was registered in 1942 and in his words, was 'his pride and joy'.

Wool was of particular interest for Ian. He aimed to produce fleece with the long, lustrous features of the Leicester, and for it to have the unique soft handle. He had many hand spinning customers waiting on shearing time.

English Leicester breeders will recall the 1992 Royal Show when they celebrated Ian's '50 years as a shepherd'.

I.J. Morrish flock rams. Australian Farm & Home, February 1949.

As an only child, he was reared on the family farm at Smeaton. When his father needed him full time on the farm, he gave up his engineering studies and shared the tasks with sheep, cattle, dairying, cropping and seed cleaning. But it was in the blacksmith shop where he was happiest, repairing and making farm machinery.

A gentle, happy family man and a thorough, hard working farmer, keen on his sport and his community. He was a committee member of the Ballarat Agricultural Show Society for many years.

In 1984 he was elected foundation president when the English Leicester breeders re-started their own association (ELAA) and in 1988 was elected to the committee of the Australian Society of Breeders of British Sheep (ASBBS).

Ian and Marion enjoyed exhibiting and promoting their Leicesters over many years. In the later years, Ian did more judging than showing. Over the years there were two visits to New Zealand and in 1988 to England judging and visiting farms and shows.[163]

Colin 'The Colonel' Taylor, 'Koenarl' F355

Colin is one of those breeders that endears himself to everyone he meets and exhibits against. For many years he has exhibited at many shows in Victoria, often being the only exhibitor of the breed. He has been very helpful to every other breeder, particularly new exhibitors that he is showing against.

In rolling up to the annual Australian Sheep and Wool Show at Bendigo each July we all think that we have arrived with well grown and pretty good looking sheep. That is until Colin arrives with his trailer full of 'very ordinary' sheep. We all anxiously await until he starts unloading and then we know that we should have all stayed at home because his 'very ordinary' sheep make everyone else's look positively inferior. 'What have you been feeding them?' we ask, only to be told nothing special. 'They have only been running in front of the dairy cows all year', he then lets on. We know that dairy farmers have the best pastures and Colin certainly takes advantage of his son's good pasture management.

'See you next year' we say to Colin as we help him load his sheep and the numerous ribbons he has won at the end of the Bendigo Show. 'I'm not showing any more' he says every year, 'besides I have sold all my sheep' he continues. We know he will be back and sure enough the next year he arrives with 'the few I just found'. We would be very disappointed if he did not find a few to show us how to do it properly.

Colin originally started a Southdown stud, not thinking he would ever have other breeds.

He used to go to Tasmania to buy Southdowns, often staying with the well respected and noted sheep breeder, the late Norm Badcock. They used to go down to the paddock to look at the Southdowns, and on the other side of the fence were those very friendly and inquisitive sheep. As usual he asked questions and was told they were English Leicesters.

Every time thereafter, when he saw Norm Badcock, he would ask 'how are the English Leicester sheep going?'. One time Norm asked, 'would I like to buy some ewes?' Hence the start of his English Leicester flock.

He purchased 10 ewes from the Badcock family and a ram from Lewis Lee, Mole Creek, Tasmania at the end of 1976. The stud was registered in 1977.

Later he bought 10 more ewes from the Badcock family. Then in the 1980's purchased ewes from Ravenswood and Beechwood studs from New Zealand. Even to this day his flock traces back to those families as no more ewes have been added. Many New Zealand sires were used, and they improved all aspects of the sheep.

Colin has been a great exhibitor of his sheep at many shows over many years and has won numerous championships. A highlight was when he won the Supreme Champion English Leicester Ram at the World Sheep and Wool Congress in Tasmania. Later semen from that ram was exported to the United Kingdom.

Bendigo Sheep Show 2018, judge Pam Tait, NZ, with Colin Taylor. Brenton Heazlewood.

Colin Taylor. Melbourne Sheep Show 1981. Dale Badcock.

Colin has also been very involved in the English Leicester Association of Australia, being the only current member who was present at the first meeting of the re-formed association.

Today Victoria is still the state with the most registered studs, having about 70% of those registered but only about 55% of the ewe population.

Millar Bros. (Vic) ewes. Australian Farm & Home.

THE HISTORY OF ENGLISH LEICESTER SHEEP IN AUSTRALIA

J.G. Habel & Sons, Yulecart, Vic. Group of three. Winners of the Centenary Breeders' Cup, Melbourne Royal Show 1937. Flock Book for British Breeds of Sheep in Australia. Vol. 30, Section 1, 1938.

Alex Habel Champion Ewe Melbourne Show 1948.

Kevin & Leslie Rorke 1988. The Muster, September 1988.

Geelong Show 2009 Group. *The Muster*, December 2009.

7
Queensland

'If the test of the value of a domestic animal be the numbers on the preservation of which human care is bestowed, and on the extent of the habitable globe over which mankind has diffused the species, then the sheep takes the first rank.'[164]

Today we do not think of English Leicesters having a place in the southern Queensland pastoral industry, but 100 years ago they were there in significant numbers. Not only was this breed that we usually associate with the cooler, temperate climate of southern Australia being used in the hotter northern pastoral areas, but a large number of them were being imported from New Zealand.

One hundred years ago the English Leicester and Lincoln, were at their peak in popularity in both Australia and New Zealand, with their numbers being second only to the Merino in Australia. They were the first choice as the sire for both the first cross ewe and the prime lamb itself. A common practice of the time was to put the English Leicester ram back over the English Leicester/Merino cross ewes to produce the prime lamb. It was for this reason, the good offspring that they produced when mated to the Merino, that they were used in the southern parts of Queensland, mainly on the Darling Downs.

Free settlement was not permitted in Queensland until 1842, and it was during the 1840's that there started to be a big push of sheep north from New South Wales, initially into the Darling Downs area. The first person to open up the stock route from northern New South Wales and take a large number of stock into the Darling Downs area was Patrick Leslie. This first stock expedition into Queensland, which started from a New England station, then the most northerly settled district of New South Wales, consisted of:

> '400 breeding ewes, 100 ewe hoggets, 1,000 wedder [wether] hoggets, 100 rams and 500 wedders 3 and 4 years old. We

> had two teams of bullocks (24 in all) and 2 drays, a team of horses and a dray, and 10 saddle horses. We had 22 men, all ticket-of-leave or convicts, as good and game a lot of men as ever existed. And who never gave us a moment's trouble-worth any 40 men I have ever had since.'

After three weeks of cutting tracks in advance over the rough country, Leslie formed the first station at Canning Downs in June 1840.[165] The rush of squatters soon followed, as when Governor Gipps arrived to see the new squatting stronghold in May 1842, he found 45 stations in existence which in that year shipped some 1,500 bales of wool, all in territory officially closed to free settlement until three months before!

The sheep numbers must have increased significantly as in 1847 the port of Maryborough was opened as a wool port. Whether English Leicesters accompanied their Merino cousins in the opening up of the Queensland pastoral industry is hard to say, but it could be safely assumed that they would have soon followed.

In 1858, Mr Broughton of Ipswich was advertising *'a number of pure improved Leicester sheep'* for sale. This would indicate that Leicesters had arrived in Queensland earlier, rather than later, and most likely these early sheep would have been taken north from either New South Wales or Victoria. Mr F.J.C. Wildash of Canning Downs, near Warwick was one of the first to engage in cross-breeding in Queensland. In 1869 he used a pure English Leicester ram over Merino ewes with success. A few years later George Davenport of Headington Hill, near Clifton also used English Leicester rams for the same purpose.[166] Certainly by the early 1870's

Leicesters were being imported into Queensland directly from England. In July 1871, the *Ramsey* from London brought 40 purebred Leicester sheep to Moreton Bay, and in November of the same year the Hon. James Taylor imported 2 Leicester rams and 5 ewes to Toowoomba at a cost of two hundred and fifty pounds. Sir Joshua Peter Bell, of Jimbour, near Dalby was also an early importer of English Leiceters, importing several rams and ewes from England and using them in his flock.[167]

Both Lincolns and Leicesters were exhibited at the Royal Agricultural Society of Queensland show which was held in Toowoomba in 1873. C.B. Fisher won first prize in the class for the '*Best pen of two Lincoln, Leicester or any other long-woolled rams*', with his Leicesters. It is interesting to note that there were also classes for the best pen of two Lincoln, Leicester or any other long-woolled *Colonial* ram hoggets and *Colonial* ewe hoggets. The ram hogget class was won by the Cotswolds, a breed that is sadly no longer in Australia.[168]

In 1876 the sheep population of Queensland was about 7.6 million. Of these about 5,000 were Leicesters or other long-wool breeds, about 150,000 were cross-breeds, principally Leicester-Merino and the remainder Merinos.[169] Queensland's sheep numbers rose quickly to 21.7 million in 1892, but these numbers fluctuated depending on droughts and good seasons. In 1916 these numbers had dropped back to 15.5 million. Queensland was the state of large pastoral stations. In 1916 there were 19 stations in Queensland with over 80,000 sheep each.

In 1906 Mr J.H.W. Mules imported 85 Leicester sheep from New Zealand from which he hoped to breed a good type of ram for producing crossbred lambs from Merino ewes. Although he was living at Ideraway (NSW) at the time, he exhibited the English Leicesters that he had imported at the Toowoomba and Brisbane shows in July and August of 1906 so those who were interested, *'would have an excellent opportunity of examining the new importations'*.[170] Mr Mules was the developer of what we now call the mulesing operation.

F.G.G. Couper

The *'foremost breeder of English Leicesters in Queensland'*,[171] Mr F.G.G. Couper of Westbrook, writing in 1908, had great praise for the English Leicester as a ram to be crossed with the Merino ewe, and he believed better still, the second cross for producing fat lambs. Mr Couper founded his stud in 1905 with the purchase of ewes from New Zealand. This was the first English Leicester registered stud in Queensland.

Mr Couper started his experiment with English Leicesters in 1901 when he imported two rams and two ewes from New Zealand, as well as six Southdowns. His aim was to produce a lamb suitable for the frozen meat trade. Mr Couper divided his flock of 1,600 Merino and crossbred ewes, mating a third each with English Leicester, Southdown and Lincoln rams as a way to find out by experiment the very best class of lamb for export.

In 1905 Mr Couper sent 2,000 frozen lamb carcasses to England. I would think it is safe to presume that these carcasses would have been from an English Leicester cross as by this time he was a great user of the English Leicester. These carcasses weighed 34 lb (15 kg) and returned a net of 11 shillings, 5 3/5 pence per carcass. It is interesting to note that at the same time Mr Couper was sending a trial shipment of frozen hares to England.

The hares were frozen with the entrails in and packed twelve per crate. These returned a net of about seven and a half pence per hare.[172]

Mr Couper was also a regular exhibitor of his English Leicesters and Southdowns at agricultural shows from as early as 1901. At the 1907 Toowoomba Show his results were, 1st ram 2 ½ or over, 2nd ram 18 months to 2 ½, 2nd ram under 12 months, 2nd ewe 2 ½ or over, 1st ewe 18 months to 2 ½, 1st ewe under 12 months, Champion ewe, in the fat sheep section 2nd pen of 5 Merino wethers, 1st pen of 5 crossbred ewes, 1st pen of 5 crossbred wethers under 18 months, 1st pen of 5 crossbred ewes under 18 months, 1st & 2nd pen of 5 crossbred wethers under 30 months suitable for export.[173]

THE HISTORY OF ENGLISH LEICESTER SHEEP IN AUSTRALIA

F.G.G. Couper advertisement 1907. The Pastoralists' Review.

F.G.G. Couper's English Leicester rams 1907. The Pastoralists' Review.

F.G.G. Couper's English Leicester lambs, Westbrook, Qld 1907. The Pastoralists' Review.

A.C. Thompson

Mr A.C. Thompson, of Fairfield Stud Farm near Dalby made a major importation of English Leicesters from New Zealand in 1912. Again his aim was to supply the fat lamb market and he selected the Leicester because it was the foundation of the *'Canterbury lamb, which has been described as the finest in the world'*. He could see no reason why they should not do well on the Downs, the annual rainfall he said was the same in Dalby as Christchurch, and although the climate was a little warmer he did not think that would affect them much, as *'Leicesters were the best fitted of the longwools to stand the heat.'*[174]

Mr Thompson travelled to New Zealand and purchased 54 rams and 64 ewes. The ewes and the best rams he said he would keep as a stud flock (although it appears that a stud was never registered), the remainder of the rams being used to cross with Merino ewes. The cross ewe lambs he planned to cross again with the Leicester ram *'and anticipates being able to turn out lambs for export equal to any sent out by any other country'*.

Six years later, in 1918, Mr Thompson wrote an article that was published in The Sydney Stock and Station Journal in which he gave a report on his use of the English Leicester as a basis for his fat lamb enterprise. He reported that along with some other lamb breeders, he had sent to London a shipment of fat lambs and had received a report from Borthwick that they, *'were equal to, if not better than prime Canterbury lamb'*. Mr Thompson also reported that he had had considerable success in the show ring with his Queensland bred rams.[175]

A year later Mr Thompson is again extolling the virtues of the English Leicester when he writes:

'If you want the best sheep under the sun for crossing with the Merino to grow fat lambs and crossbred mutton I claim it is the English Leicester for the following reasons:- The English Leicester is a fine mutton sheep. The meat is nicely mixed, it has a nice thick meaty leg, with short fine bone, and is a first class butcher's sheep. It was this sheep that made the name for Canterbury mutton in New Zealand. Canterbury was the home of English Leicesters at the time the export of mutton began in the Dominion. At last year's Ram and Ewe Fair in Christchurch there were more English Leicesters entered for sale than all the other breeds put together. That goes to show which is the popular sheep there.

Second, it is a heavy woolled sheep. Even its head and legs are covered with a dense fleece of fine lustrous wool. That is very necessary in a country like this, where we get dry times, and are sometimes compelled to hold over our lambs and young sheep. A good crop of half-bred wool will always compare favourably with Merino. The half-breds from English Leicester and Merino ewes are a hardy class of sheep. They come through drought times perhaps better than any other class of sheep, and fatten up quickly on good feed. In my opinion the English Leicester ram can be used on the half-bred and even on the three-quarter-bred with the best results.' [176]

Large numbers of English Leicester rams were being purchased by Queensland graziers about this time. In 1909 Mr Chas Binnie of the 4.D.Estate at Quirindi, shipped from Melbourne 180 English Leicester rams for his Liverpool Plains property. Mr Binnie, after trials with other long-woolled rams had decided that the English Leicester gave the best results for fat lamb raising.

Of course, the Merino was by far the dominant breed in Queensland with, in 1918, 97% of the state's wool production being obtained from the Merino and only 3% from crossbreds and British breeds.[177] The Lincoln was also being used, and certainly more heavily than the Leicester, as a crossing ram to go over the Merino. Some producers were putting the Lincoln over the Merino to produce the first cross ewe and then putting the English Leicester over this ewe to produce the fat lamb.

The use of New Zealand English Leicester bloodlines was not restricted to Queensland, with importations being made by breeders in other states, particularly South Australia.

New Zealand was of course not the only source of English Leicesters for the Queensland ram breeders. In 1910, my great grandfather, R.G. Heazlewood, sent by the steamer *Gabo* to Brisbane two English Leicester rams and several ewes which had been purchased by Mr C.H. Grove of Kelvin Grove, Nanango. It would be fairly safe to assume that other breeders in Tasmania, and particularly those from the closer states of Victoria and New South Wales would have supplied quantities of English Leicesters to this northern market.

The history of registered English Leicester flocks in Queensland is a sorry one. As already mentioned, Mr F.G.G. Couper of Westbrook, was the first Queenslander to register a stud, that being flock number 27 (no prefix is recorded). This stud was founded in 1905 but only appears in Volumes I, II and III (1910, 1911 and 1912) of the Flock Book for The Longwool Breeds of Sheep in Australia. It ceased to be registered after 1912. We then have to wait another 71 years for the next English Leicester stud to be registered in Queensland. In 1983 R.M. Thompson of Pimpama founded flock number 353, prefix Cottage Hill. This stud only remained registered for 10 years.

As far as I can tell these are the only two registered studs to represent the English Leicester breed from Queensland up until 2015, when Victor Gorring, from Wellcamp near Toowoomba registered a stud (F425, 'Tarleea') based on sheep from Ethel Stephenson's stud. After the death of Victor the flock was purchased by George and Paul Willows, and made the long trip to Tasmania to be part of George's flock.

8

The Leicester and the Merino

> 'The principal value of the New Leicester breed consists in the improvement which it has affected in almost every variety of sheep that it has crossed.'[178]

The English Leicester and the Lincoln (and later their replacement the Border Leicester) can very much owe their success in Australia to the Merino, and likewise the Merino can owe much of its success to the Leicester and the Lincoln. These breeds have been intertwined since the early 1800's.

The influence that the Leicester had on the development of the modern Merino started in France during the early 1800's at the national farm of Rambouillet. The Rambouillet breed had its origins among the Moors of North Africa during the 14th century. Distant ancestors of today's Rambouillet accompanied Moorish conquerors to Spain, and their descendants were left behind when the Spaniards drove the Moors out.

Spain's resulting Merinos were a valuable legacy of that otherwise unhappy period. The quality of the Merino wool allowed the Spaniards to dominate the European wool trade, and to maintain that dominance, the government strictly forbade sheep from being exported. France was a major buyer of the fine wool from Spain and became heavily dependent on this stable supply to keep its own mills in operation. By the mid-eighteenth century, the French began to fear that increasing Spanish industrialisation might lead to an embargo on the Merino wool as well as the Merino sheep.

This may have been one of the reasons that Louis XVI established the national farm at Rambouillet. It was advertised as a place where examples of the choicest plants and animals from around the world could be studied. The Spanish Merino was most likely very high on the list. In 1786, as a gift to his cousin Louis, the King of Spain ordered that a small flock of the finest Merinos be sent to Rambouillet. On 12 October of that year a flock of 318 ewes, 41 rams and 7 wethers arrived at Rambouillet. This was the first significant release of Merinos to the outside world and except for one small addition provided the basis of the eventual Rambouillet breed.

Course woolled breeds were being used as early as 1808 at Rambouillet. On 24 May 1808, an official advertisement appeared in the *Moniteur*, giving notice of a sale of 220 ewes and rams of the finest woolled Spanish breed. This was part of the flock on the national farm of Rambouillet, as well as 1,300 lbs of wool, the produce of the mixed breeds of sheep kept on the menagerie at Versailles. This advertisement was accompanied by a notice from Lucien Bonaparte, Minister for the Interior which states:

> '... Sheep of the ordinary course woolled breed, when crossed by a Spanish ram, produce fleeces double the weight, and far more valuable, than those of their dams; and if the cross is carefully continued, by supplying rams of the pure Spanish blood, the wool of the third or fourth generation is scarcely distinguishable from the original Spanish wool.'[179]

This would indicate that they were putting a Spanish ram over course woolled ewes, most likely Leicesters as we know that they were in France at this time as 100 'Dishley' rams and 30 ewes had been imported in 1794, rather than the expected approach of putting the course woolled ram over the Spanish ewes.

Most of the early imports of Leicesters into France were a failure, due to poor management. One exception was that of M. Duverger, near Versailles who managed to breed a high class flock. The French government was at this time becoming more concerned with the increasing amounts of long wool being imported from England. In 1833 this amounted to 1.42 million lb (645 tonnes). On hearing of the success of the Duverger flock the government sent the director of their Alfort agricultural school, to England. He was to purchase a flock of Bakewell's 'New Leicesters' which were to be used to improve the French wool, as well as study the production of long-wool sheep. In 1833, 110 ewes and 12 rams were imported and proved a success.[180]

Besides the establishment at Rambouillet, flocks were kept at Perpignan, Pampadour and Alfort[181] as well as some influential daughter flocks. One of these daughter flocks was the Videville flock which used Leicesters as part of the 'New' or 'Improved French' Merino move. Videville influenced its parent flock by example, and also exported sheep to South America and to a lesser extent North America and Australia between 1840 and 1870.[182]

At the other government farm, Rambouillet, the pure Spanish flock of Merinos which Louis XVI had imported was going backwards. The Merinos there were described as:

> '*The Royal flock ... which, for years, attracted all the sheep masters of Europe to its annual auction sale, bred the fleece so fine, and the animals so delicate, that they could no longer attract attention; and ... they changed the plan, and now sell (when they can) at private sale. The sheep have no wool on the head or legs, and but little on the belly. They are ruined by high breeding. The wool is short and fine.*'[183]

This assessment of the Rambouillet flock is also confirmed by Joshua Trimmer when he visits Ramboillet in 1827 and describes the ewes being shown to him by the head shepherd as:

> '*In his selection he evidently was wholly governed by the fineness of the fleece, without regard to the form of the sheep, some of those shown being very plain.*'[184]

The 1830's introduction of the Leicesters and their impressive impact when crossed with the Merinos was occurring as the Rambouillet sheep were experiencing poor demand and low prices. In 1834 M. Bourgeois was appointed as director at Rambouillet. He must have been aware of the developments at the Alfort and other flocks, and must have wanted to make changes:

> *'... the new management wanted a change in the direction of mutton. Several English breeds had been imported into France exclusively for mutton, and their cross on the Merinos, then distributed widely in some of the districts, attract attention, and outside pressure, if we so term it, was felt at the Rambouillet directory. They sought to have the Merinos approximate the type of races characterised by largeness of body, beauty of form, and regularity of proportions, and to attain this object as the result of selection and high feeding. There was no question about the importance of the fleece, but they sought after good forms and large weight of carcass.'*[185]

Despite this the Rambouillet sheep were always described as 'pure Spanish'. M. Bernardin, the Rambouillet director in 1880 did not admit to any other breeds being introduced to the Merinos. He said that just by the use of selection from around 1840 the Rambouillet Merino had been transformed into a large, early maturing, meat-carcase-type animal, where the fleece had ceased to be the entirely predominating consideration. This breeding resulted in a sheep that lacked some of the true Merino characteristics, so corrections were made which resulted in the extremely influential Rambouillet Merino of the 1850's and 1860's, which combined a robust and large carcass with a productive combing-wool fleece.[186]

The Rambouillet flock statistics confirm this change. While carcase and body weight were being concentrated on, the fleece weights did not improve, but when the emphasis was once again on wool, there was an improvement in the wool weights. This was due to their being a bigger carcass and more strong wool follicle traits to work with. Perhaps the most dramatic change that occurred was with body weights, there being a nearly 50% difference between those of around 1850 and the 1880's. To achieve this dramatic increase some introduced genetics must have been used as it is very unlikely that through selection alone this would have been possible. The Leicester is the most likely candidate to have helped achieve this.[187]

During this same period there were private Rambouillet-Leicester cross flocks in France that were leading the government flock by several generations. After their corrections, the Rambouillet flock achieved similar results in regard to wool.

When Australian breeders imported Rambouillet sheep from the late 1850's through the 1860's, they were importing a strong boned and large carcassed sheep with a long stapled combing wool, which at the time was regarded as a strong Merino wool. This supported the American beliefs about a Leicester cross in the large combing wool sheep that they had imported about the same time. They were not short and fine-woolled Spanish Merinos.

This view conflicted with French historians as well as practically every director of Rambouillet since the 1820's, who have emphatically denied that the Rambouillet Spanish Merions were anything but pure. Spanish purity at Rambouillet had even been demanded by government decrees since 1786. In 1834 the First Secretary of the Office of Agriculture wrote to Rambouillet's incumbent director, Bourgeois, and told him:

> *'The government can only furnish for reproduction subjects of undisputed purity of breed, and certainly no private establishment will inspire so much confidence in the public as those belonging to the state, especially the Rambouillet flock: the reputation of which must be maintained like that of Caesar's wife...'*[188]

This did not stop the rumours about a Leicester cross at Rambouillet. Special treatises on the 'non-intervention' of the 'Dishley' were issued by its defenders, with frequent denials by directors. However, every year from 1786 Rambouillet have kept wool samples from their rams and ewes and these samples show fleeces never stronger than a superfine 74's to 90's count, which backs up what the directors always maintained.

The answer seems to be in the fact that Charles Germain Bourgeois II, the second director at Rambouillet had a farm at Perray on which he kept Rambouillet ewes, some of which were crossed with *'at least one English ram – a Dishley'*. As previously stated there were also flocks at Perpignan, Pampadour and Afort, giving plenty of opportunity for cross breeding or experimenting to be done out of sight. This is the same period of time that the sheep at Rambouillet underwent quick changes to their type, supposedly only due to selection and diet.

It therefore seems that the Rambouillet flock was kept 'as pure as Caesar's wife' to keep the officials happy and supply their French clients, but a secondary flock or flocks was kept which had the Dishley influence. This more robust type of sheep was in greater demand in the overseas market.

As far as the Australian breeders were concerned, therefore, by the 1860's the directors at Rambouillet had achieved, in one or more of their flocks, through careful selection and astute crossing a combination of the best traits from both stock breeds, which was so ideally suited to their harsh economic and physical environment. The Leicester's larger, squarer, deeper carcass, longer body, greater efficiency at converting grass to protein in carcass, and a longer, freer wool staple, with greater capacity to produce raw wool fibre, had combined with the Merino's ability to produce wool of good fineness and tensile strength in proportion to that fineness, with density and the ability to produce a lot of wool on a productive skin. In the words of their proud though somewhat biassed director, Bernardin, between about 1855 and 1875 the Rambouillet sheep were then, 'the only Merinos in existence that are noted for their prodigious size, rapid growth, great hardiness, and a dense fleece of great bulk, length of staple, freedom from excessive grease, and unsurpassed fineness, equally suited for the carder or comber'.[189]

The desire to put Merino wool on a Leicester carcass has a long history. Lord Western, who was the most famous and influential of the 'Anglo Merino' breeders, in a letter to Earl Spencer stated that:

> *'My object, then, may be familiarly stated to be the placing Merino wool upon a Leicester carcass; perhaps not exactly*

resembling the short finest clothing wool of Saxony, but a fine combing wool, superior to any that has heretofore been grown.'[190]

Western then inbred the result and set his type and called his Anglo-Merinos 'pure'. Throughout the 1820's, 1830's and 1840's many of these sheep were sold in Australia.

It was in the growing sheep population of Australia that the Leicester influence was to express itself in the development of the Australian Merino.

The Australian Agricultural Company (AA Co.) was perhaps the first Merino based grazing operation to import Leicesters and use its genetics on their sheep to aid in the creation of an Australian Merino. By the late 1830's the AA Co. had become the largest pastoral enterprise in New South Wales and was running 85,000 sheep and 5,000 cattle. Three Leicester rams and eleven ewes were imported in 1837 to cross with the Merino's in an attempt to get a more robust sheep with a heavier cutting, longer stapled fleece.

'The experiment of crossing the Leicester rams imported in 1837 with the Merino flocks was being watched with some care in London; and Mr. J.T.Simes [an early agent and broker influential in Australia] ... reported that at the last sales the crossbred wool had sold at better prices than their Merino wool, but that nevertheless considered the Leicester blood too robust a type and the cross too violent ... [Simes'] advice was followed,

and the crossing was thenceforward restricted to one-third of the number of ewes stationed at the Peel.'[191]

Massy states that this must rate as one of the first influential long-wool infusions involved in the creation of an Australian Merino.[192]

John Aitken, who arrived in Van Diemen's Land about 1825 and was one of the first to follow Batman across Bass Strait with sheep to farm in Victoria, also used Leicesters to set the type of Merino he was after. Aitken would have outcrossed with the Leicester initially for staple length, larger body and the early maturity this cross would have given, and then corrected severely with fine wool infusions to fix a type with the desired combination of body size, staple length and fineness. The presence of the Leicester genes in the Aitken sheep would have been unable to be removed, and given that his sheep had a strong influence it further demonstrates the multiple genetic nature of Australia's Merino base.[193]

Perhaps the strongest influence that the Leicester and Lincoln had on the Australian Merino was in the Peppin strain. The Leicester infusion and transforming influence over the early small, Spanish Rambouillet sheep expressed itself in one of the early and influential sires used by the Peppins. 'Emperor', who Massy describes as *'undoubtedly one of the most influential rams ever used in Australia,'*[194] was a ram with strong bone, large size and barrel-like body which displayed the Leicester influence. His wool weights were almost double the then best in Australia, again coming from the Leicester influence. The Peppin's gathered genetic material from many countries and succeeded in combining and balancing these often opposite traits to develop their strain of Merino which has proven to be so suited to Australian conditions. One of these traits was the large body frame and its associated coarser and longer wool which was derived partly from an English long-wool base via the Leicester in the Rambouillet and other French sheep and also possibly by direct use of Lincolns.

The development of the South Australian Merino is also influenced by the infusion of the Leicester genetics. The South Australian Company was the early leader in South Australian sheep breeding. While they had a base breeding stock of Van Diemen's Land ewes, they also ran hairy Cape sheep, English Leicesters and Southdowns. The Leicesters had been brought out from England for the Company by John Brown. By 1844 the Company was running nearly 32,000 sheep, but the wool price slump of 1848-9 led the English shareholders to sell off the flocks and land in 1849 and 1850.

In 1840 W.H. Dutton and Captain John Finnis arranged to have overlanded into South Australia two large mobs of sheep. In the second flock to arrive were a large draft of half-bred Leicesters. Some of these were used by the Dutton's and, in view of later developments in South Australia, were to have a significant effect on the South Australian sheep evolution. These Dutton Leicester sheep were to influence J.B. Hughes and G.C. Hawker, two early key sheep men in South Australia. John Hughes was to later purchase some of these half-bred Leicesters from Dutton. In 1842 he was again to receive the Leicester influence, this time indirectly through him importing some of Lord Western's breed of Anglo-Merinos. Hughes who had large runs which extended into semi arid bluebush and saltbush country where sheep had to walk long distances to water, noticed that the best producing, most fertile and hardiest sheep that still produced a respectable fleece at shearing were the larger framed and stronger woolled ones, traits that came from non-Merino genes already in the background of most Australian sheep flocks by the mid 1840's.[195]

2012/86/1 Engraving, 'Prize combing Merino rams for 1873 and 1874, bred by John Murray, Mount Crawford', an illustrated page from 'South Australia: History, Resources and Productions', written by William Harcus, paper / ink, Australia, 1876 Collection: Museum of Applied Arts and Sciences. Purchased 2012.

Hughes's medium to strong woolled sheep on large frames were the prototype, if not the foundation source of the South Australian Merino. Significantly, one of the major genetic influences on the evolution of later South Australian and key New South Wales flocks – the Rambouillets – were essentially built on the same genetic foundation, of an English long-wool cross (Leicester) over Spanish Merinos. Thus English long-wool genes played a significant role in the evolution of the Australian Merino.[196]

During the 1850's the English worsted industry started to experience a steady increase in demand which gained pace during the 1860's. This change was brought about by technological advances in combing and spinning which led to the availability of cheaper fabrics which thus resulted in greater consumer demand. Between 1850 and 1871 the number of worsted spinning spindles more than doubled in the UK. English exports of worsted yarn rose by 50 per cent between 1857 and 1869 and exported fabrics doubled. The English wool producer could not supply this increased demand, so the manufactures looked to colonial supplies, for the longer types of wool.

To help overcome this supply shortage, in 1859 the Bradford Chamber of Commerce established a 'Wool Supply Association', whose role was to popularise the long-woolled breeds of sheep in Australia and New Zealand. The Association was wanting 'a large supply of deep-combing wool', and as part of its promotional work it published a paper in 1865 titled *The Sheep – Long-woolled as well as Short-woolled for Victoria, Tasmania and New Zealand* by 'Omega'. This was purely an advertisement for the use of Leicesters and Cotswolds in Australian flocks.

> 'The long lustrous wool required for combing and for worsted manufacturers ... owing to the limited supply of that article, commands even higher prices at the present time than the fine short wools of the Merino sheep, the single exception to this being the rare, choice, small supplies of the fine wool of the carefully tended and improved flocks of the Saxon Merino ... if our flock-owners apply themselves to the production of long-stapled wool,

in addition to the short fine-wool yielded by the Merino, they will succeed even beyond their most sanguine expectations in that branch of industry ... [The production] of long-stapled lustrous wool for combing purposes .... Would be accomplished by the introduction of the improved Leicester breed of sheep, or, better still, of the most perfect type of that improved breed, known as the New Oxfordshire, sometimes called the New Cotswold sheep.'[197]

The main South Australian flocks were already down the worsted track with the use of Leicesters and Cotswolds and this increased demand for the worsted in the 1850's and 1860's confirmed their direction. The advent of the improved Rambouillets in the late 1850's and 1860's meant that here was a Merino (with a Leicester base) that had long, lustrous and deep-crimping combing wool of a far better quality than that of the English long-wools.

Massy sums up the positive influence that the Leicester and other English long-wool breeds had on the development of the Australian Merino as:

'... Thus from the English long-wool breeds huge increases in carcass weight, body surface area and staple length were gained, giving constitutional adaptability as well as enormous gains in wool and meat productivity.'[198]

9
Cross-breeding with Leicesters

'Were I to define the art of breeding, I should say, that it consisted in the selection of males and females intended to breed together, in reference to each other's merits and defects. The breeder's success will depend entirely upon the degree in which he may happen to possess this particular talent.'[199]

While the Leicester in its own right is a true dual purpose breed of sheep, it was in its use as a sire to put over the Merino to produce the cross-bred ewe that cemented its place and influence in the early sheep industry in this country. It was the Merino breeders that introduced the Leicester into Australia to put them over their Merino ewes. The Bryant's in Van Diemen's Land, the Australian Agricultural Company in New South Wales and the South Australian Company in South Australia, were all more interested in producing wool from Merino's, than running a strong woolled British breed, but they knew the benefit of the Leicester to their operation. Not only did the Leicester help improve the Merino itself, but its main asset was its suitability in crossing with the Merino. The Lincoln was also used for this purpose and for about 50 years between 1870 and 1920 was more popular than the Leicester for this purpose. It was not until the Border Leicester became popular in the 1930's for this purpose that the English Leicester was superseded. The English Leicester can be proud of the fact that it was superseded by a breed directly descended from itself.

As early as 1836 newspaper articles were appearing which extolled the virtues of crossing the Merino with the Leicester. *Bents News* (Hobart) of April 1836 listed the seven advantages of doing this cross as:

> '1st, a larger sheep is obtained. 2nd, a greater number of lambs reared. 3rd, not so subject to scab. 4th, a greater quantity of wool, which will bring in more money. 5th, the sheep are better and quieter feeders. 6th, the cross will feed to a greater weight, in proportion to the Merino, from a lean sheep. 7th, a preference will, consequently, be given by the butcher.'

One of the early importers of Leicesters into Van Diemen's Land, W.J.T. Clarke, was also perhaps one of the greatest uses of this breed and its crosses that has ever been known anywhere in the world. This is due to the preference Clarke had for the Leicester and the sheer number of sheep that he was running. In his peak he was running sheep on 650,000 acres. His large framed Leicester sheep produced strong wool which had length of staple and almost doubled the yield of the Merino types which were so popular at that time. Clarke embarked on a breeding programme which was to make him one of the largest and most successful woolgrowers in Australia. His aim was to produce a dual-purpose sheep with a weightier carcase and a heavier fleece, without letting the wool become too coarse. But he was not without his critics, particularly from the Merino breeders.

> *'Big' Clarke's large, cross-bred Leicester sheep produced far more wool and meat than the smaller fine-woolled merinos which most sheepmen regarded fervently as the Colony's greatest asset. The advocates of the merino even went so far as to trenchantly criticise Clarke in a series of letters in the* Argus, *for debasing Victoria's reputation for growing fine wool.*[200]

In an article critical of Clarke, the *Argus*, 20 September 1858 wrote:

> *It is not surprising that the wool of a flock of these so-called Leicesters should be uneven, for Mr Clarke's flocks have been raised in the most haphazard manner, without care of selection, and from pure, half or three-quarter bred Leicester rams. Still, no catarrh or disease has yet broken out, and we have Mr Clarke's word for it that during thirty years he has found these mongrels to be the hardiest of sheep, and we have occasionally the evidence of our own senses that they yield the fattest of mutton.*[201]

Clarke replied to the *Argus* defending his cherished Leicesters.

BRENTON HEAZLEWOOD

Sir,

In the constant warfare going on with respect to sheep breeding, my name has been brought forward on many occasions. If you think my opinion worth a space in your valuable columns, you will please insert the following.

I have been an extensive sheep breeder for upwards of 30 years, have experimented on every breed in the colonies, with the exception of Cotswolds, and from what I have seen of them think favourably. I can say I have crossed more extensively with Leicesters than any other man in the colonies, and I have never had occasion to regret it, my fleeces being increased fully 30 per cent in weight, and from the length and elasticity of the wool the price per lb. was not reduced more than ten per cent; in many cases my wool sold at higher prices than the average of Port Phillip wools.

I am also certain that no sane man will dispute the value of the carcase. My sheep, before the boiling down took place, on all occasions brought in the Melbourne market nearly double the price of fine-wooled sheep, and in boiling down they produced on an average 40lb. of tallow of superior quality to that produced from fine woolled sheep.

I have further to add, that I have bred both kinds, fine wooled and Leicesters in Van Diemen's Land, with an advantage in favour of the Leicesters equal to the results in Victoria. I have also found that cross-bred sheep thrive well on the poor arid land at the Pyrenees as well as on the rich lands. Crossbreds are hardier, better mothers for rearing lambs, and I believe consume but little or no more food than fine sheep, as coarse sheep will eat and thrive on such grass as fine wooled sheep would starve on. The coarse sheep are also much less liable to footrot.

In my opinion the activity of the fine-wooled sheep is not an advantage, but the contrary, as coarse sheep will eat coarse food, and rest, which tends to produce mutton, when the Merino or Saxon are travelling the whole day, and destroying more food with their feet than they consume; and from the constant exercise they take, I believe require more grass to keep them, even in good store condition, than would fatten coarse sheep of nearly double their weight.

I am, Sir,

Wm. J.T. Clarke.[202]

The quality and evenness of Clarke's Leicester-Merino cross sheep is confirmed by Harry Peck, a noted Melbourne stock agent and auctioneer.

'For many years the wethers and cast-for-age ewes of English Leicester-Merino crosses, from the Clarke stations travelled in on the hoof in mobs of 500 to 1000, week after week, right through the winters, and topped Newmarket for quality and price. Never since have we seen the like in such numbers for evenness of type and quality combined.'[203]

But not everyone agreed with Mr Clarke in relation to the advantage of putting the Leicester over the Merino ewe. One of the strongest critics of the practice was Mr John Graham, an eminent sheep classer of the 1860's. In his book 'A Treatise on The Australian Merino', he slates the practice saying that it nearly ruined the colony and the squatters themselves.

'All so far was doing well, but the would-be sheep-breeders thought they could do better, if they could only succeed in placing the merino fleece upon the Leicester carcass (they might as well have tried to grow it on a bullock)...

... As regards the wool of the cross-breeds, it is neither one thing or the other; it retains the good qualities of neither Leicester nor Merino; it is light, tender, and destitute of character. In effect I am of opinion that the introduction of the Leicester sheep has done more injury to the entire wool-growing interests of the colony than scab, catarrh, foot-rot, or all the other ills which sheep are heirs to-at least in Australia.[204]

He does however go on to say that the Leicester is suited to the rich succulent pastures on which Mr Clarke runs a lot of his sheep, particularly Dowling Forest near Ballarat.

The pure Leicester has always been known as having 'a great disposition for laying on fat', but in Australia where there is so much light grazing land, 'the Leicester turns out with capital results'. One of the great features of the Leicester has been its value for crossing purposes with the 'crosses' with almost any breed producing good carcasses.

> *It has often appeared to me that the Leicester sheep would be far more profitable in every way than the present sheep in the Colony.*
>
> *First – They produce very considerable more wool, and although it is coarser, yet the additional quantity obtained would very much more compensate for quality. In England the improved Leicester, of a fine description, of long combing wool, will clip, on an average – rams, 12 lbs.; wethers, 10 lbs.; ewes, 8 lbs.; the coarser or old Leicester, which are of an unwieldy size, will clip more, but neither the wool nor the mutton is so marketable as the 'Dishley,' or improver Leicester. By cross breeding with well selected rams from England, a fine description of combing wool might be produced, and the sheep yield on an average at least six pounds, which, if sold in Sydney as low as 8d. per pound, would produce 4s. for each sheep, which certainly would be a very great advantage over the present system. The wool is, besides, more*

open and easier washed and got up. I would leave it for those to decide who have had more practical experience than myself, whether breeding pure unmixed Leicester sheep, for the sake of the wool alone, would not be more profitable than crossing. In some instances the experiment of crossing has been tried, and when understood, and proper rams have been selected, it has answered to the great satisfaction of the growers.

About a year ago, Dr Wilson's wool, of Braidwood, crossed in this way, far exceeded the Saxon clip, and sold for an equal price in the London market; and I am quite satisfied it is capable of doing so, or at least, approaching very near it, when properly understood and properly managed. In this case the profit would be again double, and would be an advantage of 75 per cent, on an average over the system now pursued. Some who have not properly understood crossing with the Leicester, or how to select the most suitable rams, have tried it, and complained that the wool was too course. They cannot expect fine wool; the object in view is to obtain long combing wool of as fine quality as practicable, which I am of opinion would be far more profitable, even though sold at a much lower rate, than the finer sorts. Probably the expense of shearing and conveyance of wool from the bush might be greater; but where the other advantages preponderate so much, there is room for an allowance on this head.

The next points are the carcase, and produce of lambs. The

> carcase of the improver Leicester wether will average, in fair condition, say 40 lbs. a-quarter; the old coarse Leicester, which is now only bred in Lincolnshire, will weigh nearly half as much more; but, as I have stated before, neither the wool nor mutton gives satisfaction.[205]

Robert Clarke of Bolinda Vale also preferred the Leicester rather than the Lincoln for crossing with the Merino. He also put the Merino ram over the Leicester ewe which was not commonly done as the practice was to put the Leicester ram over the Merino ewe, mainly due to the sheer number of Merino ewes.

> I tried the Lincolns on Bolinda Vale, but soon discarded them, and I am quite sure that the Leicesters on light country will clip as much wool as a Lincoln, and return far more money from the point of view of fat lambs or mutton ... The only fault you can find with them is they get too fat, and why? Because their constitutions are so strong, and they have such an aptitude to fatten ... In breeding crossbreds you never want to use the Merino or Leicester ram more than twice in succession. The way I am working now, and it is the simplest, is, when your ewes begin to get too fine, put them with a pure English Leicester ram, and when you find them getting too strong, put them to the Merino ram, but never go more than the two crosses either way.[206]

Today the English Leicester has been superseded by the Border Leicester for use in crossing purposes. The quicker maturing Border with its leaner carcass, produces a more fashionable first cross ewe for use as a prime lamb dam when mated to a terminal sire.

The English Leicester can be proud of the role it played in helping to establish the Australian fat lamb and then prime lamb industry. The English Leicester's great carcass attributes, and its ability to successfully cross with any breed, helped many early sheep breeders stay afloat when wool was in decline and helped lay the solid foundation for our important sheep meat industry.

10

Dual-Purpose Breeds

Corriedale

The Corriedale is a true dual-purpose breed. It is a large framed, plain bodied, polled sheep, capable of producing a heavy carcass. They produce a heavy cutting bright fleece, with good style, length and handle.

The Corriedale was not a development of one single breeder, but of several, both here and in New Zealand. It is often regarded as having originated in New Zealand, but the breed began in both countries simultaneously, although it expanded very much more quickly in New Zealand. The early Australian studs also developed in isolation from each other until early in the twentieth century, whilst most of the New Zealand breeders exchanged bloodlines from their earliest beginnings.

In Australia, Henry Corbett established in 1882, what was to become the most influential of all the foundation Corriedale studs. He used Merino rams over Lincoln ewes.

While most of the foundation breeders of the Corriedale used Lincoln rams over Merino ewes, or in some cases, Merino rams over Lincoln ewes, one used the English Leicester in preference to the Lincoln.

In Norman Nicolson's account of the establishment of the Egleston Corriedale stud in Tasmania he states:

> '... *Other people in Tasmania were now breeding half bred rams to half bred ewes and half bred rams were in demand at from two to three guineas. Our first step in this direction was to breed our stud Merino ewes to stud (English) Leicester rams. The ram lambs from this cross sold readily – more readily, in fact, than the Merino rams had been selling. The next step was to import six ewes and a ram of the Corriedale breed from C.H. Ensor of White Rock, New Zealand. We had gone purposefully for a Leicester type because our experience had shown that this cross grew a much finer fleece and were better doers than the Lincoln-Merino cross – although the latter were generally bigger in the bone and had a heavier fleece.*'

It is interesting to note that the White Rock sheep imported from New Zealand were also based on the mating of English Leicester rams over Merino ewes.[207]

The fact that some Corriedale studs were based on the use of English Leicester's, either locally or from New Zealand, rather than the more common use of the Lincoln did cause a serious division amongst the breeders. Some breeders believed that only those studs founded on the use of the Lincoln should be classified as Corriedales. While this division was eventually resolved, it does indicate the difficulties in firmly establishing a new breed.

11

Boiling Down and the Frozen Meat Trade

'Humanity has never been involved in a relationship with any other creature that was more pivotal to its development and ultimate success than the one it shares with sheep.'[208]

Even in adversity the English Leicester, and particularly its crosses, proved to have some advantage over most other breeds of sheep. The first major crisis for Australia's pastoral industry, which started in the early 1840's, but had its conception in the preceding decades, was caused by two market forces.

Firstly, there was a drop in the price paid for wool. By 1842, the return from wool had dropped from three shillings to less than one shilling per pound.

Secondly the natural increase of sheep in the colony far exceeded the local consumption of mutton due to the relatively slow population growth.

The Committee of Council on Immigration came to the unanimous decision at the end of 1842 that 'one grand cause of the depression in the pastoral interests of New South Wales, was the want of sale for the surplus stock'. In the preceding decades there had been a dramatic increase in sheep numbers in the colony. For example, in New South Wales sheep numbers had increased from 50,000 in 1813 to 600,000 in 1829, 2.75 million in 1838 and 7 million in 1851.[209] All the sheep and cattle had to be consumed by the local population as it was not until after 1880 when the first refrigerated ships started to take frozen meat back to England that the export trade started.

Sheep prices reached their highest point in 1837 when ewes were sold for three pounds a head. A large part of these high prices was the result of speculation. J.R. Graham, a grazier of a later period summed it up rather bluntly:

> *'From about 1835 to 1840, a perfect mania for sheep-farming had taken possession of military officers, gentleman of the merchant navy, lawyers, clergymen, merchants, and others who had capital to invest, or credit to float a set of bills. Sheep and their wool were the all-absorbing topic of the period, and immense numbers of sheep and stations became the property of gentlemen who had not the slightest idea of stations, sheep, or their management.'*[210]

With very low wool prices and no outlet for their surplus stock, sheep prices fell from 60 shillings to 1 shilling and even 6 pence per head. The squatter Threlkeld sold his 9,000 sheep, previously worth 27 thousand pounds for 37 pounds 10 shillings. The crisis deepened through 1842 and 1843 as most credit institutions failed, and traders, squatters and bankers went to the wall. Those squatters who had moved onto their land after 1837, which was the majority, virtually all went to the wall.[211]

It was an era when 'cash was king' and sheep, cattle and land were worth virtually nothing. There were many squatters who had plenty of each, but still went broke because they did not have any cash to their name.

The slow road to recovery started in 1843 when the idea of boiling down sheep carcasses to salvage the tallow started. Tallow was worth two to three pounds a hundredweight in London where it was used in the making of soap and candles.

But even in the very early days of boiling down there were those against the process. *'Save your sheep, do not kill the goose that lays the golden eggs,'* implored Rupert Kirk in a letter to the Sydney Morning Herald in September 1843. He suggested that the Chinese wool market should be looked at due to the low wool prices received from London.[212] We would now say that he was a very forward-thinking man even though at that time there was no Chinese wool manufacturing industry.

The Presbyterian minister, J.D. Lang, described the boiling down process as being *'... peculiarly offensive in the sight of Heaven',*[213] due to the fact that there were many people on the brink of starvation in Great Britain at this time.

Henry O'Brien of Yass is usually credited with starting and popularising the process of boiling down. O'Brien published an article in the *Sydney*

Morning Herald in May 1843 in which he showed that worthless sheep could realise 6 shillings per head from being boiled down.

While O'Brien may have popularised the practice of boiling down worthless sheep, he was not the first to start the practice. Tallow, which was surplus fat from sheep which had been slaughtered for mutton, had been exported as early as 1813, when 41 casks had been shipped on the *Minstrel*. WJT Clarke had commenced boiling down his own sheep on Dowling Forest in 1842[214], and continued the practice for many years. In 1849 it was reported that Clarke sent a flock of 3,000 wethers to be boiled down at the establishment of Messrs Philpott. These sheep were nearly pure Leicester and it continued the debate as to, 'whether the wool bale has not held too sovereign a supremacy over the tallow cask'. The Leicester breed was not popular with most of the 'wool men' and they did not like the idea, or a hint of the possibility that the Leicester may be more profitable than the Merino, but the Leicester's big carcass and tendency to put on plenty of fat was proving an advantage in this case.

The boiling down process was a gory and stinking process, 'peculiarly offensive in the sight of Heaven'. The sheep were slaughtered and skinned, the hind legs were removed and either sold at eightpence each or else cured and sent to Van Diemen's Land as mutton hams. The carcasses were packed tightly into wooden boilers strongly clasped with iron. Steam was then introduced and after a few hours the tallow was run off into casks. The refuse which remained in the boiler was then placed in a screw press and subject to high pressure until all the tallow had been squeezed out. This was then put into an iron boiler and refined before being put into casks. The wool was stripped off the skins, washed and packed for export.

The trotters were boiled for their oil and the bones exported to make hafts for knives.[215]

The pastoral industry was lucky that there was a ready demand in England for the tallow which Australia was to produce. In 1841 England imported 1,241,278 cwts. (63,000 tons) of tallow, the majority of which came from Russia, all of which was kept for home consumption. At this time, it was estimated that England's domestic production of tallow was approximately 120,000 tons. Australia was also fortunate in that their tallow was only subject to a duty of 3d. per cwt., whereas that imported from Russia was subject to 3s. 2d. per cwt.

The process became popular very quickly, being described as the salvation of the Colonies. Sheep that were practically unsaleable in April and May of 1843, were worth five to eight shillings two months later due to the stability that boiling down had put into the market. Whereas wool used to be the main security against bank overdrafts, tallow was now taking that place, even if at a much reduced value. John Watson and E.B. Wright, who had started as general agents in Market Square, Melbourne, survived the mercantile crisis and established beside the Yarra below Batman's Hill a melting down works that during 1844 produced 650 tons of tallow and 25 tons of lard from 56,000 sheep and 2,500 cattle.[216] The boiling down of stock was the only remedy to keep stock numbers down, and as a result the security that could be obtained against sheep and cattle was only at the estimated boiling down price.

At this time, the candle market was expanding because the supply of whale oil for domestic lighting was declining. By 1849 Sydney factories were producing 60 tons of candles per year and 1,145 tons of soap. It took 2-3 tons of tallow to produce 1 ton of soap.

Large stock owners sent big numbers of sheep to be melted down. Between 10 December 1849 and the 6 February 1850, the Clyde Company put 8,500 sheep through the Barwon Melting Establishment.[217] On the 6 December 1850, they sent 4,000 fat wethers to the melting establishment[218], January 1851 another 5,000 sheep and 5,000 more in February.[219]

The capacity of the boiling down establishments to process stock increased rapidly with an estimated 20,000 sheep per week being processed in the Port Phillip area alone. In 1844, 217,000 sheep were boiled down in New South Wales and in 1845 there were 56 boiling-down establishments. In 1850 there were 110 establishments which boiled down nearly 800,000 sheep. In seven years, some 3.6 million sheep were boiled down in New South Wales.[220] From 1844 to 1849 the number of sheep slaughtered in New South Wales and Port Phillip (Victoria) was 1,565,752 and horned cattle 184,064, producing 440,186 cwt. (22,361 tonnes) of tallow.[221] The boiling down process continued for the rest of the century while there remained depressed sheep prices.

Livestock and its fat have been an important part of mankind's history. In biblical times offerings of fattened livestock were considered important and significant. In Genesis 4, verse 4, *'Abel brought of the firstlings of his flock and of their fat portions'* to the Lord as an offering *'and the Lord had regard for Abel and his offering'*. Again, in the account of the reaction of the father to the prodigal son (Luke 15, verse 23), the slaves were ordered to *'bring the fatted calf and kill it'* for the celebration.[222]

The importance and use of livestock fat perhaps reached its peak at the same time as Bakewell was breeding his, as we would describe them today, extremely fat livestock.

Today we have the opposite view about how much fat there should be on our livestock than was held at the time of Bakewell. We are looking for the lean carcass with just a small covering of fat, although we do like some marbling within the meat.

In Bakewell's time animal fat was important for reasons other than to be eaten. It was used in the making of candles and soap, and it also had important uses in the fast growing industries by providing lubricants and artificial light. Tallow candles were the common household candle for Europeans, and by the thirteenth century candle making had become a guild craft in England and France.

The New Leicester would fatten ready for market by two years of age, whereas other breeds would take at least three years to be marketable. There was a demand at that time for what we would today describe as grossly over fat stock. The New Leicester was described as being never too fat for such hardworking people as keelmen and pitmen.[223]

'A Breeder of the Coal-Heavers Mutton' (ie a breeder of the New Leicester), writing in 1803 states the value of the fat as a good food source.

> *'The fat is frequently cut off from the surface of these animals, but not by the butchers before they bring it to the market, for the purpose of making soap and candles, but is taken off from the fattest parts by these said manufacturers, then shred and*

put into dumplings, a most nutritive, excellent, and, I am persuaded, wholesome dish, of which I have eaten many a time with great pleasure, and with a natural, unvitiated appetite, produced by hard work. Nay, I can tell him more, that the writer, when hungry from the plough, has frequently dipped toasted bread into the drippings of this delicious mutton, whilst roasting at the fire, and then eat the bread with pleasure and avidity!'.[224]

Marshall states that it was common for Bakewell's breed to have such a projection of fat on the ribs that it could be easily gathered up in the hand, as on the flank of a fat bullock. It was thus termed the foreflank, '*a point which a modern breeder never fails to touch, in judging of the quality of this breed of sheep.*'[225]

Tallow prices were, during Bakewell's working life, c.1750-76, higher than the price paid for mutton, and it would therefore seem reasonable to presume that the butchers would pay a higher price for any breed that produced large quantities of loose fat. But this was not necessarily the case as they tended to pay in relation to the quantity of meat on the carcass, the tallow, hides and bone which they did sell gave little financial return to the grazier.[226] While Bakewell's main interest was in feeding the population with his faster maturing breed of sheep, this extra price obtained for the fat must have had some influence on his decision to favour the fatter sheep, or at least not discriminate against them.

These excessively fat sheep remained popular in the marketplace well after Bakewell's death. In 1820 it was reported that, '*the fattest and completest animal ever killed in this kingdom'* was killed in Birmingham market which had eight and a half inches of fat upon its back. It was obviously of some age as it had a live weight of 280 lb (127 kg) and a carcass weight of 200 lb (90 kg).[227] This very high dressed out weight can perhaps be put down to the large amount of fat on the carcass.

With the sheep of Bakewell's time, the consideration of its ability to quickly fatten started with the rams to be used. Marshall stated that *'the fattest rams are of course the best,'* because fattening quality was the one thing needed in grazing flocks, and that it was found to be to some considerable degree hereditary.[228]

It was not rare for New Leicester rams to be, *'cracked on the back; that is, to be cloven along the top of the chine, in the manner fat sheep generally are upon the rump. This mark is considered as an evidence of the best blood'.*[229]

One of the complaints raised by farmers of Bakewell's quick fattening breed of sheep was a reduction of the internal hard fat (tallow) that they produced. Tallow was a valuable material for making candles and some kinds of soap. *'As a farmer, who must pay rent for, and support a family, on my farm, I must say, that the small quantity of tallow is an objection to them, especially to what is called the very highest bred stock. The butchers, as judges, I set aside, as they always buy whatever kind of stock they have their profit on, and the farmer should know pretty nearly what tallow stock bred on his farm will carry'.*[230] The New Leicester laid a great quantity of fat upon the bones and was seldom considered by the butcher's to 'die well'. Marshall stated that *'Tallow is a kind of boon which, if not forthcoming, incurs a disappointment the butcher cannot brook'.*[231] The prices obtained for mutton and tallow were quoted separately with tallow often selling at a higher price than mutton, for example mutton 4d. per lb and tallow 5d. per lb.[232]

The amount of tallow that a sheep produced depended on the age at which it was butchered and its breed. If the New Leicester was kept to three years of age their amount of tallow would increase. The Norfolk breed (a horned, black faced breed) of sheep, if butchered at two years of age was remarkable for its amount of tallow.[233]

While they were looking for sheep that put on the largest amount of fat the quickest, they did realise, as we do today, that the over fat ewes did not get in lamb as readily and the rams did not have the vigour if too fat. As a breeder of the modern day Leicester, I can say from experience that the breeds ability to become over fat is one thing that has to be closely watched, especially close to mating time as the ewes that are carrying too much condition will not conceive. This is one of the characteristics that Bakewell so strongly set in the breed, and we have to watch for it in our management some 200 years later.

So why were the Leicester and its crosses the subject of so much debate at this time of very low prices for both wool and sheep meat? The Leicester with its big frame and carcass and its ability to put on large amounts of fat if run under good conditions meant that even in the situation of last resort, that is boiling down, it proved to be more profitable than the smaller framed Merino sheep. This fact did not go down too well with the fine wool Merino breeders of the time as most of them did not even like the idea of a Merino Leicester cross.

THE HISTORY OF ENGLISH LEICESTER SHEEP IN AUSTRALIA

As early as 1843 there were those for and against the Leicester. 'Carnaby', writing to the Sydney Morning Herald in June 1843, states that had the settlers run more of the larger framed sheep such as the Leicester and Southdown, they would have felt very little of the depression of the times. He goes on to state:

> *'... that the Leicester sheep, in every respect, is very considerably more profitable to the grazier than the Saxon or Merino, both as to wool and carcass: and now that the idea of boiling down prevails, its advantages are greater than ever, and brings my former observations to a stricter test. I am quite persuaded that the Leicester sheep is at least 200 per cent more profitable than the Saxon, or crosses of the Merino.'*[234]

The debate regarding the value of the Leicester was still going on at the end of the 1840's, some seven years after boiling down had taken off. While there was a lot of country to which the pure Leicester was not suited, its crosses were certainly suited to a large area but the sway which the Merino and its fine wool held was very hard to break even if there was a more profitable alternative. This prejudice against the strong-woolled, large framed Leicester is summed up in the following article.

> *A flock of 3000 wethers, the property of Mr WJT Clarke, was last week boiled down at the establishment of Messrs. Philpot, and yielded an average of 40 and one eights lbs. each. There is something in this which should be suggestive to our*

squatters. It is almost unnecessary to say, that these wethers were nearly pure Leicester, a breed for which Mr Clarke has long been celebrated. At this time, when the price of wool is putting our settlers so severely to their shifts, we think the question might fairly be argued, whether a very large portion of their number have not long been on the wrong track altogether; whether the wool bale has not held too sovereign a supremacy over the tallow cask. We know that the Leicester cry is an unpopular one; that the fine wool rivalry (wholesome enough in its way) has led to an utter intolerance among the settlers of any mention of a cross; and a hint of the superior profitableness of the improved Leicester, as compared with the pure Merino, is almost universally denounced as downright heresay.[235]

Boiling down did enable management practices to be improved. While the wool prices had remained good there was an incentive to keep every sheep regardless of quality or disease. The boiling down process was the excuse to cull heavily, thus improving the overall quality of the country's flock. It also provided an outlet for diseased sheep, they were not worth treating, just fattened and sent to the melting works. The Clyde Company gives a good example of this type of management.

> *About three weeks ago a flock of 2,052 young Ewes (2 Tooth) got infected with Scab from some Rams that were purchased as clean and put to them; as they were in tolerable condition, and to save the expense of cleaning them, as well as to avoid the risk*

of infecting the other flocks, I considered it the most advisable mode to melt them for their Tallow also.[236]

The boiling down of surplus sheep continued strongly into the 1890's. While the freezing works were an outlet for the better quality sheep, there were still large numbers being boiled down in the 1890's. New boiling down plants were still being built. In October 1893, 70,000 sheep were waiting to be processed through the Albury plant and at Nyngan, NSW, two plants had some 200,000 sheep booked in.[237]

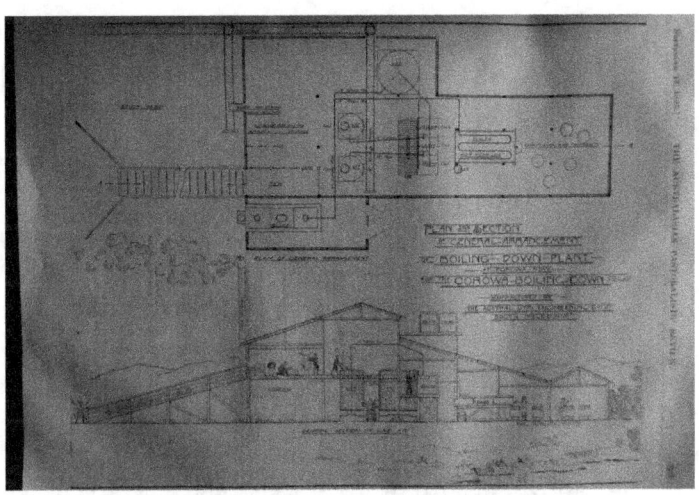

Plans for the Corowa Boiling Down Works.

Boiling down plant.

Despite all the adversity thrown at the pastoral industry from the 1840's, the sheep population of the eastern colonies grew from about 5 million in the early 1840's to over 90 million by 1892.

While we may today look at the boiling down era as just another episode of survival in the Australian pastoral industry, it did perhaps have a greater positive impact that was to last for many years. It initiated changes from which a stronger and more soundly based industry went forward until the next over expansion and depression occurred in the 1890's.

While the boiling down of surplus stock gave some relief to farmers it was not a permanent solution to the fact that Australia was producing ever greater numbers of sheep and cattle with no market for them. Australia's isolation from the consuming countries of the world became more and more apparent to all stock owners.

It was on 6 December 1879, that the steamer *Strathleven*, left Melbourne carrying the first cargo of frozen meat bound for the London market. While some freezing works had already been established in Australia, the first in the world being established in 1861 by T.S. Mort at Darling Harbour in Sydney, no cold storage facilities had before been fitted to a vessel trading between Australia and England. In 1877 the *Paraguay* took a cargo of frozen meat from Buenos Aires to Marseilles, this being the first entirely successful frozen meat shipment in the world.

Just before the *Strathleven* sailed from Williamstown she was visited by a number of pastoralists and leading Melbourne men to inspect the Bell-Coleman preserving system installed in the vessel. Part of the hold in the after portion of the ship was used for the purpose. The meat chamber was sealed by means of boarded sides and caulked to prevent the escape of air. The entrance to the freezing chamber was a porch with two doors, one to be shut before the other was opened. The vessel had been partly loaded in Sydney where 400 sheep and 55 head of cattle were placed on board. All the sheep were Merinos with some of them being un-skun. An additional 75 sheep and 2 lambs were obtained from Mr Bennett of Melbourne. Altogether there were 26 tons of meat and a few kegs of butter on the *Strathleven* when she left for London.[238] On arrival the London Central Markets reported favourably on the consignment and forecast a ready sale for similar supplies.

The trade in frozen meat was slow to take off. Ten years after the *Strathleven* sailed, New South Wales still earned more by exporting cases of tinned meat than carcasses of frozen meat, but by the last ten years of the century frozen exports were becoming a valuable source of Australian wealth.

Geoffrey Blainey sums up the beneficial effects of the advent of this outlet for Australia's surplus stock as:

Refrigerated ships raised the standard of living in millions of European homes and provided a new source of wealth to thousands of Australian farmers and pastoralists. Few innovations in the nineteenth century did more to improve health in one part of the world and to relieve waste in another.[239]

The English Leicester was one of the breeds to benefit from the frozen meat trade. With the explosion in sheep numbers (except for the periods of drought), particularly in Queensland and New South Wales, British breeds were required to complement the numerically dominant Merino. The English Leicester, along with the Lincoln, was the breed of choice to put over the Merino to produce the first cross ewe. The Southdown then became a popular ram to mate with this first cross ewe, producing the export carcass. The flock ram breeders in the southern states had a ready market for their rams, and this is reflected in the increase in stud numbers during the first half of the 20th century.

The English Leicester showed its versatility and dual purpose ability during the period when sheep were basically worthless, as they proved to be more profitable in the boiling down process, and again when European markets became available their crossing ability was sought after.

British Breeds lamb advertisement 1949. Australian Farm & Home, 1949.

12
Quarantine

Australia's first quarantine law was enacted in New South Wales in 1861 to prevent the importation of animals from countries in which known infectious diseases were present. In 1866, due to the fear of the introduction of disease, cattle, sheep and swine were included under the provisions of the Cattle Disease prevention Act (30 Victoria No. 11) of New South Wales. The introduction of animal stalls, fittings, fodder, litter, manure and other objects used in the working with animals on board ship, was prohibited unless it was proved that the animals had been in the Australian colonies for at least the previous six months. The Governor could exclude animals from the colony and ship masters had to sign a declaration that no overseas animals were on board.

The real beginning of animal quarantine in the colonies began with an Act in New South Wales to prevent the introduction of contagious and infectious diseases of cattle in 1871. The Act stated that all stock arriving by sea, not the produce of Australian colonies, were imported stock. It proclaimed the diseases of cattle plague (rinderpest), foot-and-mouth disease, sheep pox, and any other diseases which may be proclaimed.

Animal attendants who had looked after the livestock during the voyage had to remain on board until the inspector gave clearance and the ship's master had to declare if any stock were on board or if any had been carried by the ship. No stock or fittings could be landed without a permit, and no fodder could be landed. All stock went directly into quarantine.[240]

This Act was soon put into use as in 1873 a two-year ban was placed on the importation of sheep and cattle from Europe. Some British authorities were not pleased with the importation ban as they felt there was:

> *'No fear of disease being carried around the world by a few carefully selected animals, kept for three months on board ship, well tended, separately lodged in airy deck apartments, and subject on landing to further quarantine under the supervision of the colonial inspectors. ... Mouth-and-foot disease would show itself in a week after the animals were on board, cattle plague within a fortnight, and pleuro-pneumonia before half the voyage was over; whilst a very casual inspection should discover any scab which might have appeared amongst the sheep.'*[241]

There is at least one early record of sheep being destroyed on arrival in Australia. In 1886 the Victorian Government's Stock Department slaughtered the Leicester sheep which Mr Harper had imported into Victoria. The newspaper reports of the incident raised questions as to whether this was a disease based decision, or one brought, *'at the instigation of some influential Melbourne gentlemen who are interested in pastoral pursuits.* [242]

A conference of Chief Inspectors of Stock of New South Wales, Victoria, Queensland, South Australia and Tasmania was held in Sydney in the mid 1870's to consider a cooperative approach or other means of dealing with infectious and contagious diseases of stock. The conference recommended the continuation of a ban on the importation of stock, with the exception of horses, and an extension of the ban until certain diseases that were not in Australia had been eradicated from Britain. Another recommendation was for the Chief Inspectors to communicate with one another by the telegraph. Needless to say, the attempt to have uniform legislation failed because of intercolonial differences.

At the time of Federation in 1901 each state had separate legislation to keep out unwanted animal diseases and all had differing regulations to police their Acts. In 1909, in anticipation of the declaration of a Commonwealth Quarantine Act coming into force to replace the numerous State Acts and Regulations, a conference of Chief Veterinary Officers met to draft a uniform code. The Commonwealth legislation came into operation on 1 July 1909.[243]

Even before the Commonwealth Act came into force all sheep coming into Australia had to be quarantined in one of the quarantine stations in Sydney or Melbourne. While the Sydney quarantine station was situated on Shark Island in Port Jackson, separated from the mainland, the early Melbourne quarantine station was in Jones's bonded stores on the south bank of the Yarra River, an undrained flat close to the centre of the city. Tasmania's first quarantine station was established in 1901 at Nubeena, but it was felt by some importers that the accommodation and landing facilities left a little to be desired. This fact was the reason that George Simonds had his 1907 importations quarantined in Melbourne rather than Nubeena.

Live sheep imports into Australia were banned from all countries except New Zealand in May 1952. In 2016 imports of live sheep, semen and embryos were also banned from New Zealand due to the fear of Scrapie being introduced through these possible imports.

13

England Imports

> 'Nothing is of greater importance [to Australia] than the improvement of our sheep. The climate and soil both seem to agree well with sheep.'[244]

England has always been considered the premium place to purchase stud livestock from.

Phrases such as 'they can be traced back through a long line to Bakewell's flock', 'imported from the most famous Leicester flocks in England', or 'imported at considerable expense', are used frequently in the flock histories of many of the early Australian flocks. In some cases we can trace back fairly accurately to Bakewell's flock, and there is no doubt that a lot of imports were purchased from famous English flocks, but in other cases it is very difficult to trace the ancestry of the sheep concerned. When the breeders of a lot of the early imported sheep are mentioned, it is usually only their surname, no Christian name, that is given, with no address or place of farming. As there were often several breeders with the same surname, as has been the case here in Australia for many years, working out which person is being referred to, is difficult.

My assumption is that a lot of the early importers purchased their sheep from breeders that were in their local area, had had some association with, or were recommended by someone who purchased on their behalf. I would guess that the Bryant's would have purchased from the Somerset area and that Edward Sparke would have purchased from Devon. We can be fairly certain that Toosey would have purchased on behalf of The Cressy Company from Charles Champion due to his farming apprenticeship with him. During the 1900's a lot of purchases were made by agents on behalf of Australian breeders. Mostly these were Australian agents that often spent up to a year in England purchasing all types of stud stock for export to Australia. It is interesting to note the consistency with which some of the English breeders' names keep appearing. This could be due to their willingness to sell for export, their good stock, their reputation with the English livestock agent, or all of these.

After Bakewell's death his flock passed to the care of his nephew, Mr Honeybourne, who also took over the tenancy of 'Dishley Grange'. William Smith took over from Honeybourne in about 1825 and seems to have stayed there until his death. Philip Oakden stated that the sheep he and Henry Reed imported in 1837 (52 rams and ewes) were direct descendants from the Dishley flock. This may be correct as it is recorded that these sheep came from Richard Smith of Dishley, Richard Smith of Swarthstone (perhaps no relation), a Mr Coke, Joseph Burgess and John Simpson. Joseph Burgess's flock came from Nathaniel Stubbins flock who had also purchased sheep from both Honeybourne and Richard

Smith, so a double connection back to Bakewell's flock. The Richard Smith of Dishley, which Philip Oakden mentions, is a son of William Smith. Richard was born in 1803, and he was at Dishley in 1835 when his daughter Henrietta was born so it is quite possible that he was there in 1836 or 1837 when Oakden would have purchased his sheep.

It must be remembered that the British system of land ownership and farming was very much a system of land owner and tenant. While this system was practised in some places in early Australia, particularly in Van Diemen's Land, most of the land here was held freehold. In Britain at that time most of the land was held in large estates with sections being leased out to tenant farmers. These tenant farmers would often move to a new tenancy when their lease expired so it is sometimes hard to trace where they moved to, or from and when.

Honeybourne and Smith sold sheep to Stubbins of Stone Barford, Paget of Elman and Philip Skipworth the elder. The Skipworth flock was the foundation of the Aylesbury flock, and as far as I can tell there were not any imports from this flock. But from the Stubbins sheep we can trace imports. From the Stubbins flock was descended the once celebrated stock which was in 1814, divided between the nephews of Mr Nathaniel Stubbins, Joseph and Robert Burgess.[245]

Joseph Burgess was succeeded at Holmepierrpoint in 1834 by Mr Sandy. Sir Charles Sladen imported from Mr Sandy. While Mr Sandy Jr sold the Leicesters off in preference to Shropshires, the Holmepierrepoint blood lines were kept through Sir Tatton Sykes's flock as he commenced operations by the purchase of ten ewes from Holmepierrepoint.[246] John Crooks imported one ram from Sir Tatton Sykes in approximately 1868.

During my research I have continually been surprised to read about the numerous trips that were made back to England to purchase livestock. Both breeders and agents went there to make purchases, some of them making several return trips to purchase stock, and also, undoubtedly do other business while in England.

Edmund Bryant, Edward Sparke, W.J.T. Clarke, Philip Oakden, George Simonds, Sir Richard Dry and E.W. Oliver were some of those who made one, or sometimes two trips to England to purchase livestock. These would have been long, and we can presume, expensive trips with a considerable amount of money being spent on procuring high quality animals. Then there was always the high risk that some of the animals would die on the long voyage out to Australia, particularly in the early days when sailing ships were the only means of conveyance. Some of these people would have purchased sheep solely for themselves, or perhaps one or two others as well.

Alfred Mansell & Co, of Shrewsbury, were the largest exporters of pedigree livestock in Britain. While they concentrated on Shropshire sheep, they exported all types of livestock to all parts of the world. They were the agents used by George Simonds when he purchased 254 sheep, 189 of them being Shropshires, for export back to Australia in 1906.

George Simonds was a company director and manager of the livestock activities of Hobart stock and station agents Roberts & Co.. His 1906 buying trip to England was his second trip there buying sheep, cattle, horses and poultry for clients in New South Wales and Victoria, but mostly for Tasmanians. Mr Simonds left Hobart on 1 November 1905, and arrived back in Melbourne on 16 December 1906, a 13-month buying trip. In total he purchased 254 sheep comprising of 180 Shropshire ewes and 9 rams, 37 Leicester ewes and 2 rams and 26 Lincolns, plus 3 bulls, 5 racehorses and a number of pigeons and fowls.

Leicester ewes were purchased from Mr George Harrison, Gainsford Hall. Ten for Mr J.B. Rowe of Noble Farm, Sorell, ten for himself, ten for Mr Button of Oatlands and six for Mr Lester. He also purchased two ewes from Mr Jordan for my great grandfather, R.G. Heazlewood. All the ewes were supposed to be in lamb.

> *I bought from Mr Harrison because I had previously purchased sheep from him which turned out marvellously well, notably the ram with which Mr Heazlewood took the championship in Melbourne last year.*

The Leicester Sheep-Breeders Association, UK transfer. Brenton Heazlewood.

Winning championships was not a novel experience for my great grandfather, Robert Heazlewood. Between 1896 and 1920 he won at least 18 at Royal Shows in Melbourne and Sydney.

Mr Simonds also purchased two rams from Mr Harding of Gainsford Hall. He does make the comment that, *'it is no easy matter to get good Leicesters, or good Hampshire Downs, as these two breeds of sheep are in comparatively few hands.'*[247]

Mr E.W. Oliver, Chudleigh, Tasmania, was also in England at this time purchasing sheep for himself and others. It was Henry Reed (the 2nd) who had arranged for Mr Oliver to go to England, and if possible, purchase the two best rams in England for him. It is obvious that Mr Oliver and George Simonds must have co-operated in their purchases as all the sheep were transported back to Tasmania together, after having completed a quarantine period in Melbourne. Some of the purchases are also mentioned in both reports.

> *It will be remembered by breeders that Mr E.W. Oliver the well-known Chudleigh breeder of Leicesters, recently returned from England where he purchased a number of stud Lincoln, Leicester, Southdown and Cotswold sheep for various breeders in this state. His selection arrived by the Loongana from Melbourne yesterday. The whole consignment, numbering 30, arrived in first class order, not a sick or sorry animal being among them. A number of the leading breeders and judges inspected the sheep on their arrival, and it was freely remarked that they comprised the best consignment of the various breeds*

ever imported from the old country to Tasmania. Mr Oliver had no limit to bind him in his purchases, but was instructed to bring the best sheep obtainable, and it is only fair to say that he has justified the confidence placed in him by his clients, as among the consignment are to be found some of the cream of English flocks. During his visit Mr Oliver was promised by the English Leicester Sheepbreeders Association a cup valued at five guineas, to be competed for at the Launceston show, and to be won three times by the same exhibitor. This will be welcome news to our Leicester breeders, and competition should be exceedingly keen as there is very little to choose between a number of the sheep imported. A few particulars as to the breeding etc, of the sheep will doubtless be interesting to readers.

Leicesters

Mr Oliver purchased for Mr Henry Reed, of Logan, Evandale, two rams, one two-tooth, bred by Mr G. Harrison – Unionist, sire Gainsford Exchange, dam champion ewe. This ram has never been exhibited, but was used by Mr Harrison in this stud. Mr Oliver thinks this is the best woolled Leicester he has ever seen. The other is a ram lamb, bred by Mr E.F. Jordon (breeder of the winner Royal Show 1906), and was in the prize pen of three rams at the Royal Yorkshire and Driffield shows. This ram was the pick of the flock of ram lambs. He is a very promising youngster, and should be heard of in our show pens this year. He also purchased for Mr Reed 20 ewes from Mr

Jordons flock including 13 two-tooths and 2 four-tooths. These are in lamb to selected rams, some to Royal winners. The other five are ewe lambs.

Mr R.G. Heazlewood of Glenore has two rams and two ewes. One of the rams, Pilmoor Champion, was bred by Messrs. Simpson Bros., the oldest flock in England, being handed down for generations from father to son. He was sired by Neptune. This ram took second prize at the Yorkshire Show, and in some judges' opinion is equal to the Royal winner. He was wanted for New Zealand but the owners would not sell until they had used him in their flock. The second ram, Humanity Compact, was bred by Mr John Cranswick, and was sired by Gaily. He is an exceptionally fine sheep. The two ewes were both bred by Mr E.F. Jordan. They are in lamb to Royal Park, the Royal winner, and should drop something exceptionally prime.

Mr Oliver's purchase for Mr F.T. Holmes was one ram and three ewes. The ram was bred by Mr E.F. Jordan – Royal Colonist II, sired by Royal Colonist I. This ram was third at the Royal Show, and Mr Jordan thought he should have beaten his champion ram. The three ewes were bred by Mr John Cranswick. These were the reserved number at the Royal Show, which speaks for itself, and are in lamb to a selected stud ram. One ram, bought for Mr John Chugg, of White Hills was bred by Mr E.F. Jordon. He was never exhibited, but was used by Mr Jordon in his stud. Mr Oliver considers this a typical

Leicester, and one that should stand well in the show ring in any company. Mr Oliver purchased for Messrs. John Badcock and Son, of Glenore, a ram lamb, Pilmoor Leader, bred by Messrs. Simpson Bros. this sheep was the pick of the flock of ram lambs, and he had great trouble in striking a bargain. One ewe, bred by Mr G. Harrison, was in the second prize pen at the Royal Show, and is in lamb to Unionist, the ram purchased for Mr Reed. Mr Harrison considered this ewe one of the best he had ever handled. Messrs. A. Oliver and Sons secured Pilmoor Selected, bred by Messrs. Simpson Bros., but purchased and used by Mr E.F. Jordon in his flock. He is a fine animal and a typical Leicester, but has never been exhibited.[248]

The Leicester Sheep-Breeders' Association.

CERTIFICATE OF THE BREEDING OF A LEICESTER ~~ram~~ ewe EXPORTED, OR SOLD FOR EXPORTATION, FROM THE UNITED KINGDOM.

Name and Number *Lady Seamer* 401 (Ear Tag) 58   Lambed 1903
Breeder E. F. Jordan
Sire Pilmoor Exchange 674
Grand Sire Seamer No. 2, 474
G.G. Sire Rupee 331

BREEDER'S CERTIFICATE.

I hereby declare that I bred the ~~ram~~ ewe above referred to, named 401 (Ear Tag 58) and that the foregoing pedigree and particulars are correctly stated.

Signature E. F. Jordan
Address Eastburn, Driffield
Dated Nov. 14th 1906

VENDOR'S CERTIFICATE.

I hereby declare that the Animal above referred to has been sold by me to Mr. R. G. Heazlewood of Tasmania for exportation to Tasmania.

Signature E. F. Jordan
Address Eastburn, Driffield
Dated Nov. 14th 1906

CERTIFICATE OF THE ASSOCIATION.

I hereby certify that I have examined the particulars above stated with the entries in the Flock Book of the Association, and that I have found them to be correct.

DRIFFIELD, Yorks.
November, 1906.   W. A. Brown

Jordan ewe transfer to RGH. Brenton Heazlewood.

Simpson ram transfer to RGH. Brenton Heazlewood.

Mr Oliver also purchased Lincoln, Southdown and Cotswold sheep for Tasmanian breeders. The whole consignment consisted of 34 Leicesters, 9 Lincolns, 6 Southdowns and 1 Cotswold. In relation to Leicesters it is interesting to note that Mr Oliver made the comment that the English style of sheep had changed greatly during the previous 20 years, with the breeders getting away from the extremely fine head and bones, and are going in for robustness.

The export trade in British stud sheep had been increasing for some years prior to these exports to Australia. In 1894, 4,638 stud sheep were exported from Britain at an average value of eight pounds, ten shillings and five pence. By 1897 this had increased to 11,569 sheep at an average value of twelve pounds, four shillings and eleven pence.[249]

After Mr George Simonds's trip to England during 1906 he summarised this British export trade.

> *'I was greatly impressed with the enormous extent of the export trade in stud stock from England to all parts of the world. All foreign countries send to England for stud animals of every kind, from birds to horses. They all go to England for the best. There is no doubt they have got the finest sheep and cattle of all standard breeds in the world.'*[250]

The Leicester breeders in England realised the value of this trade as they comment on it in their annual report presented at their Annual Meeting held on 7 February 1907.

'The expansion of the foreign trade in Leicesters was alluded to in the report for 1905. The sales for export this year have exceeded all previous records, 120 certificates having been issued for sheep of both sexes for Tasmania, New Zealand, New South Wales, the Argentine Republic, and Spain.'[251]

To compile an accurate list of all Leicester sheep imports into Australia is impossible. The list below is obtained from Flock Books and other sources but is by no means complete. I would guess that this list is only a small percentage of the sheep that were actually imported.

Also, the list of English breeders from whom the sheep came has only been compiled from information I have been able to source. I have tried to give some history of these English flocks, particularly if there is any relationship going back to Bakewell's flock.

THE HISTORY OF ENGLISH LEICESTER SHEEP IN AUSTRALIA

Importer	Date	Number	Breeder
Bryant Bros., TAS	1825	3 Rams & 2 Ewes	Thomas Ewer
Cressy Company, TAS	1826		Charles Champion
Edward Sparke, NSW	1827	60 sheep	
W.J.T. Clarke, TAS	1829	20	
VDL Company, TAS	1830	?	?
George Stokell, TAS	1838?	?	Mr W. Sandy
			Mr Hodgson of Gainsford
			Mr Clarke of Old Crough
			Mr Carter of Scales, Yorkshire
			Mr Mason of Hopper
Robert Jones & Samuel Blackwell, TAS	?	?	Mr Burgess
			Mr Stubbins
			Mr Stone
South Australian Co.	1836		
Australian Agricultural Company, NSW	1837	3 Rams & 12 Ewes	John Wilkinson
Philip Oakden & Henry Reed, TAS	1837	52 Rams & Ewes	Richard Smith of Dishley
			Richard Smith of Swarthstone Coke
			J. Burgess
			John Simpson
Bryant Bros.	1838?	8 Rams	John Ewer, W. Shepperd?
T. Robinson, NSW	1840	4 Rams	
AACo.	1841	3 Rams	Burgess
		2 Rams	Pratt
Hugh Wallace, NSW	1841	2 Rams	Wilkin of Dumfriesshire
		4 Ewes	Watson of Perthshire
William Field, F24, Enfield, TAS	1847	Ewes	R. Forrester, Raeselands, Penrith
		Ram	Bell of Brompton
		Ram	Bell of Scale Hill
		Ram	Robinson of Leakby
		Ram	Palace
		Ram	Beatie of Neeby House
			Duke of Devonshire

BRENTON HEAZLEWOOD

John Stokell, TAS	1856-58	Sheep	Hodgson, Roppleny Carter, Michuion Mason, Hoppen of Marton
Sir Charles Sladen, VIC	1872	2 Rams	Hutchinson
Thomas Gibson, Eskvale, TAS	1863	2 Rams 14 Ewes	George Turner
John Trethewie, TAS	1865?	?	Turner Tremain
John Crooks	1868?	1 Ram 'Sir Tatton' 1 Ram	Sir Tatton Sykes Joshua Bates
?	1871	40 Sheep to Moreton Bay	?
Hon. James Taylor, Toowoomba, QLD	1871	2 Rams 5 Ewes	? Bolton
Oliver Bros., F7, Chudleigh, TAS	?	1 Ram 'Gainford Victor'	G. Harrison
Mr Woods, Dennistoun, TAS	1880	3 Rams 1 Ram 'No 3'	John Barton, Malton, Yorkshire
McArthur Bros., F4, Hillside, Vic	1905	2 Rams	G. Harrison
R.G. Heazlewood, F6, Whitemore, TAS	1905	1 Ram 'Lord Sheffield No.18'	G. Harrison
James Woods, F22, Almond, VIC	1906	1 Ram	G. Harrison 'Gainsford Eclipse'
R.G. Heazlewood, F6	1907	1 Ram 'Pillmoor Champion' 1 Ram 'Eastburn No. 183' 1 Ram 'Hunmanby Compact' 2 Ewes	Simpson Bros. E.F. Jordan John Cranswick E.F. Jordan
John Badcock, F9, Whitemore, TAS	1907	1 Ewe 1 Ram	Harrison Simpson
Henry Reed, F11, Wesley Dale, TAS	1907	2 Rams, 20 Ewes	E.F. Jordan ('Pastoralist & Eastburn Champion')
Percy Hart, F19, Westbury, TAS	1907	1 Ram, 6 Ewes	E.F. Jordan
G. Holmes, White Hills, TAS	1907	1 Ram 3 Ewes	E.F. Jordan J. Cranswick
J.B. Row, Sorell, TAS	1907	10 Ewes	G. Harrison
G. Simonds, Stonor, TAS	1907	10 Ewes	G. Harrison
Button, Oatlands, TAS	1907	10 Ewes	G. Harrison
Lester, TAS	1907	6 Ewes	G. Harrison & W. Marshall
J.H. Fairchild, F8, Lang Lang, VIC	1907	20 Ewes 2 Rams	G. Harrison

J.R. & A.W. Carter, F7, Clunes, VIC	1910	1 Ram	G. Harrison
J.H. Fairchild, F8	1910	2 Rams	G. Harrison
J.H. Fairchild, F8	1911	1 Ram	G. Harrison
		6 Ewes	E.F. Jordan
J.H. Fairchild, F8	1912	2 Rams, 10 Ewes	??
J.H. Fairchild, F8	1915	1 Ram	G. Harrison
W.S. Kelly, F29, Giles Corner, SA	1923	1 Ram 'Driffield'	E.F. Jordan
McArthur Bros., F25, Lindenow South, VIC	1923	1 Ram 'Gainford Connie'	G. Harrison
McArthur Bros., F25	1929	2 Rams 'Garton No1 & No2'	R. Megginson
H.B. Slaney, F66, Mornington, VIC	1929	2 Rams 'Garton No3 & No4'	R. Megginson
H.H. Shillabeer, F77	1935	1 Ram 'Garton Royal Jubilee'	R. Megginson
T.G. Stancombe, F157	1939	5 Ewes	Exors. R.H. Stocks
W.G. Spencer, F69, Grass Valley, WA	1945	2 Rams 'Haywold Resolution', 'Haywold Victory'	Sir William Prince Smith
L.J. & A.L. Graves, F209, Mansfield, VIC	1945	2 Rams 'Garton David', 'Garton Winston'	Megginson
		3 Ewes	Megginson
E.P. Hart, F262, Mole Creek, TAS	1946	6 Ewes	C.E. Simpson
R.K. Heazlewood, F36, Whitemore, TAS	1946	1 Ram 'Speeton Sample'	W.A. Coleman & Sons
		1 Ewe	C.E. Simpson
Nicholas Cole, F12, Camperdown, VIC	1950	1 Ram 'Haywold No33'	Southburn Estates
F.N. Everitt, F140, Duranillin, WA	1951	1 Ram 'Taranaki'	T.A. Stephenson & Son

There is not a lot of information regarding the purchase price of sheep in England, or of the freight and extra costs involved in bringing the sheep to Australia.

In 1837 the AA Co paid four pounds for ewes which equates to approximately $840 in 2020 values, a high price for ewes.

The cost associated with the 1837 purchase of 52 sheep by Henry Reed and Philip Oakden were recorded by Oakden as follows.

Account of Bulls & Sheep purchased in joint account of Henry Reed Esq. & Philip Oakden.

1837

Oct 9	To Sundry expenses on Sheep	4	18	6
30	" Cash paid Richard Smith Dishley for Sheep	44		
" " "	Richard Smith Swarthstone for sheep	43	10	
" " "	Fox for Cokes Sheep	13	9	
" " "	Allowance to Elly	2		
	" Sundry expenses on Sheep 5 4 10 Turnips 20/-	16	4	10
Nov 2	" Cash paid J Burgess for Sheep	66		
	" Cash paid Hon. John Simpson for Bulls and Sheep	91		
7	" Cash Freight sheep to Liverpool	11	1	11
9	" Sundry expenses in bringing sheep to Liverpool	8	1	10
	" Allowance off the freight	2	12	10
10	" Cash sundries for sheep on board	1	9	4
	" Do--------Do-----------Do------	1	6	6
	" Preserved meats for Walker	3	6	6
11	" Insurance on 680 pounds & duty less brokerage & discount	13	7	6
	" Freight on 52 sheep 5 pounds each	260		
	" Freight on 2 Bull calves 22 ½ pounds each	45		
		609	10	4
	Mr Reid's share of the above	304	15	2
	Interest from 7th Nov 1837 to 7th April 1838 @ 5%	6	7	0
		311	2	2
	Interest from 7th April 1838 to 10 August @ 10%	10	13	0
		321	15	2

Philip Oakden and Henry Reed had neighbouring properties at Chudleigh, Oakden had 'Bentley', and Reed 'Wesley Dale'. They not only shared an obvious interest in agriculture with Leicester sheep being part of that, but both were successful merchants, both were evangelical Methodists sharing a common interest in the establishment of such Launceston bodies as the Mechanics Institute, the Benevolent Society, Hospital and Sunday Schools.

Oakden does record the freight cost for the sheep, Liverpool to Launceston, as five pounds each (approx. $1,050 in 2020 value), but not the actual price per sheep nor the numbers purchased from each vendor.

THE HISTORY OF ENGLISH LEICESTER SHEEP IN AUSTRALIA

If we can guess that four pounds (approx. $840 in 2020 value) was paid for each sheep, then these two costs alone amount to approximately $1,900 (2020 value) per head, plus all the other extra costs.

The one cost not accounted for is the deaths that occurred in freight. In this case they lost 20 sheep, landing 32 alive. While insurance may have perhaps covered the initial purchase price, they would still have been considerably out of pocket for all the other costs.

When George Stokell purchased eight sheep in about 1856 it is recorded that he paid ninety three pounds for them. We do not know how many rams or ewes there were in this purchase, but as this equates to approximately $2,400 per head in 2020 values, my guess is that they were all rams at this price.

The most detailed breakdown of the costs involved in procuring English sheep that I have, comes from a detailed invoice to my grandfather, Roy K. Heazlewood, from the Launceston agents, Allan Stewart Pty Ltd. This relates to his 1946 purchase of the ram 'Speeton Sample', from WA Coleman & Son, and one ewe from Charles Simpson. The ram cost 57 pounds, 15 shillings, in 2020 value being approximately $4,000. The ewe cost 21 guineas, in 2020 value being approximately $1,500. These two sheep were shipped to Tasmania with six other Leicester ewes.

The English charges in relation to their export were (I have stated the costs per sheep in approximate $AUS as in 2020[252]):

- Freight: $1,170

- Carriage to quarantine: $209

- Pens: $570

- Fodder and bedding: $210

- British quarantine: $209

- Attendant's passage: $210

- Marine insurance: $325

- Exchange and charges: $1,225

- Bill of Lading, Entry, Port dues etc.: $26

- Post & Petties: $3

- Agency (Shipping): $58

- Commission to New Zealand Loan: $27

The Australian charges were:

- Wharfage paid Melbourne Harbor Trust: $41

- Inspection fee to Dept of Ag.: $66

- Stacking charges on crates: $13

- Shearing sheep in quarantine: $5.75

- Disinfection fees on crates: $8.50

- Green lucerne and cartage: $5.75

- Cartage quarantine station to Taroona & Fodder for 3 weeks in quarantine: $75

- Bill of Lading and Stamp shipment: $0.25

- Freight Melbourne to Beauty Point: $58

- Cooper's Livestock Transport, transporting Sheep to boat and feeding: $148

The Launceston charges were for wharfage, Inspection, gratuity and agency: $12

This amounts to a total of approximately $4,675 per sheep, more than the cost of the ram and more than three times the purchase price of the ewe.

Melton Vale imported ewe, Australia. Australian Farm & Home, April 1949.

Melton Vale ewes by the imported ram Speeton Sample.
Australian Farm & Home, April 1949.

Notes

George Stokell imports were from:

- Mr W. Sandy of Holme Pierre-Point

- Mr Hodgson of Gainsford, Yorkshire

- Mr Clarke of Old Crough

- Mr Carter of Scales, Yorkshire

- Mr Mason of Hopper, Yorkshire (Same Mr Mason as below?)

John Stokell imports:

- Feb 16, 1857 letter. From Newbiggin, Sadberge, Darlington. 8 Leicester sheep cost 93 pounds & signed by H. A. Harrison.

- Sept 27, 1857 letter. Same address & mentions Leicester sheep purchased from:

- Mr Hodgson, Roppleny in Guilford, Yorkshire

- Mr Carter, Michuion, Yorkshire

- Mr Mason, Hoppen of Marton, Yorkshire

- Signed by H A Harrison of 'Newbiggin House'

Simpson, Pillmoor House, Hunmanby, Yorkshire

Flock No. 2 'Pillmoor'

The history of this flock is recorded in Vol.1, Improved Leicester Flock Book, 1893. It is in the name of John Jordon Simpson. Ewes put to the ram, 165.

> *This flock has been in the Owner's possession 39 years, taken from his late father, Mr. John Simpson, Hunmanby, and the late Mr. G Simpson, Marton. It had been in existence over 30 years* [established approx. 1824]. *Rams from the Improved Leicester type have been used in the Flock from the late Mr. G Walmsley, Rudston; also from Mr. T C Dixon, Brandesburton Barff, Mr. H P Robinson, Carnaby House, Mr. R Fisher, Leckonfield, and Mr T Spink, Hunmanby.*

John Cranswick, Field House, Hunmanby

Flock No. 3 'Hunmanby'

The history of this flock is recorded in Vol.1, Improved Leicester Flock Book, 1893. It is in the name of John Cranswick. Ewes put to the ram, 300.

> *This flock has been in the owners possession 13 years, being established from Ewes bought of the late Mr Reaston, Carnaby, Mr Keith, Carnaby, and Mr Etherington, Spring Dale, also from Mr T Lee, Field House, Hunmanby. Those from Mr Lee had been bred by the late Mr J Simpson, Hunmanby, that flock being in existence nearly 70 years.*

Charles Champion was a breeder of prized Durham cattle as well as Leicester sheep. When Champion commissioned Thomas Weaver to paint his *Blyth Comet Ox* in 1818, Weaver included four Leicester wethers in the foreground. The inclusion of the New Leicesters in the front of the ox gave the picture more interest, as well as promoting two breeds that were becoming popular.[253]

George Turner, Beacon Downs, Exeter. He sold sheep to Thomas Gibson, 'Eskvale', and supplied a certificate which is dated 'Beacon Downs, Exeter, August 1863,' and states: 'The fourteen ewes and two rams, sold to Messrs. Thornton (evidently the agents through whom the purchase was made) are pure English Leicesters, and have been bred for the last seventy years from the pure flocks of Messrs. Stubbins, Burgess, Stowe and Langdale. They have won more prizes at the Royal Agricultural Shows than any other flock in Great Britain.' The certificate is signed George Turner.[254]

Mr **Paget** of Ibstock, was an associate of Mr Bakewell, and an ardent and successful breeder of the new Leicester sheep. His stock of ewes was sold by auction on Nov 16, 1793. Buyers with perhaps connections to early imports.

S Stone, 5 shearlings @ 30 Guineas each

    5 shearlings @ 22 Guineas each[255]

The **Burgess** flock has been purely bred from the original stock of Bakewell for upwards of fifty years.[256] In 1809 Burgess farming at Hugglescote, Leicestershire.

The original Dishley stock went to the relatives of Bakewell, Mr Smith and Mr Honeybourne. These two sold to Stubbins of Stone Barford, Paget of Elman and Philip Skipworth the elder.

The Skipworth flock was the foundation of the Aylesbury flock.

From Stubbins sheep was descended the once celebrated stock which was in 1814 divided between the nephews of Mr Nathaniel Stubbins, Joseph and Robert Burgess.

Joseph Burgess was succeeded at Holmepierpoint in 1834 by Mr Sandy.[257] Mr Sandy retired from sheep breeding in 1862 and sold all his Leicester sheep in three sales held in 1862 and 1863.

English breeders from whom known sheep were purchased

- **Barton John**, Barton House, Malton Yorkshire

- **Bates Joshua**

- **Beatie**, Neeby House

- **Bell** of Brompton

- **Bell William**, Seale (Scale) Hill, Cumberland

- **Bolton**

- **Burgess Joseph**

- **Burgess Robert**

- **Burgess W,** near Homepierpoint, Notts, (mentioned in Oakden letter to AA Co)

- **Champion Charles**, Blyth, Nottinghamshire

- **Clarke**, Old Grough

- **Carter**, Scales Yorkshire

- **Carter**, Michuion

- **Cranswick John**

- **Coleman WA & Sons**

- **Coke**

- **Duke of Devonshire**

- **Ewer Thomas**

- **Forrester** of Raiselands, Penrith

- **Fox**

- **Harrison George**, Gainford Hall, Darlington, Durham

- **Harrison HA**, Newbiggin, Sadberge, Darlington

- **Hodgson**, Roppleny in Guilford, Yorkshire

- **Hodgson**, Gainsford, Yorkshire

- **Hutchinson**

- **Jordan EF**, Eastburn, Driffield, Yorkshire

- **Mason**, Hopper of Marton, Yorkshire

- **Megginson Robert**, Riplingham House, Brough, East Yorkshire

- **Pratt**, Caldwell

- **Robinson**, Leakby Palace

- **Sandy W**, Holme Pierre-Point

- **Stubbins**

- **Stone**

- **Spencer Earl**

- **Simpson**, Pillmoor House, Hunmanby, Yorkshire

- **Smith Richard** of Dishley (W.R. Smith in Philip Oakden letter to AA Co)

- **Smith Richard**, Swarthstone, Derbyshire. (W. Smith in Oakden letter to AA Co)

- **Smith Sir William Prince**, Haywold, North Dalton

- **Southburn Estates**

- **Stephenson TA & Sons**, Arras, Sancton, Yorkshire

- **Sykes Sir Tatton**, farmed at Sledmere

- **Turner George**, Beacon Downs, Exeter

- **Tremain**

- **Watson**, Perthshire

THE HISTORY OF ENGLISH LEICESTER SHEEP IN AUSTRALIA

- **Wilkin**, Dumfries-shire

14

New Zealand Imports

'One may consider the ancestors as pure only when all the progeny are pure.'[258]

While we may tend to think of most of the imports of stud sheep as coming from England, the majority actually came from New Zealand. While J. MacGregor (F22) did import six rams from New Zealand in 1887, the steady stream of imports from there did not start until 1903, and has continued, with a few breaks, until the present time. The big years for large numbers of rams and ewes being purchased in New Zealand occurred in the ten years starting about 1906. There was an 18-year gap of no imports from 1959 until 1977, and then a 25-year gap of no imports from 1988 to 2013.

The flock books record that while 256 stud rams have been imported from New Zealand, there have been 2,012 stud ewes imported. This large number of sheep imported may be due to the fact that the stud breeders were looking for different blood lines, but perhaps the main reason was that the number of ewes in particular, were not available to be purchased here in Australia. It would have also been a lot cheaper and easier to import sheep from New Zealand than England.

There were many breeders who regularly imported from New Zealand, but perhaps the most consistent was the stud of W. & W.T. Grant (F25), Curlewis, NSW, who imported regularly over a 29-year period from 1913 to 1942.

W.T. Grant's (NSW) New Zealand imported rams.

The Inglewood stud of Philip C. Threlkeld, Flaxton, Canterbury was a major supplier of sheep to Australian sheep breeders, not only studs but also commercial breeders. This, which was the largest flock in New Zealand running 2,000 ewes in 1893, was started in 1865 with a draft of ewes and rams from the English flocks of Jefferson of Preston Hows and Bell of Scale Hill.

Some of the ewes were in lamb to a ram by Sandy of Holm-Perriepont. Later additions were made from George Mann of Scorsby Hall, John

Nicholson of Kirbythorpe Hall, Jefferson and Hutchinson. Indirectly the blood lines of Bugess, Stone and Turner were also introduced.[259]

The Reid flock was founded in 1893 with the purchase of ewes from P.C. Threlkeld. No direct English imports were used in this stud.[260]

The following list is taken from the flock books and lists those introductions recorded. While it is as accurate as I can get it, it must be remembered that breeders did not always submit flock returns. As with those imports listed from England, this list would only be a small percentage of the sheep that actually came from New Zealand. Large numbers of flock rams would have crossed the Tasman to be used on Merino properties as perhaps the local studs could not supply the number of rams needed.

The initial auction sale of New Zealand Leicester's in Australia was in 1893 at the Sydney sheep sale.

BRENTON HEAZLEWOOD

Importer	Date	Number	Breeder
Henry Beattie, Mount Aitken, VIC	1870?	?	?
J. McGregor, F22, Dalmore, VIC	1887	6 Rams	R. Grieve
Rupert Clarke, VIC	1899	10 ewes 2 Rams	D. Grant & E. Menlove
F.G.G Couper, F27, Westbrook, Qld	1901	2 Rams 2 Ewes	S.R. Lancaster P.C. Threlkeld
A.S. Fotheringham, F2, Dashwoods Gully, SA	1903 1904	2 Rams, 6 Ewes 40 Ewes	A.J. Murray, Kaikoura A.J. Murray, Kaikoura
W.J.R. Clarke, Mt Gambia, SA	1904	90 Rams	
A.S. Fotheringham, F2	1905	3 Rams, 50 Ewes	P.C. Threlkeld
F.G.G. Couper, F27	1905	43 Ewes 1 Ram	F. Moore Perry
H.M. Osborne, F19, Poowong, VIC	1905	1 Ram, 2 Ewes 12 Ewes	P.C. Threlkeld J. Reid
John Langham, F21, Woodside, VIC	1905	13 Ewes 2 Rams	J. Reid P.C. Threlkeld
?	1905	150 Rams	P.C. Threlkeld
Henry Reed, F11, Wesley Dale, TAS	1906	20 Ewes	F.H. Wright
R.G. Heazlewood, Whitemore, TAS	1906	1 Ram, 5 Ewes	
C.W. Pye, Windsor NSW	1906	5 Rams 200 Ewes	
J.H.W Mules	1906	85 sheep	D. Grant
A.S. Fotheringham, F2	1906	1 Ram, 200 Ewes	W. Grant
	1906	1 Ram	P.C. Threlkeld
	1906	2 Rams	A.W. Rutherford
	1907	1 Ram, 24 Ewes	P.C. Threlkeld
	1907	1 Ram, 4 Ewes	W. Nixon
	1907	1 Ram	R. & J. Reid
F.G.G. Couper, F27, Westbrook, Qld	1907	46 Ewes 3 Rams	Mr Eagle jnr. William Perry
A.L. Bennett, F1, The Oaks, NSW	1907	2 Rams 41 Ewes	W. Nixon, Killinchy
J. Fleming Douglas, F4, Curlewis, NSW	1907	28 Ewes 25 Ewes 10 Ewes 1 Ram	D. Grant W. Nixon P.C. Threlkeld Little Bros.
H. Val Wright, Cambooya, Qld	1907	30 Rams	Thomas Gilber
John Badcock, F9, Whitemore, TAS	1907	1 Ram	John McCrostie
W.S. Kelly, F29, Giles Corner, SA	1907	10 Ewes 1 Ram	P.C. Threlkeld Lincoln College
J.G. Habel, F9, Hamilton, VIC	1907	10 Ewes	W. Grant
D. Porter, F11, Tallarook, VIC	1908	10 Ewes 1 Ram, 11 Ewes	W. Nixon C.M. Threlkeld
A.L. Bennett, F1	1908	8 Ewes	P.C. Threlkeld
John Beattie, F13, Gisborne, VIC	1908	6 Ewes 1 Ram	W Nixon R. & J. Reid

THE HISTORY OF ENGLISH LEICESTER SHEEP IN AUSTRALIA

Kybyolite Experimental Farm, F74, SA	1909	2 Rams, 40 Ewes	F.C. Murray
Joseph Gill, F32, Breadalbane, NSW	1910	4 Rams, 450 Ewes	John Reid
Fraser Churchill, F33, Manildra, NSW	1910	1 Ram, 26 Ewes	W. Nixon
NSW Govt. Farm, F34, Wagga Wagga, NSW	1910	6 Ewes	J. Nixon
		2 Ewes	J. Kelland
		1 Ram	W. Nixon
John Miller, F39, Gunnedah, NSW	1911	10 Ewes	C.M. Threlkeld
John Habel, F9	1911	1 Ram	C.P. Threlkeld
Sir R.T.H. Clarke, F15, Bolinda Vale, VIC	1912?	13 Rams	W.T. Williams
		104 Ewes	W.T. Williams
Sir R.T.H. Clarke, F16, Somerset Farm, VIC	1912?	1 Ram	D. Grant
		2 Rams	F.C. Murray
		2 Rams	P.C. Threlkeld
		2 Rams	R. & J. Reid
Humphrey Dixon, F49, Gisborne, VIC	1912	1 Ram	R. & J. Reid
F.G.G Couper, F27	1912	1 Ram	R. & J. Reid
A.C. Thompson	1912	54 Rams, 64 Ewes	?
John Miller, F39	1912	2 Rams, 70 Ewes	F.C. Murray
	1913	1 Ram, 77 Ewes	E. Ruddock
A. Francis, Oakbank, SA	1913	8 Rams	C.M. Threlkeld
		4 Ewes	C.M. Threlkeld
McMillan Bros., F18, Caldermeade, VIC	1913	2 Rams	D. Grant
		10 Ewes	D. Grant
John Beattie, F13	1913	1 Ram	D. Grant
		5 Ewes	D. Grant
W. & W.T. Grant, F25, Curlewis, NSW	1913	2 Rams	D. Grant
W. & W.T. Grant, F25	1914	1 Ram, 113 Ewes	D. Grant
Fraser Churchill, F33	1914	5 Ewes	W. Nixon
S.E. Roberts, F44, Barwang, NSW	1914	1 Ram, 50 Ewes	W.T. Williams
F.A. Webb, F45, Cudal, NSW	1914	1 Ram, 2 Ewes	J. Nixon
Robert Howard, F23, Tangie, NSW	1914	1 Ram, 1 Ewe	J. Nixon
Hyland Bros. F40, Araluen, NSW	1914	1 Ram	J. Nixon
Robert Elliott, F42, Cumnock, NSW	1914	1 Ram, 1 Ewe	D. Grant
W. & W.T. Grant, F25	1915	2 Rams	D. Grant
H.G.M. Thackeray, F50, Young, NSW	1915	1 Ram, 50 Ewes	W.T. Williams
G.C. Brunskill, F53, Wagga Wagga, NSW	1915	15 Ewes	D. Grant
AJ Simpson, F32	1915	1 Ram, 25 Ewes	W.T. Williams
Fraser Churchill, F33	1916	3 Ewes	J. Nixon
		10 Ewes	D. Grant

BRENTON HEAZLEWOOD

John Miller, F39	1916	35 Ewes	D. Grant
F.A. Webb, F45	1916	22 Ewes	W. Nixon & J. Nixon
W. & W.T. Grant, F25	1917	3 Rams	D. Grant & E. Kelland
F.A. Webb, F45	1917	3 Rams, 3 Ewes	J. Nixon
G.C. Brunskill, F53	1917	1 Ram	D. Grant
W. & W.T. Grant, F25	1918	3 Rams	D. Grant
G.C. Brunskill, F53	1918	2 Rams	J. Nixon
F.A. Webb, F45	1919	4 Ewes	J. Nixon
Moffat Bros., F58, Duri, NSW	1919	1 Ram	J. Nixon
Sir R.T.H. Clarke, F15	1919	3 Rams	E. Kelland
John Beattie, F13	1920	1 Ram	R. & J. Reid
Sir R.T.H. Clarke, F15	1921	10 Ewes	J.C. Kelland
W. & W.T. Grant, F25	1923	2 Rams	L.J. Grant
James Woods, F22, St. James, VIC	1923	1 Ram, 3 Ewes	T.A. Stephens
Arnold Caddy, F33a, Tylden, VIC	1924	2 Rams	Sir R. Heaton Rhodes
		1 Ram	I. Andrew
W.J. Clark, F64, Sunbury, VIC	1924	1 Ram, 20 Ewes	Sir R. Heaton Rhodes
John Beattie, F13	1924	1 Ram	Canterbury Agricultural College
Sir R.T.H. Clarke, F15	1924	2 Rams	R. & J. Reid
W. & W.T. Grant, F25	1924	3 Rams	J.E.P. Cameron
W. & W.T. Grant, F25	1925	4 Rams	J.E.P. Cameron
G.C. Brunskill, F53	1925	1 Ram	J. Nixon
H.K. Nock, F59, Nelungaloo, NSW	1925	1 Ram	J. Nixon
R.F. Taylor, F70, Gundary, NSW	1926	1 Ram, 3 Ewes	J. Nixon
		1 Ram, 1 Ewe	RJ Law
W. & W.T. Grant, F25	1927	6 Rams	J.E.P. Cameron
H.K. Nock, F59	1927	4 Rams	J. Nixon
W.L. Armstrong, F78	1927	40 Ewes	H.F. Wright
F.W. Moffat, F58	1928	2 Rams, 3 Ewes	W Nixon
W. & W.T. Grant, F25	1928	10 Rams	J.E.P. Cameron & WJ Kelland
Augustus Scott, F72, Bungendore, NSW	1928	1 Ram, 11 Ewes	JB Laidlaw
		4 Ewes	James Reid
		10 Ewes	W.J. Jenkins
W.J. Clark, F64	1929	1 Ram	James Reid
Arnold Caddy, F33a	1929	1 Ram	W.J. Clark
Arnold Caddy, F33a	1930	1 Ram	W.J. Clark
W. & W.T. Grant, F25	1930	3 Rams	J.E.P. Cameron
F.W. Moffatt, F58	1930	1 Ram, 6 Ewes	J. Reid
		8 Ewes	J.O. Coop
R.F. Taylor, F70	1930	13 Ewes	J.O. Coop
W.J. Clark, F64	1930	1 Ram	Sir R. Heaton Rhodes
W. & W.T. Grant, F25	1931	5 Rams	J.E.P. Cameron
Arnold Caddy, F33a	1931	2 Rams	W.J. Clark
Arnold Caddy, F33a	1932	1 Ram	A.S. Elworth
W. & W.T. Grant, F25	1933	5 Rams	J.E.P. Cameron
John Williamson, F51, Carisbrook, VIC	1933	1 Ram, 2 Ewes	Canterbury Agricultural College
W. & W.T. Grant, F25	1934	6 Rams	J.E.P. Cameron

THE HISTORY OF ENGLISH LEICESTER SHEEP IN AUSTRALIA

Buyer	Year	Quantity	Seller
W.J. Clark, F64	1934	1 Ram	James Reid
A.G. Williams, F101, Rutherglen, VIC	1934	1 Ram	Canterbury Agricultural College
W. & W.T. Grant, F25	1935	5 Rams	J.E.P. Cameron
Est. R.H. Dugdale, F73	1935	1 Ram	James Reid
J.G. Habel & Sons, F8, Yelecart, VIC	1936	2 Rams	Sir R. Heaton Rhodes
W. & W.T. Grant, F25	1936	3 Rams	J.E.P. Cameron
Arnold Caddy, F33a	1936	1 Ram	Sir R. Heaton Rhodes
W.J. Clark, F64	1936	1 Ram	Sir R. Heaton Rhodes
W.J. Jenkins, F114, Mt Gambier, SA	1936	1 Ram	ES Taylor
W. & W.T. Grant, F25	1937	4 Rams	J.E.P. Cameron
John Williamson, F51	1937	1 Ram	Canterbury Agricultural College
W.J. Scott, F128, Henty, NSW	1937	1 Ram 35 Ewes	R.J. Low James Reid
J. MacGregor & Son, F115, Baringhup, VIC	1937	1 Ram, 2 Ewes	Sir R. Heaton Rhodes
R.O. Zander, F120, Angaston, SA	1937	1 Ram	James Reid
R.O. Zander, F120	1938	1 Ram, 1 Ewe	H. Bushell
W. & W.T. Grant, F25	1938	2 Rams	J.E.P. Cameron
J.B. Stephenson, F119, Sevenhills, SA	1938	15 Ewes	Canterbury Agricultural College
Williamson Bros, F51	1938	1 Ram	Canterbury Agricultural College
W. & W.T. Grant, F25	1939	2 Rams	J.E.P. Cameron
W. & W.T. Grant, F25	1940	3 Rams	J.E.P. Cameron
Arnold Caddy, F33	1940	1 Ram	Sir R. Heaton Rhodes
Williamson Bros, F51	1940	1 Ram	Canterbury Agricultural College
W.J. Clark, F64	1940	1 Ram 20 Ewes	E.S. Taylor Sir R. Heaton Rhodes
Armstrong & Co, F121, Deniliquin, NSW	1940	1 Ram	Canterbury Agricultural College
Richard Ball, F29, Colac, VIC	1940	2 Rams	G. Lill & E. Taylor
William Johnston, F35, Seymour, VIC	1940	1 Ram	G. & A. Aynsley
Sharp & Taylor, F129, Tallarook, VIC	1940	5 Rams 14 Ewes	G. & A. Aynsley A.S. Elworth
Keith Campbell, F146, Broadmeadows, VIC	1940	1 Ram	G. Lill
John Badcock, F188, Whitemore, TAS	1940	1 Ram	G. Lill
J. Watkins, F178, Trafalgar South, VIC	1941	1 Ram	G. Lill
John Beattie, F13	1941	20 Ewes	A. Elworth & Sir R. Heaton Rhodes

BRENTON HEAZLEWOOD

J.W. Anderson, F136, Bamawm, VIC	1941	1 Ram	Sir R. Heaton Rhodes
Keith Campbell, F146	1941	1 Ram	Sir R. Heaton Rhodes
W. & W.T. Grant, F25	1942	1 Ram	J.E.P. Cameron
Williamson Bros, F51	1943	2 Rams	Canterbury Agricultural College
D. & M. MacGregor, F115	1943	1 Ram	Sir R. Heaton Rhodes
Keith Campbell, F146	1943	39 Ewes	Sir R. Heaton Rhodes & G Woodhouse
W. & J. Cockbill, F94, Footscray, VIC	1943	1 Ram	Canterbury Agricultural College
Williamson Bros, F51	1944	1 Ram	E.S. Taylor
L.J. & A.L. Graves, F209	1944	6 Ewes	James Reid
W.J. Jenkin, F114, Yahl, SA	1946	1 Ram	James Reid
D. & M. MacGregor, F115	1946	1 Ram	R.J. Low
J. Hamilton-Smith, F231, Tallangatta, VIC	1947	3 Rams	R.J. Low
Williamson Bros, F51	1947	2 Rams	J. Reid
J. Hamilton-Smith, F231	1948	10 Ewes	Est. J. Reid
W.J.A. Higham, F278, Williams, WA	1950	1 Ram	E.S. Taylor
D. & M. MacGregor, F115	1951	1 Ram	R.J. Low
W.G. Burges, F117, Burges Sidling, WA	1953	1 Ram	W.J. Symes
		1 Ram, 5 Ewes	D.W. Graham
		3 Ewes	R.J. Low
W.J.A. Higham & Co, F278	1955	1 Ram	Est. E.S. Taylor
A.G. Berryman, F195, Moama, NSW	1956	1 Ram	Est. J. Reid
I.J. Morrish, F181, Blampied, VIC	1956	1 Ram	Est. Sir RH Rhodes
		1 Ram	Est. J Reid
A.G. Berryman, F195	1957	1 Ram	Est. J Reid
D. & M. MacGregor, F115	1957	1 Ram	Est. Sir RH Rhodes
Williamson Bros, F51	1958	1 Ram	R.J. Low & Sons
		1 Ram	Est. J. Reid
Lewis Lee, F63, Mole Creek, TAS	1977	1 Ram	Robert Reid & Sons Ltd
Ian J. Morrish, F181	1978	1 Ram	Ravenswood Stud
Colin R. Taylor, F335, Timboon, VIC	1982	1 Ram	Ravenswood Stud
Colin R. Taylor, F335	1984	1 Ram, 2 Ewes	R.J. Burrows, 'Beechwood'
		2 Ewes	Ravenswood Stud
P. & E. Stephenson, F352, Benalla, VIC	1986	1 Ram	H. Bennett, 'Riverside'
Bellbrook Ptn., F347, Richmond, TAS	1986	1 Ram	R.J. & M.A. Burrows, 'Beechwood'
Gliksten Pastoral, F361, Adelong, NSW	1987	2 Rams	J. Reid 'Riversleigh' & H. Bennett 'Riverside'
Colin R. Taylor, F335	1988	1 Ram	R.J. & M.A. Burrows, 'Beechwood'
Ian J. Morrish, F181	1988	1 Ram	Ravenswood Stud
B.P Heazlewood, F36	2013	Semen	P. Tait, 'Seaview'

R. Cohen & S Miller, F357, Heidelberg Heights, VIC	2013	1 Ram	R.W. & M.A. Manson 'Ellesmere'
E. Stephenson, F352	2013	Semen	B.L. & D.J. McCloy 'Karendale'
R. Cohen & S. Miller, F357	2013	1 Ram	R.W. & M.A. Manson, 'Ellesmere'

15

Transport of Sheep

'The two Sheep pen's were swept away from their fastenings.'[261]

The transport of livestock to the new Colony was by no means new in relation to livestock being transported on long sea voyages. Since Roman times animals have been transported over long distances for purposes such as the improvement of livestock production, food supply, scientific interest, public entertainment, war and other purposes. While this long distance transportation was originally limited to the Mediterranean area, it did extend to the rest of Europe during the Middle Ages. The first major transport of large numbers of animals across the oceans occurred with the conquest of the New World.

Sheep, along with other small animals such as pigs, goats, ducks and chooks were carried on ocean voyages as a fresh food supply well before they were carried for commercial purposes.

The transportation of sheep and other animals to Australia began with the first fleet. While the fleet was at the Cape of Good Hope purchases of livestock and poultry were made. The Colonial Secretary listed the animals that were purchased as: horses: 1 stallion, 3 mares and 3 colts (foals); cattle: 2 bulls and 6 cows, 44 sheep, 4 goats, 28 pigs, poultry etc. In total there were 500 animals of various species on the ships. Surprisingly only 9 stock were lost during the 77-day voyage to Botany Bay.[262]

Livestock continued to be purchased at the Cape of Good Hope by sailing ships on their voyage to the Colony, but most of the early sheep and cattle came from India. Up until 1793 most of the cattle were lost on the voyage to New South Wales. It was Captain William Bampton, the master of the *Shah Hormuzear* who devised a management system that enabled cattle, and later horses, to be shipped to the Colony from India without these disastrous losses. He prepared the cattle for the voyage and sailed at the time of the most suitable winds. He arranged the cattle on each side of the gun deck, fore and aft, and not confined in separate stalls so that they were a support for each other. They were also supplied with mats, well fed and watered, and kept well aired.[263]

Livestock were not only lost on the voyage to Australia, but also to America. In the frantic trading rush of 1810 and 1811, of the estimated twenty-six thousand Spanish Merinos that were shipped from Spain to America, it is estimated that only about nineteen thousand survived the thirty to forty day passage. Taking into account the considerable number that died soon after arrival, it is doubtful if more than fifteen thousand actually survived to be useful breeding sheep, a loss of over forty percent.[264]

While a lot of livestock were transported long distances to the Colony, and short distances within the Colony by sea, there is not a lot of information as to the conditions of their transport. The voyages by sailing ship from

England to the Colony often took four to five months to complete, during which they would experience very rough weather. Livestock losses during these voyages were often very high if good preparations had not been made.

Most of the small vessels that made the long voyage to the new colony were between 3 and 4 hundred tons, which meant that their length varied from about 80 to 120 feet, and their width in the vicinity of 25 feet. Most likely the livestock pens were not permanent structures on the vessels, but constructed as required depending on the make-up of the livestock cargo.

The confinement of livestock indoors for long periods of time was not new to British farmers as they were, and still are, used to housing animals over winter. In those days the cities were full of stables and the stables were full of horses. It is estimated that there were more horses in London before World War I than in the whole of Australia. While the transfer of the land-based stable to the maritime situation was not perhaps such a big step, the estimation of how much feed and water to take with them did present some problems due to the fact that travel times could only be estimated. The following details of livestock being transported to the colony that I have been able to locate give some information regarding the fodder that was secured for the voyage as well as the trouble encountered in keeping fresh water available for the livestock.

On 17 November 1825, James Denton Toosey embarked on board the *Albion* of 320 tons with 28 thoroughbred English, Flemish and Cleveland horses; 20 Hereford and improved Durham cattle; 97 Merino, Leicester and Southdown sheep, as well as several dogs and pigs. Toosey was coming to Van Diemen's Land on this initial shipment as a manager for The New South Wales & Van Diemen's Land Establishment (later known as the

Cressy Company) which was formed in London in 1825 with a capital of fifty thousand pounds.

At Cape Town a further 7 horses, 8 Merino sheep and 2 pigs were purchased. During the voyage 2 calves and 39 lambs were born.

Toosey described the voyage as 'nothing but misery and misfortune' with 23 horses, 9 cattle, 34 sheep, 32 lambs and 1 calf being lost en route. This was a result of '147 head of stock squeezed into a vessel ill calculated to hold half the quantity: hay of a bad description, partly stowed upon deck, with no covering to protect it either from above or below, and partly between decks and in the hold, so as to allow the poor animals but just space enough to breath the breath of life; a long tedious passage of nearly six months, attended with much rough weather'. He describes some of the horses as 'expiring by inches and others dropping momentarily from thorough exhaustion or acute inflammation'. Those that 'survived the voyage were landed in a very emaciated state'.[265]

The loss of expensive livestock on such voyages only added to the value of those that did survive, even if they did take some time to recover once landed. Insurance was available to those that could afford it as Toosey states that, 'three of the best [horses] were insured to their utmost value (two of which died), the remainder only against the risks of the sea'.

In 1865 Robert Linton got a job to help look after a shipment of Lincoln and Border Leicester sheep that were being brought out to a station in the Western District of Victoria. In preparation for the voyage the sheep had been housed at Grays, a village on the Thames, where they had been getting used to being hand fed. The 100 sheep were put on a barge and floated down river to the ship. On board they were all housed on deck in a very confined space with many of the alley-ways being so low and narrow that Robert had to crawl on hands and knees to feed and water the sheep. The sheep did very well until they got to the tropics, being fed plenty of roots, but on reaching the warmer zone what was left of the roots rotted, leaving only dry fodder. The agreement was that the sheep-owners would supply the fodder and the ship the water in sufficient quantity. The Captain's idea of the volume of water required by the sheep was very limited and he restricted them to one pint a day. Even as the sheep started to die he could not be persuaded to increase this amount. One or two sheep continued to die each day, with the situation only improving when the ship left the tropics. They reached Melbourne with about 75% of the sheep still alive.[266]

In 1829 when W.J.T. Clarke decided to relocate to Van Diemen's Land, he went to a lot of trouble regarding the transport of his livestock. He had heard of high losses during the long voyage and was determined that he was going to do all he could to prevent the death of any of his valuable animals, in which he had invested hundreds of pounds.

The account of his voyage is interesting as it gives us insight into the perils of the long sea voyage and the preparation and work that was required to successfully transport the animals on a sailing ship.

Clarke chose the *Deveron* because she was cheap and for her size, 271 tons, her hold had an extraordinary capacity considering her small size. Also she was still largely empty of cargo and had sufficient room in which he could build the stalls and pens for his livestock.

Clarke intended taking his favourite saddle-mare, Jessica; a draught stallion named Champion; two cart-mares recently in foal; a red bull called Comet and some good cows already in calf. He also bought some pure Leicester sheep, a breed recommended for early fattening qualities in a mild climate, and for their docility.

It was still the practice to shear all sheep before sailing and it was therefore essential to rug them warmly in flannel. So it was that instead of shopping and taking in the sights of London, Eliza, the curates daughter with whom he had just eloped and married, spent her honeymoon busily fashioning twenty flannel coats for her husband's sheep.

Clarke built the stalls in the hold to his own design. Each of the horses and cattle was provided with an individual compartment four feet wide and seven to eight feet long. Strong, moveable, well-padded bars were fitted to receive the chest and hindquarters, and strong padding twelve inches wide and three inches thick put along the sides at the height of the ribs. To prevent the animals from being thrown down in rough weather, wide canvas belly-bands were fixed by ring bolts to the deck above and other canvas bands placed across the horses' necks so they could not rear up and crack their polls, for the stock stood on shingle ballast six inches deep and Jessica's and Champion's heads would be very close to the roof.

He then had to cart and stow on board the barrels of oats, barley, linseed and beans, the hogsheads of carrots and turnips, and the trusses of hay. He also loaded his dismembered farm cart, his plough, the harness and assorted tools.

He did all this before paying any attention to his wife. Eliza was horrified at the prospect of having to spend five long months in their cabin which was bare except for two narrow wooden bunks. But Clarke in his usual efficiency installed chests of drawers, put up shelves, screwed hooks into the walls and nailed everything into position so that nothing could come adrift in a storm. He even provided a candleholder so that his wife could read in bed, and by the time a rug had been procured for the floor and Eliza had made curtains for the porthole and the doorway, their accommodation was as good as could be attained on the *Deveron*.

The supply of water caused Clarke much concern even though he had installed water-casks and tanks for his own use, and a day or two before sailing loaded five thousand gallons of water. The Thames was so polluted by the hundreds of sewers discharging their nauseous contents into it that water drawn from the river often became putrid after a short time at sea. Sheep, in particular, refused to drink it and had to be drenched by bottle. Even if this tedious operation kept them from dying of thirst, the contaminated water would cause severe scouring that led to their eventually dying.

Once the voyage was underway Clarke spent most of his time in the hold, milking the cows, mucking out the stalls, feeding out trusses of hay and buckets of oats and barley, carrying flagons of water to the horses and cattle and drenching the reluctant sheep, for to his dismay, the water was already developing the feared evil taste and smell.

Eliza had to coax him up for meals, as he had little appetite, but she had no objection to him sleeping with the stock at night. Indeed, he was not a very welcome visitor to her cabin. His clothes smelt strongly of the stable and he could be seldom persuaded to change them or to even have a wash. His constant pipe-smoking she always found most unpleasant and the reek of coarse tobacco in the small and stuffy cabin with its port closed against the sea, was most revolting.

Some two weeks and 1,215 miles out, the masthead lookout sighted Madeira and soon after the *Deveron* anchored off the town of Funchal where pure spring-water could be obtained. The foul Thames water was thrown overboard and the crew helped Clarke refill his water casks. This greatly assisted the health of the animals and played a vital part in their safe arrival in Van Diemen's Land. Oranges, lemons and figs, potatoes and onions and wine were in plentiful supply and taken on, so that the *Deveron* was well provisioned when, after a brief stay, she sailed again.

Only a couple of weeks later they called at the island of St. Jago, anchoring of Porto Praya. Once again they filled up with water and bought tropical fruit, papaws, coconuts, dates and pineapples which Eliza saw for the first time.

Unfortunately, the blissful conditions were not to continue. Passing along the African coast, the brig was sighted by a pirate vessel and for three days the *Deveron* was forced to run for its life. The ancient carriage-guns were loaded and crew and passengers alike were issued with cutlasses, muskets and pistols. It came closer and every man on the *Deveron* was on deck, armed and prepared to fight to the death when out of the blue an English Frigate suddenly appeared in sight and the pirates fled.

Apart from this alarming episode the long stretch southward from St. Jago to the Cape of Good Hope proved dull indeed. The north-easterly breeze died away and for many stifling days the *Deveron* lay becalmed in the doldrums.

In that bout of sweltering tropical weather, the heat in the hold was proving intensely distressing to the livestock and they required far more to drink than Clarke had envisaged. The water became tainted again and much time had to be spent purifying it. He hung buckets on the ship's rail so that the sea-air might remove the smell; he filtered it through charcoal and then added chloride of lime and magnesia. It was hard and tedious work; at least there was little opportunity of boredom. When he happened to have an unoccupied half-hour, Eliza would come down to join him, reading aloud from a work that had been published just before they sailed. It was Henry Widowson's *Present State of Van Diemen's Land* and its advice and practical information was to prove a great assistance to Clarke later on.

The *Deveron* arrived in Table Bay on 31 October, 107 days, or three-and-a-half months out from London. While there Clarke purchased more fodder and water. While ashore during a week's stay at the Cape, Clarke formed a very poor opinion of the local sheep which had long deer-like legs and enormously fat tails that were melted down for tallow.

After setting sail for Hobart Town, the *Deveron* spent the next six weeks being driven and buffeted by stormy westerlies that whipped up mountainous seas. Clarke now found it difficult and strenuous work to protect his animals from injury. Particularly, he feared the dark nights when it was almost impossible to see what was happening in the dimly-lit hold.

After five months and seven days the *Deveron* arrived in Hobart Town. The fact that Clarke had not lost a single animal during the voyage speaks volumes for the skill and care he bestowed upon them during the long and trying voyage.[267]

Records of the Australian Agricultural Company (AAC) also give an insight into the preparations for the transport of three Leicester rams to New South Wales. H.T. Ebsworth (London Secretary of the AAC) wrote to Robert Williamson, who was to be in charge of the sheep during the voyage with the following instructions.

> *On Monday next will be shipped on account of this Company, the three Leicester ram, on board the Hero of Malown now lying in St Katherine's Docks. You will be pleased to accompany them on board and to take charge of them during the voyage.*
>
> *You will feed them with water, hay and fodder, oil cake, bran & oats according to the annexed scale – these articles, however, are not to be considered as a regular issue, but to be varied as circumstances may suggest. You must be guided by the health and strength of the animals, the weather and other circumstances as to diminishing or increasing the daily issue. Water & provisions for the sheep have been put on board for 150 days consumption. In cold weather they will not require so much as three pints of water per day, in warm weather they will require, perhaps, more.*

You are entitled to draw the quantities of water and provisions named in the annexed scale – should you on some occasions not require so much and take less, on other occasions, you can take such quantities as may be wanted but not to exceed the average daily and weekly issue.

It may be of service to wash the mouths of the sheep occasionally with salt water, and you will, of course, at all times keep the pens as clean as possible, both for the health of them comfort of themselves, the passengers & crew.

Should their health required it, a little salt may be administered; in the hotter climates it may be requisite to shear them. A little rock salt should occasionally be put in their pens.

The following articles have been put on board for their purposes

1 bucket
3 birch brooms
1 pair sheep shears
5 lbs salt
6 lbs rock salt
2 tin pannikins for giving water.

Mr William Graham will assist you in keeping a daily account of the quantities of water & food issued to avoid the possibility of the average daily issue being exceeded.

On arrival at Sydney, Mr Graham will report to Messrs Hunter & Co, the Company's Agents, who will assist in providing for the animals during their stay at Sydney. They are not, however, to go out of your care or management until you have delivered them to Port Stephens, the Company's Principal Settlement. Messrs Hunter & Co will arrange for the shipment of the sheep for Port Stephens, and you will act in accordance with the instructions of those gentlemen regarding your accompanying them on board such vessel as they may select.

When the sheep are delivered at Port Stephens, you will receive instructions regarding your journey to Newcastle, where you will enter upon your duties at the Company's Colliery.

Wishing you and your companions a safe & pleasant voyage, and prosperity in the Colony...

Scale of water & provender to be given to the three rams on board the Hero of Malown

3 pints of water each per day
1½ lbs of hay each per day
1½ lbs of fodder each per day
1½ lbs of oilcake each per day
12 pints of bran each per week
12 pints of oats each per week[268]

The diary of John Brown, who sailed out on the *John Pirie* gives an interesting insight into the voyage to South Australia, particularly as John Brown appears to have been in charge of looking after the livestock on board. The *John Pirie* was one of the at least nine ships that were employed by the South Australian Company in 1836. The ships began sailing from England in 1836, from January to about June, and arrived on the South Australian coast from July to December. The *John Pirie* departed on 22 February 1836 and arrived at Nepean Bay, Kangaroo Island on 16 August.

Monday 29th February

... also one of the Ewes has had a very bad Cold ever since leaving London, and for the last 3 Days has not taken anything, except what we have given her from a Bottle.

Tuesday March 1st

... have had a rough Night, the Wind continuing to blow hard from the S,W, & has raised a very heavy Sea which knocks the Vessel about tremen – diously – the Ewe died this morning at 5 O'Clock and should we not get better Weather, I'm afraid these which remain will stand but a poor chance of surviving as they are nearly overhead in Water & tumbled about most dreadfully, indeed the Wind has been a complete Hurricane all this Day – and at 3 O'Clock.

PM, a tremendous Sea struck the Vessel, which it is most wonderful has not swept every thing of the Deck – but enough mischief is done – two Ewes in the lee Pen are Drown'd and the remainder are more Dead than Alive – likewise 6 Bags of Fodder are washed overboard with the, Top galt Qr boards, &c – We have removed all the Sheep, into the weather Pen, and fill'd it half full of Hay, where they have a little more comfort, than the lee One, and are not in so much danger of being drowned.

March 2nd

We have had another most awful rough Night, and during which, I am sorry to say four more of our Ewes perished, likewise 3 Rabbits, and three Turkeys have suffered the same melancholy fate – those poor Animals which still survive, are in the most miserable plight, they consist of 11 Sheep, 2 Turkeys, and 2 Rabbits, – but the Swine, have weathered the Gale, very well indeed – not one of them having died – and at Day break this Morng the Wind became more moderate, so that all the Men, who are not Sick, have been very busily engaged to Day clearing away and making the live Stock, as comfortable as circumstances will admit of – but the Sea is still tremendiously heavy, which knocks them about most cruelly, in their weak state.

March 10th

... this Morng One of our Sows died, while another has been off her meat for a few Days, and is not able to stand upon her Legs this Eveng I cannot assign any other reason for their Ilness, than the very confined situation, in which they are placed, preventing them getting, the least possible exercise – During this Day, the Capt sent on board, six young Sheep, likewise a lot of Sweedish Turnips, and some Hay.

March 15th

... The Weather has now become quite moderate, but two of our Sheep, have caught very bad Cold's, and are removed to the Hospital, (a place we have partitioned off from the others,) where they can be better attended too, and made more comfortable, than being amongst those, that are healthy – We have also taken the Troughs, from the inside of the Pens, and hung them outside, which does not only make a great deal more room within, but will likewise prevent the Sheep, from cutting their Legs which they did do most dreadfully, against the edges of them things, when the Vessel roll'd about at Sea, besides making it far more handy to clean them Out and feed – During the Afternoon two small Pigs, belonging to the Capt, have been in several Fits, which I have no doubt is caused, by the wet and Cold they have caught, in their confined Births, indeed it is impossible to keep any of the Pigs dry, in heavy Rains, besides their being cramped to

Death, for the want of room – but with all these difficulties, I am glad to say, that the large Sow, is recovering again.

March 16th

The Weather has contd moderate, all last Night and to Day, except a few Squalls, and showers of Hail – one of the little Pigs (a Boar) that was so very Ill in Fits, Yestdy died during the Night, the other looks a great deal better.

March 17th

The Wind continues Squally, from the W,S,W,accpd with Showers of Hail and Rain – Two more of our Capts little Pigs, have had the staggers all Day, and One of them has often been in Fits, We have had them on Deck, about 8, Hours, for exercise, thinking the want of that, is the principle cause of their disease.

March 18th

The Wind has been blowing fresh all Night from S,W, but became quite moderate this Morng – and at Noon was nearly Calm – In the Afternoon a gentle Air sprang up from the Southward, which in the Eveng became more Easterly – This morng another of Capt Martin's Pigs was found Dead, and the remaining two, of the same litter, are very Ill indeed continually taking Fits, and staggering about the Decks.

March 19th

There being a fine Breeze, from S,E, this Morng we got under weigh, at Daylight, and proceeded to Sea, in company with several other, outward bound Vessels, – The two little Pigs, being left upon the Deck, last Night – one of them was missing this Morng it had most probably staggered overboard, and the other poor thing, not seeming at all likely to recover, was killed in the Afternoon.

March 20th Sunday

There has been very little Wind from the Southward during the Night, untill 4, AM, and at Mch 20th 10,A.M. veer'd to West, where it remained all Day, but sometimes being nearly a Calm – Two of the Ewes have had very bad Legs for a Week past, but which are now covered with Sores, resembling in appearance, large ripe Strawberry's, We have put them into the Hospital, those which were there, having greatly recover'd from their Cold's.

March 22nd

... but a strong Sea, still coming from N,W, causes the Vessel to labour very much, and has prevented the Sheep getting any rest in their Pen's since Yestdy morng, besides making several of the Passenger's squeamish.

March 24th

There has been a tremendous high and cross Sea, all Night, with heavy Squalls of Wind, from N,W, by N, accompd by pelting showers of Hail, and Rain, – All our live Stock seem very much distress'd, for the want of rest, and one of the Rabbits brought forth six young One's, during the Night, all of which were dead.

March 27th

The Gale has contd since Thursday, from the Westward, and without the least do 27th, intermission, or abatement, but at 3, PM, of this day it veer'd to S,W, and increased to a perfect Hurricane, raising the Sea, to the greatest possible pitch of Madness, and violent uproar, so that fearing everything would be washed off the Deck's, we bore away, right be–the Wind, at 4, PM, hoping by this means, to save them, from destruction, but the Weather has contd (to the end of this Day) so truly awful, as to baffle all description, indeed the

Elements, seem to be engaged in the most dreadful Warfare, with each other, and violence is the order of the Day, in which the Rain likewise takes a good share, for it is pouring down in Torrents – At 10, PM, the Wind backed round to N,W, and I think (if possible) it blows more terrifickly than ever.

March 28th

At 2, P,M, a most tremendious Sea, overlap'd the Vessel, and giving her such a violent Shock, as caused both the Capt and every Soul on board, to suppose She must founder, being for a time completely buried under Water, however after a few Moments, of the most horrible suspense the little Vessel again arose out of the angry Deep, when both Pumps were set to work, and which to our unutterable satisfaction, very soon sucked her dry, but the loss sustained by that dreadful Sea is truly lamentable – The two Sheep pen's were swept away from their fastenings, and One of them dashed to pieces, when all the poor Sheep which it contained, were washed overboard, the other pen is also greatly injured, and thus were 12 of our Sheep either killed or drown'd, likewise 3 Pigs, 23 Fowls, 2 Turkeys, and 2 Rabbits, shared the same hard fate, besides 5 Sacks of Fodder, and all the Turnips also 1 Barrel of Beef, 1 Piece of Pork, the log – reel, and several other Articles, were all swept off the Decks, along with the Bulwark &c, after which Capt Martin, ordered all the Hay to be thrown overboard, deeming such a course expedient for the safety of the Vessel ...

March 30th

... During the Forenoon our white Sow brought forth 10 young Ones, all of which were dead, and the size of half grown Rats.

March 31st

Our Vessel contd to tumble about exceedingly heavy the whole of last Night and shipping a great deal of Water, while the Wind kept blowing very strong, untill 3 A,M, when it began to moderate, and at 8 A,M, had gradually lessend, to a clever Breeze, and we had once more the pleasure of seeing all the reefs shaken out, and our Vessel again under whole Sails, after having experienced one of the most severe Gales of Wind, ever witnessed, indeed I am astonished, how this little Vessel, has weathered such a violent and terrific Storm, in the awful rough Seas, of the Western Ocean, – The only live Stock which have survived its fury, are 5 young Sheep, (which were purchased at Falmouth) 2 Sows, 3 Rabbits, and 1 Turkey, besides several of the Crew being almost fatigued to Death, our Cook has not been able to come upon Deck, since Tuesday last ...

April 8th

We have had variable Weather since Tuesday, and we sent our 5 remaining Sheep, on 8th Shore this Morng to recover themselves a little from the effects of the late disastrous Weather, which has made them very sickly and weak.

[Due to the damage received in the rough weather the *John Pirie* had to return to Dartmouth for repairs before setting sail once more]

April 13th

During this Afternoon we got on board, 2 Rams, 2 Ewes, 1 Sow, 1 Boar, and 4 small Pigs....also a lot of Fowls, and 3 Turkeys, besides 3 Bags of Oats 3 Bags of barley, and a quantity of Swe – ish Turnips.

April 15th

... we got on 11 Bundles of Hay, and this Afternoon have got back the 5 Ewes, that do 15th were sent on to Shore to recruit themselves, but one of the poor Animals is in a most deplorable condition, being as lean as Wood, and so very lame'd that it can scarcely put a foot to the Ground, indeed one of its Legs is most shocking to look upon, and quite incureable.

April 16th

... In the Afternoon our Capt ordered the Ewe to be killed, which I spoke of Yestdy, as it was in extreme Misery, and at 4 P,M, one of the Turkey's, unfortunately flew overboard and was drown'd.

April 28th

There is less Wind to Day, than has been for the last few Days, but it still continues do from N,E, and the Weather feels so much warmer, that we had the Wool cliped off our Sheep this Morng.

June 29th

... Our Sheep consumed the last mouthful of hay Yesterday, They have been very much knocked about lately, by the heavy rolling and pitching, of the Vessel and are as lean as Wood, yet, they appear in excellent Health, so that I make no doubt but we shall get them safe to Kangaroo Island, should the Weather prove al all decent, during the remainder of Our Voyage.

July 9th

... During the last Week, two little Pigs belonging to our Captain, have been so very Ill, in Fits, that they were obliged to be kill'd.

July 17th

Our Sheep, have the most miserable appearance, that can be imagined, having suffer'd dreadfully, by the fury of this Gale, in which the poor Animals, have been in a manner, almost buried under Water, besides it being by far the coldest Weather, that we have had, on the South side of the Equator, yet, still we may be thankful, that the Wind was Right Aft, or they must all have perished.

July 20th

The Wind contd Squally from the S,W, accompd with cold Weather, untill last Night, when it became moderate, and veer'd round to South, but has been nearly a Calm, the whole of this Day – On Monday we got a place, contrived, down below, for the two poor Ram's, (where the Carpenter, Second Mate, and T, Waldron, live,) as there is not height enough, in the Pen's for them to stand, without chafeing their Backs against the top part of it, and during the many Gales of Wind, that we have lately experienced, they have been very much bruised, by tumbling about, upon the Deck's.

July 28th

... The Rams are uncommonly weak, nor having taken any nourishment, for the last 10 Days, except what has been given to them, by means of a Bottle, and the little Ewws, have again suffer'd most severely in the late Gale, indeed it is really astonishing, how these poor Animals survive, after enduring so many hardships being almost continually Wet, and Cold, besides when the Sea's, come rushing upon the Deck, on Gales of Wind, they are for while, almost over head in Water, and dread – fully knocked about, by the violent rolling, and pitching of the Vessel ...

August 3rd

... One of the Ram's is getting better again, and the Ewe's are still in good Health, notwithstanding all the hard – ships, they have endured.

August 7th

Both Ram's are now recovering, and have taken to their Food again, so that I yet have hope, of getting them, safely to Kangaroo Island.

August 15th

... when a gentle Air sprang up, from the Northward, that in the Eveng had Contd gradually increased into a clever little Breeze, but our Vessel was put under easy Sail, the whole Night, on account of being thought, not far from the Land, and accor – dingly at Day break this Morng, the West end of Kangaroo-Island, was distinctly seen, right ahead, bearing due East, about 20 Miles distant. – One of the Ram's has again refused, to take his Provender, and had likewise got a very bad Eye, the sight of which, I am afraid he will loss – Three of the Ewe's, have wonderfully improved within the last 10 Days, particularly the two, we purchased at Dartmouth – ... when a gentle Air sprang up, from the Northward, that in the Eveng had Contd gradually increased into a clever little Breeze, but our Vessel was put under easy Sail, the whole Night, on account of being thought, not far from the Land, and accor – dingly at Day break this Morng, the West end of Kangaroo-Island, was distinctly seen, right ahead, bearing due East, about 20 Miles distant. – One of the Ram's has again refused, to take his Provender, and had likewise got a very bad Eye, the sight of which, I am afraid he will loss – Three of the Ewe's, have wonderfully improved within the last 10 Days, particularly the two, we purchased at Dartmouth – Thos Waldron has also got considerably better ...

... At 9, A,M, we rounded Point Marsden, and had the pleasure of seeing two Barques at Anchor in "Nepean Bay", which proved to be the "Duke of York", and "Lady Mary Pelham", (they had arrd about 3 Weeks before us) – For the course of an Hour, we were visited by Saml Stephens Esqr C,M, who was saluted with three times three Cheer's, and shortly afterwards a Boat came from each of the Vessels, in one of which was Capt Morgan of the "Duke of York", who undertook to be our Pilot.

August 16th

... and at 3, P,M, we were safely Anchor'd, in a well sheltr'd Roadstead, not more than a Mile from the Shore, Contd and right abreast, of the Company's Tents, at the "New Colony" of "South Australia".[269]

Stock losses not only occurred on the long voyages from England. Perhaps some of the greatest losses were experienced on the relatively short voyages around the Australian coast, particularly to and from Tasmania.

In 1820 Lieutenant-Governor Sorell purchased 300 Merino lambs from Mr McArthur and had them shipped to Van Diemen's Land in an effort to improve the wool production of the island. During the voyage 91 died and a further 24 soon after.[270]

In October 1836, one of the Imlay Brothers' ships experienced a very slow 14-day passage from Twofold Bay to Hobart, a round trip of usually two to three weeks, during which 49 of the 50 cattle on board were lost, and 442 of the 600 sheep. The sheep and cattle were part of regular shipments that the Imlay's sent to the weekly Hobart markets from their properties around Twofold Bay.[271]

The perils that often beset stock owners in moving sheep, the relatively short distance across Bass Strait, is exemplified by one of Charles Swanston's shipments. On what was probably the *Norval's* sixth round trip from Launceston to Port Phillip Bay in January 1836 heavy stock losses were experienced. She was transporting 1,130 sheep (1,000 ewes, 80 Saxon rams and 50 wethers), which was part of the livestock Swanston (after whom the Melbourne street is named) was taking across Bass Strait to stock his land in the Port Phillip district. The voyage across Bass Strait was extremely rough and due to a severe gale from the north-west the *Norvel* was prevented from entering Port Phillip Bay. The ship was blown further east, 115 sheep were lost on board and fearing the remaining sheep would not survive it was decided to land them at Westernport. Once ashore the sheep strayed and some drank the salt water. The 800 or so surviving sheep were placed under the surveillance of two shepherds, but through the night the sheep strayed again and a search the next day revealed about 280 dead carcasses.[272]

The rough passages across Bass Strait did not only see stock losses but often resulted in uncomfortable sailings for the passengers who were in close proximity to the livestock. The Rev. Joseph Orton describes one such trip aboard the *Caledonia*, from Launceston to Melbourne in 1836.

> *'Scarcely any possibility of (staying) in the cabin owing to the steam from the livestock in the hold. I soon became sick and took my lodging for the night in a quantity of loose hay on the quarter deck.... Everything in the cabin extremely filthy in addition to the fluvial from the cattle. The unpleasantness of my situation occasions a dissipation of mind which prevents reading or profitable reflection.... Last night I laid on the deck to avoid the intolerable stench from the cattle in the hold, the nuisance finding its way into the cabin in dense, suffocating drafts. Upon examination, it was ascertained that ten head of cattle had died on the passage, four of which were hauled up out of the hold in a complete state of putrefaction; probably having been dead for many days occasioning nauseous effluvia which have endured for several days past.'*[273]

THE HISTORY OF ENGLISH LEICESTER SHEEP IN AUSTRALIA

Shipping stud sheep from Hobart 1906. The Weekly Courier, June, 1906.

Sheep arriving Hobart wharf 1906. The Weekly Courier, June, 1906.

Sheep being loaded aboard ship Hobart wharf 1906. The Weekly Courier, June, 1906.

Heavy losses were also experienced on the voyages from Launceston to Western Australia. In April 1839 the schooner *Ellen* departed with 450 sheep but only landed 190, a loss of 260. Again, in May 1840 the *Fox* lost 500 sheep and 3 heifers on a voyage from Launceston to Perth. The number that survived were 13 heifers, 4 Merino rams and 150 lambs, these being advertised for sale at Fremantle.[274]

In 1936 B.M. Badcock sent some of the first documented English Leicesters to Western Australia from Tasmania. These were most likely sent to Adelaide by ship and then by trans-continental railway to the west.

When the roll-on, roll-off ferries started operating between Devonport and Melbourne, sheep were trucked directly from Tasmania to most destinations on the mainland. In 1971 the first load of sheep was trucked from Tasmania right through to Perth, WA.

As a young boy I can remember my father air freighting show sheep to Melbourne for the Royal Show. Air freighting sheep across Bass Strait started in August 1946 by Australian National Airways. This form of transport across Bass Strait was only used for show sheep or those sheep going to mainland sales as well as stud sheep coming into Tasmania. Cattle were also air freighted in both directions. From the first uplift in 1946 there was a steady increase in numbers. In the 12 years to 1958 about 101,000 sheep, 1,200 horses, 1,000 cattle as well as 300 monkeys from Manila for Salk vaccine purposes were air freighted across Bass Strait. Sheep were also air freighted to Sydney, Brisbane and Perth. Bristol aircraft were initially used for this purpose as a double deck could be used if required.[275] The advent of the roll-on ferries would have put a stop to the expense of air freight.

Loading sheep onto a Bristol aircraft at Launceston Airport, 1958. Australian Stud & Farm Monthly, 1958.

Today sheep are transported long distances by road, and I think nothing of the show sheep and I taking the overnight ferry from Devonport to Melbourne so we can compete at the annual Bendigo Sheep Show.

16

Exports

'In the management of wool, especially if we would obtain it in a perfect state, the time of shearing though not of prime importance is a circumstance deserving of some attention.'

The history of English Leicester's in Australia is one dominated by imports, not exports. There has however been some genetic transfer to New Zealand, Canada, the United States, England and Sweden in the past 160 years.

New Zealand

While there has been a long history of New Zealand Leicester genetics coming to Australia, there has been very minimal reciprocal trade. The first sale to New Zealand that I can find occurred in 1843 when the Hobart Town Advertiser reported that:

'We have had the pleasure of seeing a small lot of prime Leicester sheep, selected from the flock of a gentleman in the interior of this colony, who stands pre-eminent as a Leicester breeder. They are about to be conveyed to New Zealand, per SISTERS. The sheep have been selected by Mr Boste, Auckland.[276]

Who these sheep were purchased from I have not been able to locate.

William Field, 'Enfield', Tasmania, sold 191 rams to New Zealand in 1864. This sale is quite something because of the large number involved and that they were all from the one breeder. In 1906, A. Oliver & Sons, Tasmania, sold a ram at the Royal Melbourne Show sales to C.H. Ensor, New Zealand for 25 guineas. My guess is that during the period that the Sydney and Melbourne sheep shows and sales were strongly attended by both Australian and New Zealand breeders that rams in particular would have been sold to some New Zealand breeders. In 1981 the Connaughtville stud of N.S. & S.B. Badcock and the Melton Vale stud of Ivan Heazlewood each sent one ram to New Zealand. In 1987 another two rams were sent from the Melton Vale stud.

In the early 1870's there was a transfer of English Leicester's owned by W.J.T. Clarke to New Zealand. In 1871, Clarke purchased the Moa Flat Station in Otago, New Zealand. This was his last major land purchase before his death in January, 1874. Part, or all of the English Leicester flock from Rupertswood, Sunbury, was transferred to the Moa Flat Station. This is not surprising as Clarke was a big user of Leicesters over his Merino ewes in Australia, and his manager there, John Fry Kitching, who he had brought over from Tasmania, was obviously staying with the same formula in New Zealand. After several years the nucleus of this New Zealand flock came back to Victoria.

These would have been some of the early English Leicester's into New Zealand, but not the first. English Leicester rams first arrived in New Zealand sometime in the 1840's, and the first English Leicester ewe in 1867.

Canada

In 1986 one ram from the Melton Vale Stud was taken to Canada as part of a consignment of sheep sent from Tasmania to the World Sheep and Wool Congress which was held in Edmonton.

Ivan Heazlewood with the "Melton Vale" ram at the World Sheep and Wool Congress, Edmonton, Canada 1986. Brenton Heazlewood.

United States of America

English Leicesters had been part of the American sheep industry from its very early days. George Washington ran 900 Leicester's on his Mount Vernon property, and they were popular and numerous for much of the 19th century.

There was at least one exportation of Leicesters to the United States in the late 1800's. In 1870 Messrs. W.C. & J.C. McGillis, who owned a large sheep station in California, purchased Leicester's, these coming from Mr Lester, Denham Court, 9 sheep; Mr Reynolds, Tocal, 16 sheep; Mr Peter Green, Windermere, 20 sheep and from Mr J.F. Doyle, Kaloudab, 11 sheep.[277]

> *'It is with great pleasure that we have to notice that our American cousins are just beginning to come to Australia to select their stud stock, to which our colonists should be somewhat proud. The Messrs M'Gillis, two American gentleman, who are large and successful breeders in California, are now in this colony, and in the market for the purchase of pure-bred Leicester sheep; and within the last few days these gentleman have paid a visit to Tocal, and have selected and purchased from Mr Charles Reynolds ten pure-bred ewes and six rams for exportation to their estates in California. Mr Reynolds has always been awarded the palm for breeding Leicesters, and when he has exhibited his sheep we believe he has never been defeated; and these sixteen sheep selected by Messrs M'Gillis, in point of size and symmetry, are not to be excelled in the colony. We have no doubt that this is only the first of many purchases which will be made in the colony on behalf of residents of California, and probably other American states, particularly when the present splendid selections are exhibited there.'* [278]

By about 1930 the breed was relatively obscure in America. The 1988 death of the last remaining ram in the United States, led to the Colonial Williamsburg Foundation importing sheep from Tasmania to re-introduce the breed back into the United States.

In 1989 nine ewes and one ram were sent from Tasmania to the Colonial Williamsburg Foundation living museum in Virginia. Four ewes were supplied by I.A. Badcock's Glen Dhu stud, N.S. & S.B. Badcock's

Connaughtville stud, and E.B. Gray's Marengo stud, with the remaining five ewes and the ram coming from Ivan Heazlewood's Melton Vale stud.

Melton Vale ram and ewes sent to Colonial Williamsburg, USA. Brenton Heazlewood.

From this initial import the number of registered Leicester ewes in the United States is now more than there are in either Australia, New Zealand or Britain. This has basically been achieved because of the strong interest in the Leicester wool for the craft industry which is very strong in that country.

Since this initial live sheep export there have been several consignments of semen sent to United States breeders.

In 2003 the Ostlers Hill stud of P. & E. Stephenson sent semen from the two year old ram 'Chequers', which had been Champion Ram and Best Woolled Sheep at the ASBA Sheep & Wool Show, Bendigo 2003.

The Melton Park stud of Brenton Heazlewood sent semen in 2008 (Melton Park 301/06), 2014 (Melton Park 225/12 'Melton') and 2015 (Melton Park 34/13 'Tasman').

Melton Park 225/2012 Melton. Brenton Heazlewood.

Melton Park 301/2006. Brenton Heazlewood.

England

In 1990 Colin Taylor's Koenarl English Leicester stud sent semen to England, along with a shipment of Southdown, Lincoln and Ryeland semen. VAB Genetics, a Division of Victorian Artificial Breeders Cooperative facilitated the collection and export of the semen.

This is, as far as I know, the only return of genetics to the breed's country of origin, and therefore must represent an important milestone for the breed in Australia.

Breeders who visited VAB Sheep Genetics to inspect the new facilities and discuss the progress of the semen collection program were (left to right) Andrew Beard, Marylyn Stevens, Colin Taylor holding his English Leicester ram, Geoff Baker, John Beard, and Des Stevens with his Ryeland.

Colin Taylor at VAB Sheep Genetics with UK semen ram. The Muster, July 1990.

Sweden

In 2016, I was asked by a member of The Swedish Leicester Sheep Association, with whom I had been corresponding with for some time, if it would be possible for Australia to send English Leicester embryos to Sweden. My initial research indicated that this would be possible and in early 2017, the Swedish Association asked the English Leicester Association of Australia if any of its members would be interested in being part of a program to send embryos to Sweden.

Jason Southwell (Langs Crossing, Flock No. 420) and myself (Melton Park, Flock No. 36) indicated that we would be prepared to supply ewes and rams for this project. As the program did not get the official request for 100 embryos from The Swedish Association until after we had put the rams out with our ewes for the 2017 mating, it was decided that everything would have to be put back until 2018.

After much discussion with Adrian Veitch (Allstock WA Pty Ltd) and Jean and Moozie Niekerk from Murray River Genetics, Moama, NSW, Jason and I delivered 20 ewes and 3 rams to Murray River Genetics, in early March 2018 to start the program. The sheep were in quarantine at Murray River Genetics for 3 months, during which time the 100 embryos were successfully collected. They were then sent to Adrian Veitch in Western Australia, where he combined them with other embryos that he was sending to Sweden.

Adrian Veitch travelled to Sweden and implanted 38 embryos in 2018, resulting with 11 ram lambs and 11 ewe lambs being born in April 2019. The breeders who purchased the embryos were very pleased with the resultant lambs, as they have a good growth rate, confirmation and are exhibiting more muscle than the Swedish lambs.

Embryo lamb Sweden. Brenton Heazlewood.

Wool

While not an export of Leicester genetics, Leicester wool has been sent overseas for exhibition on at least three occasions.

Philip Oakden sent Leicester wool to London in 1851 as part of the Tasmanian contribution for the Industry of all Nations Exhibition. This was also known as the Crystal Palace Exhibition.

In 1876, W. Murray of Brie Brie, Victoria , sent Leicester wool to the Philadelphia Exhibition (this may have been Border Leicester wool).[279] Cheviot, Lincoln and Merino wool was also sent from Australia.

Mr W.H. Chaffey, Tamworth, NSW, sent English Leicester wool to the World's Columbian Exposition held in Chicago in 1893. This was part of the New South Wales display.

In 1910 my great grandfather, R.G. Heazlewood, sent a fleece from a three-year-old ram to the Argentine Centenary Exhibition being held in Buenos Aires in June of that year. This was part of a consignment, which also consisted of Merino and Lincoln wool from breeders in Victoria and New South Wales.

In 1925 my grandfather, R.K. Heazlewood, sent wool to the British Empire Exhibition held at Wembley. This was part of the contribution of the Tasmanian branch, Australian Longwool Sheepbreeders' Association to the State Commission's exhibit.

British Empire Exhibition 1925 Certificate. Brenton Heazlewood.

17
Flock Book Notes

'Without a shepherd sheep are not a flock.'[280]

Robert Bakewell was perhaps the first sheep breeder to use pedigrees in his breeding programme even though he deliberately avoided keeping written records of what he did. Animal pedigree keeping did not really take off until the mid-eighteen hundreds as it co-emerged with the breeding and showing of prized livestock by the mostly, but not always, wealthy elite. There had been debate in the English society for some time as to whether the qualities of 'good breeding' were hereditary or could be cultivated. The record keeping of livestock breeding was a new field which was the beginning of the standardisation of livestock animal bodies. The first of this kind was the *General Stud Book*, which was a registry of thoroughbred horses published in 1793.

In 1822 George Coates published the first Volume of *The General Short-Horned Herdbook containing the pedigrees of Short-Horned bulls, cows etc of the improved Durham breed from the earliest account to the year 1822*.

Coates's Herd Book, as it was commonly known, was perhaps valued more for its advertising potential by the farmers, rather than the pedigrees it contained which were valued by the aristocracy of the day.

Today the pedigree entitles the animal to registration with their appropriate breed society, and this in theory increases their value as a sire or dam. Today the written pedigree is the most basic of data collection information, with birth weight, weaning weight, fat thickness, eye muscle area and everything that DNA testing can tell the breeder, being the main selection criteria used by a lot of breeders.

The word 'pedigree' comes from the French for 'crane's foot', from the resemblance of a crane's foot to the lines of succession on a genealogical chart.

Pedigrees are a very important tool used in the practice of animal breeding, particularly where inbreeding is being used. Bakewell was a great user of inbreeding while he was developing and stabilising the New Leicester.

Before the publication of any flock books in Australia there were two books printed which documented the flock histories of some early flocks. In 1880, G.A. Brown's *Sheepbreeding in Australia* documented many Merino studs and a few British Breed flocks. The only two Leicester flocks documented are those of E.N. Allen and Robert Jones, both in Tasmania. Detailed flock histories and purchases are documented. In the introduction Brown wrote for volume one of the Flock Book for British Breeds of Sheep in Victoria, he states that '*my object in writing the book was to preserve a record of the grand old studs of merino sheep in these colonies before all the pioneers of Australian sheep husbandry had passed away*'.

THE HISTORY OF ENGLISH LEICESTER SHEEP IN AUSTRALIA

Clarence McIvor's 1893 *History and Development of Sheep Farming* has a similar format but in much more detail. Again, while concentrating on Merino studs, 8 English Leicester flocks, all from Tasmania, are also detailed. Those studs documented are: C.W. Allen, 'Leicesterville', Westbury; Basil Archer, 'Woodside', Cressy; Miles Bennett, 'Esk Farm', Longford; George Hogarth, 'Raeburn', Breadalbane; James Ritchie, 'Mayfield', Chudleigh; J.W. Brumby, 'Ashton', Cressy; R.C. Gibson, 'Belmont' and B. Gibson, 'White Hills'.

There were calls for a flock book as early as 1893.

> *The want of good stud-sheep registers is one which is sure to be felt very shortly to a much greater extent than it is now, and it is to be deplored that efforts to compile such works have hitherto been defeated by the apathy of breeders themselves ... this registry should be a national affair ... the result would be less 'blow' and more useful information.*[281]

While the flock book as we know it is our present day 'bible' in relation to the breeding of our sheep, it did not come into existence in Australia until 1898 when volume 1 of the Flock Book for British Breeds of Sheep in Victoria was published. This was published by the Royal Agricultural Society of Victoria, who adopted a recommendation of a special committee that the work of compiling and publishing a Flock Book for pure sheep should be undertaken by the Society.

It should be noted that the first Improved Leicester Flock Book in the UK was only published a few years earlier in 1893 due to *the desirability of regulating the breed by a recognised standard,* and, *to establish a permanent record of the Improved Leicester, and to form a Society under the title of 'The Improved Leicester Sheep Breeders' Association'.* The founders of the Society rightly anticipated the increase in exports of the Improved Leicester sheep and that *Foreign and Colonial Breeders rightly recognised the value attaching to Pedigree. English Flock Masters, in order to ensure an export trade, must satisfy the requirements of their customers in this particular.*[282]

The first flock book to be produced in Australia was issued in 1894 by the South Australian Shropshire Society, when it published its first volume. During the following year the Tasmanian Shropshire breeders compiled a similar volume representing flocks in Tasmania. It is believed that the first Society of Longwool Sheepbreeders formed in Australia was inaugurated in Tasmania. It was on April 30, 1892, when an association was formed, known as the Tasmanian Longwool Sheepbreeders' Association. It was largely influenced by the well known stud stock breeder Mr C.W. Allen of Westbury. This society, however, lapsed in 1897, but in about 1906 it was re-formed, and a flock book was prepared but did not go to print as it formed the nucleus of the Australian Longwool Sheepbreeders' Association which was formed in Sydney.

Longwool Sheepbreeders.

AN AUSTRALAIN ASSOCIATION FORMED.

A MEETING of breeders of Longwool sheep was held in

Sydney on the 2nd inst. to consider a proposal to form an association of breeders of Lincoln, Leicester and Romney Marsh sheep.

Mr Charles Hebden occupied the chair.

Mr A.I. Bennett's motion that an association be formed to include British Longwools only was carried, as were motions that the headquarters of the association be in Sydney, and that it be called 'The Australian Longwool Sheepbreeders' Association'.

The following provisional committee was elected to arrange and to form the association, prepare rules, &c., and that the fee for membership in the association be one guinea:-

Romney Marsh. Messrs. E. Fuerbardt, S.A. ; G. Doyle, Molong, NSW.; W.H.Yelland, Victoria; C. Mallinson, Bathurst; S.S. Hunt, Sherring

Lincolns. Messrs. J.H. Glasson, Orange; G.A. Higgins, Manilla; J.A. Wallace, Orange; W. Richardson, SA; R.C. Forsyth, Victoria; J. Tyson, Donelly, Queensland; W. French, Tasmania.

Border Leicesters. Messrs. H.F. Marr, Moss Vale; J.A.Cochrane, Victoria; J.F. Douglass, Gunnedah; D

McGregor, Queensland; R. Thompson, Brewongle, and the manager Edgerol Station.

English Leicester. Messrs. A.L. Bennett, Camden; J. McGregor, Victoria; Oliver, Tasmania; A.S. Fotheringham, South Australia.

The committee was declared elected. Mr C.E.H. Maitland being appointed secretary.[283]

The Australian Longwool Sheepbreeders' Association receipt 1911. Brenton Heazlewood.

The Australian Longwool Sheepbreeders' Association (ALSA) also published a flock book. The ALSA was formed in Sydney in 1907 by breeder's representatives from all states except Western Australia. It was therefore more nationalistic than the rival, solely Victorian based RASV flock book. The ALSA was formed as it was felt '*necessary to maintain the purity and standard of the Longwool breeds in Australia, and this work can only be successfully carried out by an organisation of breeders themselves.'*

The Australian Longwool Sheepbreeders' Association Medal 1932. Brenton Heazlewood.

The Australian Longwool Sheepbreeders' Association Medal reverse side. Brenton Heazlewood.

There were only 2 (English) Leicester flocks registered in Flock Book Volume 1 of the British Breeds of Sheep in Victoria, those being of Mr L.R. Carter, Scale Park, Clunes who was mating 50 ewes and Mr D. MacGregor, Dalmore, Pakenham who was mating 500 ewes. Flock numbers were allocated, not in order of foundation date, but in alphabetical order. So, in the case of the two Leicester flocks, that of Mr Carter, which was started in 1883 was allotted flock number 1, while that of Mr MacGregor which was founded in 1869, was allotted flock number 2.

Despite the preface of volume 1 detailing the work that had gone into checking the purity of the sheep before being eligible to be entered into the flock book, it is very lacking on actual pedigrees. It does give extensive and descriptive backgrounds of each flock stating who the foundation and subsequently acquired sheep were purchased from, but virtually no pedigrees.

The other interesting fact regarding the establishment of the flock book is that Mr Frederick Peppin, Fernbrook, Loch, was a member of the Central Committee to oversee the establishment of the flock book. Mr Fred Peppin had with his brother founded the famous 'Peppin' strain of Merino sheep on their Riverina property 'Wanganella', but was now a Southdown breeder and member of the RASV council.

In its early years the Flock Book was not an annual production. Volume 2 covered the period up until 1901 and volume 3 for the years 1902-1906.

By 1907 when volume 3 was published the committee decided to change the name to the Flock Book for British Breeds of Sheep in Australia and invited entries from breeders in all states of the Commonwealth. It retained this name until the Centenary Edition in 1997 when its name was changed to Australian Flock Register.

Up until 1925 when the 17th Volume of the BBSA flock book was produced it had been published by the RASV. It was not until the 1924 Melbourne Royal Show that at a meeting of owners of registered flocks, *'it was decided to ask the Royal Agricultural Society to hand the book over to the control of the breeders themselves, with the object of having for all British breeds one Commonwealth organisation with branches in the various states.'* [284] Thus the Australian Society of Breeders of British Sheep (ASBBS) was formed and from 1925, Vol. 17, took over the production of the flock book.

The 1909, Vol.1 of the Australian Longwool Sheepbreeders' Association (ALSA) flock book is very similar to Vol.1 of the British Breeds of Sheep in Victoria in that it is very strong on giving the flock histories, but it does include individual registrations for 70 rams and 160 ewes.

In relation to English Leicester flocks recorded in the ALSA Flock Book Vol.1, it includes the flock of Ritchie Bros. Chudleigh, Tasmania, flock number 5. This flock is now under the management of Neon Ritchie, the great, great grandson of the founder. This stud is still operating on the same property and is by some 20 years the oldest continually registered stud in the current flock book. It also records the registration of my great grandfather's, Melton Vale flock number 6, from which my current flock is directly descended. These are the only two English Leicester flocks recorded in either the ALSA or BBSV volume 1 books that are still in existence under descendants of the original owners and still on their original properties.

There is only one flock which is recorded in both the BBSV Vol. 1 and ALSA Vol. 1, this being that of D. MacGregor, Dalmore, Pakenham. Both books contain the same flock history, word for word but in the BBSV flock book it is under the name of Mr D. McGregor (Flock No. 2) with 550 ewes put to the ram in 1897, whereas in the ALSA flock book it is under the name of J. McGregor (Flock No. 22) with the flock numbering about 900 ewes (in 1909). In 1891 Duncan turned over the management of his estate and Dalmore to his sons Donald and John[285] so perhaps the 'D' and 'J' McGregor's are the sons.

The Australian Longwool Sheepbreeders' Association amalgamated with the Australian Society of Breeders of British Sheep (ASBBS) in 1934 due to its declining numbers and the first combined flock book was published in 1935, Vol. 27.

The combining of the two flock registrations did not appear to cause any conflict with regard to flock numbers. In the last flock book published by The Australian Longwool Sheepbreeders' Association (Vol. 21) there are 17 English Leicester studs recorded. Of these only 12 continued to register their stud with the Australian Society of Breeders of British Sheep. The other 5 withdrew their registration. No reason is given for these withdrawals, but it did mean that there was not any doubling up of flock numbers. The 12 studs that kept their registration going with ASBBS all retained the flock number that they had had with ALSA. The only change that occurred with flocks already registered with the ASBBS was that the flock of J.G. Habel moved from being Flock No.9, to being Flock No.8.

From the analysis of the flock books it can be seen that the number of English Leicester ewes peaked in the 1940's with 1949 recording the highest number of over ten thousand ewes. From then the numbers declined fairly rapidly as the breed went out of favour mainly due to the rise in popularity of the quicker maturing Border Leicester. The figures also show that Victoria was the powerhouse state for numbers.

In looking at the analysis of the flock book numbers it must be remembered that the figures do not always show a truly accurate picture. As happens today when breeders forget to get their returns in on time and miss the cut-off date for publication, my reading of the early flock books showed that this must have occurred quite frequently as some flocks were sometimes absent for several years before reappearing again. So while the figures given in the table are as accurate as the flock books allow, it must be remembered that they are not necessarily giving us a truly accurate picture.

Roy K. Heazlewood ASBBS Certificate 1932. Brenton Heazlewood.

The use of a stud prefix seems to have developed more by natural means rather than regulation. Today we often know the sheep stud more readily by its prefix than the actual owner's name. In studying the early flock books and reading details of the various stud's sales, purchases etc, the sheep and stud are by default commonly referred to as the property name. When Sir R.T.H. Clarke registered two separate studs in 1913, they were recorded as 'Bolinda Vale' and 'Somerset Farm', being the property names on which the respective flocks resided. The stud of A.S. Fotheringham, Dashwoods Gully, SA, is more often referred to as the 'Hillyfields' stud than that of Mr Fotheringham.

The Australian Longwool Sheepbreeders Association does not make any mention of prefixes at any time during its existence. It is not until Vol. 17, 1925, when the Australian Society of Breeders of British Sheep came into existence that the flock book for British Breeds of Sheep in Australia include in their rules under *Stud prefix* that:

> *'Every registered owner will be required to register a separate stud prefix for its exclusive use in connection with the names of sheep bred by him. The registration fee shall be five shillings.'*

Commencing with Volume 18, the prefix for each stud has been included in the stud details listed in each flock book. In the list of Flock Numbers which I have compiled from the flock books, up until the prefix is stated, I have either used the name that the stud was known by, guessed it, or used the prefix that was registered in 1925 if the stud was still in existence then.

The ALSA does not list any flock dispersals in its flock books. The Flock Book for British Breeds of Sheep in Australia starts to list flock dispersals from Vol. 7. This is continued through to volume 67, then volumes 68, 69 and 70 do not list any dispersals, nor do volumes 83 to 88.

From 1913, the Flock Book of the Australian Longwool Sheepbreeders' Association included a list of the Grand Champion Award winners at Royal and major shows. It is interesting to note the long distance that some breeders travelled with their sheep to exhibit at these shows. New Zealand breeders are exhibitors at the early Sydney shows, no doubt due to the strong demand for New Zealand sheep in Australia about this time. Tasmanian breeders are regular exhibitors at Sydney and Melbourne, again no doubt due to the demand for Tasmanian sheep in New South Wales and Victoria. I can remember my father saying how his father and grandfather would walk the sheep to the Whitemore train sidling where they would be loaded into the sheep wagons for the train trip to either the Hobart or Launceston port, where they would board the ship for the trip to Sydney.

It would no doubt be a long, and probably expensive exercise to exhibit at these shows, with the aim of winning some broad ribbons and then selling the sheep at the subsequent sale.

ALSBA Receipt 1911. Brenton Heazlewood.

ALSBA R.G. Heazlewood transfer to Hepworth Bros. 1917. Brenton Heazlewood.

ALSBA R.K. Heazlewood transfer to G. Lee 1918. Brenton Heazlewood.

English Leicester Grand Champion Ribbon Awards

The following registered English Leicester breeders have won the Grand Champion Ribbons awarded by the Longwool Sheepbreeders' Association at various shows in the Commonwealth of Australia:

1913: **Mr DONALD GRANT**, Winchester, New Zealand. Grand Champion English Leicester Ram, NSW Sheepbreeders' Association's Show, Sydney.

1913: **Mr F.H. BADCOCK**, Hagley, Tasmania. Grand Champion English Leicester Ewe, NSW Sheepbreeders' Association's Show, Sydney.

1913: **Messrs. J. BADCOCK & SONS**, Willow Vale, Glenore, Tasmania. Grand Champion English Leicester Ram, Tasmanian Agricultural and Pastoral Society's Show, Launceston.

1913: **Mr N. HEAZLEWOOD**, Glenore, Whitemore, Tasmania. English Leicester Ewe, Tasmanian Agricultural and Pastoral Society's Show, Launceston.

1914: **Messrs. R. & A. SCOTT**, Umagarlee, Wellington, NSW. Grand Champion English Leicester Ram, NSW Sheepbreeders' Association's Show, Sydney.

1914: **Mr R.G. HEAZLEWOOD**, Glenore, Tasmania. Grand Champion English Leicester Ram, NSW Sheepbreeders' Association's Show, Sydney.

1914: **Messrs. JOHN BADCOCK & SONS**, Willow Vale, Glenore, Tasmania. Grand Champion English Leicester Ram, Tasmanian Agricultural and Pastoral Society's Show, Launceston.

THE HISTORY OF ENGLISH LEICESTER SHEEP IN AUSTRALIA

1914: Messrs. JOHN BADCOCK & SONS, Willow Vale, Glenore, Tasmania. Grand Champion English Leicester Ewe, Tasmanian Agricultural and Pastoral Society's Show, Launceston.

1915: Mr DONALD GRANT, Winchester, New Zealand. Grand Champion English Leicester Ram, NSW Sheepbreeders' Association's Show, Sydney.

1915: Mr R.G. HEAZLEWOOD, Glenore, Tasmania. Grand Champion English Leicester Ewe, NSW Sheepbreeders' Association's Show, Sydney.

1915: Messrs. JOHN BADCOCK & SONS, Willow Vale, Glenore, Tasmania. Grand Champion English Leicester Ram, Tasmanian Agricultural and Pastoral Society's Show, Launceston.

1915: Mr E.G. HALL, Allendale, Newnham, Tasmania. Grand Champion English Leicester Ewe, Tasmanian Agricultural and Pastoral Society's Show, Launceston.

1916: Mr R.G. HEAZLEWOOD, Glenore, Tasmania. Grand Champion English Leicester Ram, NSW Sheepbreeders' Association's Show, Sydney.

1916: Mr G.C. BRUNSKILL, Forest Hill, Wagga Wagga, NSW. Grand Champion English Leicester Ewe, NSW Sheepbreeders' Association's Show, Sydney.

1916: Mr J.G. HABEL, Viladale, Yulecart, Victoria. Grand Champion English Leicester Ram, The Australian Sheepbreeders' Association's Show, Melbourne.

1916: **Mr J.G. HABEL**, Viladale, Yulecart, Victoria. Grand Champion English Leicester Ewe, The Australian Sheepbreeders' Association's Show, Melbourne.

1916: **Mr R.G. HEAZLEWOOD**, Glenore, Tasmania. Grand Champion English Leicester Ram, Tasmanian Agricultural and Pastoral Society's Show, Launceston.

1916: **Messrs. JOHN BADCOCK & SONS**, Willow Vale, Glenore, Tasmania. Grand Champion English Leicester Ewe, Tasmanian Agricultural and Pastoral Society's Show, Launceston.

1917: **Mr JOHN NIXON**, Canterbury, New Zealand. . Grand Champion English Leicester Ram, NSW Sheepbreeders' Association's Show, Sydney.

1917: **Mr G.C. BRUNSKILL**, Forest Hill, Wagga Wagga, NSW. Grand Champion English Leicester Ewe, NSW Sheepbreeders' Association's Show, Sydney.

1917: **Mr A. FRANCIS**, Liverton, Oakbank, South Australia. Grand Champion English Leicester Ram, The Australian Sheepbreeders' Association's Show, Melbourne.

1917: **Mr A. FRANCIS**, Liverton, Oakbank, South Australia. Grand Champion English Leicester Ewe, The Australian Sheepbreeders' Association's Show, Melbourne.

1917: **Mr J.H. FAIRCHILD**, Fairfield, Lang Lang, Victoria. Grand Champion English Leicester Ram, Royal Agricultural Society of Victoria, Melbourne Royal Show.

1917: **Mr J.H. FAIRCHILD**, Fairfield, Lang Lang, Victoria. Grand Champion English Leicester Ewe, Royal Agricultural Society of Victoria, Melbourne Royal Show.

1917: **Messrs. JOHN BADCOCK & SONS**, Willow Vale, Glenore, Tasmania. Grand Champion English Leicester Ram, Tasmanian Agricultural and Pastoral Society's Show, Launceston.

1917: **Mr N. HEAZLEWOOD**, Glenore, Whitemore, Tasmania. Grand Champion English Leicester Ewe, Tasmanian Agricultural and Pastoral Society's Show, Launceston.

1918: **Mr ROY K. HEAZLEWOOD**, Whitemore, Tasmania. Grand Champion English Leicester Ram, NSW Sheepbreeders' Association's Show, Sydney.

1918: **Mr ROY K. HEAZLEWOOD**, Whitemore, Tasmania. Grand Champion English Leicester Ewe, NSW Sheepbreeders' Association's Show, Sydney.

1918: **Mr D. PORTER**, Riverlea, Tallarook, Victoria. Grand Champion English Leicester Ram, Australian Sheepbreeders' Association's Show, Melbourne.

1918: **Mr D. PORTER**, Riverlea, Tallarook, Victoria. Grand Champion English Leicester Ewe, Australian Sheepbreeders' Association's Show, Melbourne.

1918: **Mr ALFRED FRANCIS**, Liverton, Oakbank, South Australia. Grand Champion English Leicester Ram, Royal Agricultural Society of Victoria, Melbourne Royal Show.

1918: **Mr W.S. KELLY**, Merrindie, Giles' Corner, South Australia. Grand Champion English Leicester Ewe, Royal Agricultural Society of Victoria, Melbourne Royal Show.

1918: **Mr R.G. HEAZLEWOOD**, Glenore, Tasmania. Grand Champion English Leicester Ram, Tasmanian Agricultural and Pastoral Society's Show, Launceston.

1918: **Mr E.G. HALL**, Allendale, Newnham, Tasmania. Grand Champion English Leicester Ewe, Tasmanian Agricultural and Pastoral Society's Show, Launceston.

1919: **Mr G.C. BRUNSKILL**, Forest Hill, Wagga Wagga, NSW. Grand Champion English Leicester Ram, NSW Sheepbreeders' Association's Show, Sydney.

1919: **Mr G.C. BRUNSKILL**, Forest Hill, Wagga Wagga, NSW. Grand Champion English Leicester Ewe, NSW Sheepbreeders' Association's Show, Sydney.

1919: **Mr R.G. HEAZLEWOOD**, Glenore, Tasmania. Grand Champion English Leicester Ram, Tasmanian Agricultural and Pastoral Society's Show, Launceston.

1919: **Mr B.M. BADCOCK**, Glenore, Tasmania. Grand Champion English Leicester Ewe, Tasmanian Agricultural and Pastoral Society's Show, Launceston.

1919: **Mr J.H. FAIRCHILD**, Fairfield, Lang Lang, Victoria. Grand Champion English Leicester Ram, Royal Agricultural Society of Victoria, Melbourne Royal Show.

1919: **Mr A. FRANCIS**, Liverton, Oakbank, South Australia. Grand Champion English Leicester Ewe, Royal Agricultural Society of Victoria, Melbourne Royal Show.

1920: **Mr W. PADBURY**, Guildford, Western Australia. Grand Champion English Leicester Ram, Royal Agricultural Society of Western Australia, Perth Royal Show.

1920: **Mr W. PADBURY**, Guildford, Western Australia. Grand Champion English Leicester Ewe, Royal Agricultural Society of Western Australia, Perth Royal Show.

1920: **Mr G.C. BRUNSKILL**, Forest Hill, Wagga Wagga, NSW. Grand Champion English Leicester Ram, NSW Sheepbreeders' Association's Show, Sydney.

1920: **Mr G.C. BRUNSKILL**, Forest Hill, Wagga Wagga, NSW. Grand Champion English Leicester Ewe, NSW Sheepbreeders' Association's Show, Sydney.

1920: **Mr R.G. HEAZLEWOOD**, Glenore, Tasmania. Grand Champion English Leicester Ram, Tasmanian Agricultural and Pastoral Society's Show, Launceston.

1920: **Mr B.M. BADCOCK**, Glenore, Tasmania. Grand Champion English Leicester Ewe, Tasmanian Agricultural and Pastoral Society's Show, Launceston.

1920: **Messrs. S.O. and F.W. WOOD**, Mooroopna, Victoria. Grand Champion English Leicester Ram, Royal Agricultural Society of Victoria, Melbourne Royal Show.

1920: **Messrs. S.O. and F.W. WOOD**, Mooroopna, Victoria. Grand Champion English Leicester Ewe, Royal Agricultural Society of Victoria, Melbourne Royal Show.

1920: **Mr G.C. BRUNSKILL**, Forest Hill, Wagga Wagga, NSW. Grand Champion English Leicester Ram, NSW Sheepbreeders' Association's Show, Sydney.

1920: **Mr G.C. BRUNSKILL**, Forest Hill, Wagga Wagga, NSW. Grand Champion English Leicester Ewe, NSW Sheepbreeders' Association's Show, Sydney.

1921: **Mr W.S. KELLY**, Giles' Corner, South Australia. Grand Champion English Leicester Ram, Royal Agricultural Society of Victoria, Melbourne Royal Show.

1921: **Mr W.S. KELLY**, Giles' Corner, South Australia. Grand Champion English Leicester Ewe, Royal Agricultural Society of Victoria, Melbourne Royal Show.

1921: **Messrs. A.W. EDGAR and Co.**, Gingin, Western Australia. Grand Champion English Leicester Ram, Royal Agricultural Society of Western Australia, Perth Royal Show.

1921: **Messrs. A.W. EDGAR and Co.**, Gingin, Western Australia. Grand Champion English Leicester Ewe, Royal Agricultural Society of Western Australia, Perth Royal Show.

1921: **Mr R.G. HEAZLEWOOD**, Glenore, Tasmania. Grand Champion English Leicester Ram, National Agricultural and Pastoral Society of Tasmania, Launceston Show.

1921: **Mr NORMAN HEAZLEWOOD**, Whitemore, Tasmania. Grand Champion English Leicester Ewe, National Agricultural and Pastoral Society of Tasmania, Launceston Show.

1922: **Mr J. NIXON**, New Zealand. . Grand Champion English Leicester Ram, The NSW Sheepbreeders' Association's Show, Sydney.

1922: **Mr G.C. BRUNSKILL**, Forest Hill, Wagga Wagga, NSW. Grand Champion English Leicester Ewe, The NSW Sheepbreeders' Association's Show, Sydney.

1922: **Mr J.G. HABLE**, Hamilton, Victoria. Grand Champion English Leicester Ram and Ewe, the Australian Sheepbreeders' Association's Show, Melbourne.

1922: **Mr W.S. KELLY**, Giles' Corner, South Australia. Grand Champion English Leicester Ram, Australian Sheepbreeders' Association's Show, Melbourne.

1922: **Mr O. INGLIS**, Avoca, Victoria. Grand Champion English Leicester Ewe, Australian Sheepbreeders' Association's Show, Melbourne.

1922: **Mr W.S. KELLY**, Giles' Corner, South Australia. Grand Champion English Leicester Ram and Ewe, Royal Agricultural Society of South Australia, Adelaide Royal Show.

1922: **Messrs. A.W. EDGAR and Co.**, Gingin, Western Australia. Grand Champion English Leicester Ram, Royal Agricultural Society of Western Australia, Perth Royal Show.

1922: **Mr W. PADBURY**, Guilford, Western Australia. Grand Champion English Leicester Ewe, Royal Agricultural Society of Western Australia, Perth Royal Show.

1922: **Mr R.G. HEAZLEWOOD**, Glenore, Tasmania. Grand Champion English Leicester Ram, National Agricultural and Pastoral Society of Tasmania, Launceston Show.

1922: **Mr B.M. BADCOCK**, Whitemore, Tasmania. Grand Champion English Leicester Ewe, National Agricultural and Pastoral Society of Tasmania, Launceston Show.

1923: **Mr W.S. KELLY**, Giles' Corner, South Australia. Grand Champion English Leicester Ram, Royal Agricultural Society of Victoria, Melbourne Royal Show.

1923: **Mr J.G. HABLE**, Hamilton, Victoria. Grand Champion English Leicester Ewe, Royal Agricultural Society of Victoria, Melbourne Royal Show.

1923: **Mr W.S. KELLY**, Giles' Corner, South Australia. Grand Champion English Leicester Ram and Ewe, Royal Agricultural Society of South Australia, Adelaide Royal Show.

1923: **Mr W.G. SPENCER**, Grass Valley, Western Australia. Grand Champion English Leicester Ram and Ewe, Royal Agricultural Society of Western Australia, Perth Royal Show.

1923: **Mr B.M. BADCOCK**, Whitemore, Tasmania. Grand Champion English Leicester Ram, National Agricultural and Pastoral Society of Tasmania, Launceston Show.

1923: **Mr R.K. HEAZLEWOOD**, Glenore, Tasmania. Grand Champion English Leicester Ewe, National Agricultural and Pastoral Society of Tasmania, Launceston Show.

1923: **Sir RUPERT CLARKE**, Bolinda Vale, Victoria. Grand Champion English Leicester Ram and Ewe, Australian Sheepbreeders' Association's Show, Melbourne.

1924: **Sir RUPERT CLARKE**, Bolinda Vale, Victoria. Grand Champion English Leicester Ewe, Australian Sheepbreeders' Association's Show, Melbourne.

1924: **Dr. ARNOLD CADDY**, Chandpora, Tylden, Victoria. Grand Champion English Leicester Ram, Australian Sheepbreeders' Association's Show, Melbourne.

1924: **Mr W.G. SPENCER**, Grass Valley, Western Australia. Grand Champion English Leicester Ram and Ewe, Royal Agricultural Society of Western Australia, Perth Royal Show.

1924: **Mr ROY K. HEAZLEWOOD**, Whitemore, Tasmania. Grand Champion English Leicester Ram, National Agricultural and Pastoral Society of Tasmania, Launceston Show.

1924: **Mr E.G. HALL**, Newnham, Tasmania. Grand Champion English Leicester Ewe, National Agricultural and Pastoral Society of Tasmania, Launceston Show.

1924: **Mr W.S. KELLY**, Giles' Corner, South Australia. Grand Champion English Leicester Ram and Ewe, Royal Agricultural Society of South Australia, Adelaide Royal Show.

1924: **Mr J. NIXON**, Kellenchy, New Zealand. Grand Champion English Leicester Ram, NSW Sheepbreeders' Association's Show, Sydney.

1924: **Mr G.C. BRUNSKILL**, Forest Hill, Wagga Wagga, NSW. Grand Champion English Leicester Ewe, The NSW Sheepbreeders' Association's Show, Sydney.

1925: **Mr ROY K. HEAZLEWOOD**, Whitemore, Tasmania. Grand Champion English Leicester Ram and Ewe, National Agricultural and Pastoral Society of Tasmania, Launceston Show.

1925: **Mr W.S. KELLY**, Giles' Corner, South Australia. Grand Champion English Leicester Ram and Ewe, Royal Agricultural Society of Victoria, Melbourne Royal Show.

1925: **Dr. ARNOLD CADDY**, Chandpora, Tylden, Victoria. Grand Champion English Leicester Ram, Australian Sheepbreeders' Association's Show, Melbourne.

1925: **Mr W.S. KELLY**, Giles' Corner, South Australia. Grand Champion English Leicester Ram and Ewe, Royal Agricultural Society of South Australia, Adelaide Royal Show.

1925: **Mr ROY K. HEAZLEWOOD**, Whitemore, Tasmania. Grand Champion English Leicester Ram, Tasmanian Royal Agricultural Society, Hobart Royal Show.

1925: **Mr JOHN. NIXON**, Kellenchy, New Zealand. Grand Champion English Leicester Ram and Ewe, NSW Sheepbreeders' Association's Show, Sydney.

1926: **Mr J. NIXON**, Kellenchy, New Zealand. Grand Champion English Leicester Ram and Ewe, NSW Sheepbreeders' Association's Show, Sydney.

1926: **Mr W.S. KELLY**, Giles' Corner, South Australia. Grand Champion English Leicester Ram and Ewe, Royal Agricultural Society of South Australia, Adelaide Royal Show.

1926: **Dr. ARNOLD CADDY**, Chandpora, Tylden, Victoria. Grand Champion English Leicester Ram, Australian Sheepbreeders' Association's Show, Melbourne.

1926: **Mr W.S. KELLY**, Merrindie, Giles' Corner, South Australia. Grand Champion English Leicester Ram and Ewe, Royal Agricultural Society of Victoria, Melbourne Royal Show.

1926: **Mr B.M. BADCOCK**, Willow Vale, Whitemore, Tasmania. Grand Champion English Leicester Ram, National Agricultural and Pastoral Society of Tasmania, Launceston Show.

1926: **Mr ROY K. HEAZLEWOOD**, Whitemore, Tasmania. Grand Champion English Leicester Ewe, National Agricultural and Pastoral Society of Tasmania, Launceston Show.

1927: **Mr H.K. NOCK**, Nelungaloo, NSW. Grand Champion English Leicester Ram and Ewe, New South Wales Sheepbreeders' Association's Show, Sydney.

1927: **THE KYBYBOLITE GOVERNMENT EXPERIMENT FARM**, South Australia. Grand Champion English Leicester Ram, Royal Agricultural Society of South Australia, Adelaide Royal Show.

1927: **Mr H.A. SHILLABEER**, Oakbank, South Australia. Grand Champion English Leicester Ewe, Royal Agricultural Society of South Australia, Adelaide Royal Show.

1927: **Dr. ARNOLD CADDY**, Tylden, Victoria. Grand Champion English Leicester Ram, Australian Sheepbreeders' Association's Show, Melbourne.

1927: **Mr B.M. BADCOCK**, Willow Vale, Whitemore, Tasmania. Grand Champion English Leicester Ram, Royal Agricultural Society of Victoria, Melbourne Royal Show.

1927: **Mr E.G. HALL**, Allendale, Newnham, Tasmania. Grand Champion English Leicester Ewe, Royal Agricultural Society of Victoria, Melbourne Royal Show.

1927: **Mr B.M. BADCOCK**, Willow Vale, Whitemore, Tasmania. Grand Champion English Leicester Ram, National Agricultural and Pastoral Society of Tasmania, Launceston Show.

1927: **Mr ROY K. HEAZLEWOOD**, Whitemore, Tasmania. Grand Champion English Leicester Ewe, National Agricultural and Pastoral Society of Tasmania, Launceston Show.

1928: **Mr H.K. NOCK**, Nelungaloo, NSW. Grand Champion English Leicester Ram, New South Wales Sheepbreeders' Association's Show, Sydney.

1928: **Mr H.K. NOCK**, Nelungaloo, NSW. Grand Champion English Leicester Ewe, New South Wales Sheepbreeders' Association's Show, Sydney.

1928: **Dr. ARNOLD CADDY**, Tylden, Victoria. Grand Champion English Leicester Ram, Australian Sheepbreeders' Association's Show, Melbourne.

1928: **Mr B.M. BADCOCK**, Willow Vale, Whitemore, Tasmania. Grand Champion English Leicester Ewe, Royal Agricultural Society of Victoria, Melbourne Royal Show.

1928: **Mr H.H. SHILLABEER**, Liberton, Oakbank, South Australia. Grand Champion English Leicester Ram, Royal Agricultural Society of Victoria, Melbourne Royal Show.

1928: **Mr B.M. BADCOCK**, Willow Vale, Whitemore, Tasmania. Grand Champion English Leicester Ram, National Agricultural and Pastoral Society of Tasmania, Launceston Show.

1928: **Mr B.M. BADCOCK**, Willow Vale, Whitemore, Tasmania. Grand Champion English Leicester Ewe, National Agricultural and Pastoral Society of Tasmania, Launceston Show.

1928: **Mr B.M. BADCOCK**, Willow Vale, Whitemore, Tasmania. Grand Champion English Leicester Ram, The Royal Agricultural Society of Tasmania, Hobart Show.

1929: **Mr H.K. NOCK**, Nelungaloo, NSW. Grand Champion English Leicester Ram, New South Wales Sheepbreeders' Association's Show, Sydney.

1929: **Mr H.K. NOCK**, Nelungaloo, NSW. Grand Champion English Leicester Ewe, New South Wales Sheepbreeders' Association's Show, Sydney.

1929: **Mr J.G. HABEL**, Yulecart, Victoria. Grand Champion English Leicester Ram, Royal Agricultural Society of Victoria, Melbourne Royal Show.

1929: **Mr R.J. CLEMENT**. Grand Champion English Leicester Ewe, Royal Agricultural Society of Victoria, Melbourne Royal Show.

1929: **Dr. ARNOLD CADDY**, Chandpara, Tylden, Victoria. Grand Champion English Leicester Ram, Australian Sheepbreeders' Association's Show, Melbourne.

1929: **Mr E.G. HALL**, Allendale, Newnham, Tasmania. Grand Champion English Leicester Ram, The National Agricultural and Pastoral Society of Tasmania, Launceston Show.

1929: **Mr H.H. SHILLABEER**, Oakbank, South Australia. Grand Champion English Leicester Ram, Royal Agricultural Society of South Australia, Adelaide Royal Show.

1929: **Mr E.G. HALL**, Allendale, Newnham, Tasmania. Grand Champion English Leicester Ram, Royal Agricultural Society of Tasmania, Hobart Show.

1930: **Mr H.K. NOCK**, Nelungaloo, NSW. Grand Champion English Leicester Ram, New South Wales Sheepbreeders' Association's Show, Sydney.

1930: **Mr H.K. NOCK**, Nelungaloo, NSW. Grand Champion English Leicester Ewe, New South Wales Sheepbreeders' Association's Show, Sydney.

1930: **Mr H.H. SHILLABEER**, Oakbank, South Australia. Grand Champion English Leicester Ram, Royal Agricultural Society of Victoria, Melbourne Royal Show.

1930: **Mr H.H. SHILLABEER**, Oakbank, South Australia. Grand Champion English Leicester Ewe, Royal Agricultural Society of Victoria, Melbourne Royal Show.

1930: **Mr B.M. BADCOCK**, Willow Vale, Whitemore, Tasmania. Grand Champion English Leicester Ram, The National Agricultural and Pastoral Society of Tasmania, Launceston Show.

1930: **Mr ROY K. HEAZLEWOOD**, Whitemore, Tasmania. Grand Champion English Leicester Ewe, The National Agricultural and Pastoral Society of Tasmania, Launceston Show.

1930: **Mr H.H. SHILLABEER**, Oakbank, South Australia. Grand Champion English Leicester Ram and Ewe, Royal Agricultural Society of South Australia, Adelaide Royal Show.

1930: **Mr B.M. BADCOCK**, Whitemore, Tasmania. Grand Champion English Leicester Ram, Royal Agricultural Society of Tasmania, Hobart Show.

1931: **Mr H.H. SHILLABEER**, Oakbank, South Australia. Grand Champion English Leicester Ram, Royal Agricultural Society of Victoria, Melbourne Royal Show.

1931: **Mr H.H. SHILLABEER**, Oakbank, South Australia. Grand Champion English Leicester Ewe, Royal Agricultural Society of Victoria, Melbourne Royal Show.

1931: **Dr. ARNOLD CADDY**, Chandpara, Tylden, Victoria. Grand Champion English Leicester Ram, Australian Sheepbreeders' Association's Show, Melbourne.

1931: **Mr NORMAN HEAZLEWOOD**, Whitemore, Tasmania. Grand Champion English Leicester Ram, The National Agricultural and Pastoral Society of Tasmania, Launceston Show.

1931: **Mr ROY K. HEAZLEWOOD**, Whitemore, Tasmania. Grand Champion English Leicester Ewe, The National Agricultural and Pastoral Society of Tasmania, Launceston Show.

1931: **Mr H.H. SHILLABEER**, Oakbank, South Australia. Grand Champion English Leicester Ram, Royal Agricultural Society of South Australia, Adelaide Royal Show.

1931: **Mr ROY K. HEAZLEWOOD**, Whitemore, Tasmania. Grand Champion English Leicester Ram, Royal Agricultural Society of Tasmania, Hobart Show.

The Flock Book has had a couple of name changes during its existence to reflect the changing times. Volumes I and II of the Royal Agricultural Society of Victoria publication were called the *Flock Book for British Breeds of Sheep in Victoria*. Volume III was changed from Victoria to Australia to no doubt try and present itself as a more national publication.

With the formation of The Australian Society of Breeders of British Sheep (ASBBS) in 1925, the name of the flock book stayed the same, the only difference being that it was now published by the ASBBS. In 1997 with volume 89, the name was changed to *Australian Flock Register*, the name it retains today.

The only other change with the flock book has been when ASSBA had a name change. Volume 93, 2001 was published by ASBBS but volume 94, 2002 was published by the Australian Stud Sheep Breeders Association (ASSBA).

18
English Leicester Association of Australia

'Live as if you expect to die tomorrow; farm and breed stock as if you expect to live forever.'[286]

While the present day English Leicester Association of Australia (ELAA) was not re-formed until 1982, there was talk of forming an association as early as 1905.

> *Leicester breeders are talking of forming an association to watch the interests of that type of sheep. As the Southdown and Shropshire breeders have associations, the Leicester people think they ought to start one also. It is held that it would give an impetus to the breeding of this class of sheep, which is being utilised so largely in the freezing industry.*[287]

I have not found any evidence to say that an association was formed at this time, but obviously the breeders were then thinking about the benefits of such an association.

Australian English Leicester Breeders' Association. Brenton Heazlewood.

The first such association that I have been able to locate was The Australian English Leicester Breeders' Association which was formed in late 1947, with plans for the general advancement of the breed.[288] I do not know how long this association was in existence for. This first association must have lapsed at some stage as it was reformed in 1982 as the English Leicester Association of Australia.

First meeting of reformed Association

The inaugural meeting of members of the English Leicester Association of Australia was held in the Administrative Building, Royal Show Grounds, Epsom Road, Ascot Vale at 3.30pm on Sunday, 19 September, 1982, pursuant to notice.

PRESENT:

Mr D.S. MacGregor in the Chair.

Messrs M. Gadd. T. Taylor, C.R. Taylor, I.J. Morrish, M.M. MacFarlane, A. Habel, A.F. Williamson, R.D. Morrish, R. Scott, Mrs J. Gadd, Mrs C. Stuart, Mrs D. Taylor, Miss P. Taylor, Mrs D. MacFarlane and Mrs B.B. Scott. Mr Stuart.

Mr MacGregor welcomed breeders to the meeting which was called with a view to re-establishing the English Leicester breeders Association. He stated that a search for the records of the previous Association had been unsuccessful but a copy of the Constitution may be available.

REFORMING OF ASSOCIATION:

After discussion it was decided on the motion of Mr Habel, seconded by Mr Williamson that an Association of English Leicester breeders be reformed.

ELECTION OF PRESIDENT:

The Chairman then called for nominations for President.

Mr C.R. Taylor nominated Mr I.J. Morrish and Mr I.J. Morrish nominated Mr D.S. MacGregor. Mr MacGregor declined nomination but stated his willingness to assist the Association in any way possible.

As there were no further nominations Mr I.J. Morrish was declared elected.

APPOINTMENT OF HONORARY SECRETARY:

The Chairman called for nominations for the appointment of Honorary Secretary.

Mr C.R. Taylor was appointed on the nomination of Mr I.J. Morrish.

APPOINTMENT OF HONORARY TREASURER:

The Chairman called for nominations for the appointment of Honorary Treasurer.

Mrs D. MacFarlane was appointed on the nomination of Mr Williamson.

ELECTION OF COMMITTEE:

It was decided on the motion of Mr I.J. Morrish, seconded by Mrs Scott that a committee of four members be appointed. The Chairman then called for nominations.

Mrs MacFarlane nominated Mr A. Habel, Mr Habel nominated Mr D.S. MacGregor who declined nomination. Mrs Scott nominated Mr A.F. Williamson, Mr Taylor nominated Mr M. Gadd and Mr Williamson nominated Mrs B.B. Scott.

As there were no further nominations the following members were declared elected for the ensuing year:- Messrs A. Habel, A.F. Williamson, M. Gadd and Mrs B.B. Scott.

ANNUAL SUBSCRIPTION:

The meeting then discussed annual membership subscriptions and it was decided on the motion of Mr Habel, seconded by Mr Gadd that the subscription for the ensuing year be $10.00 per member.

Mr MacGregor informed the meeting that he had received $10 in donations from Dr W.J. Granger and Mr E.B. Gray. This money was handed to Mrs MacFarlane, the Honorary Treasurer to be deposited into the Association's account.

VOTE OF THANKS:

Mr C.R. Taylor then moved a vote of thanks for the work carried out by Mr MacGregor towards the re-establishment of the Association, and for supporting the breed. The motion was carried with acclamation.

Mr MacGregor thanked members for their comments and for attending the meeting. Mr MacGregor then left the meeting.

Mr I.J. Morrish occupied the chair.

CIRCULATION OF MINUTES:

After discussion it was decided that all English Leicester breeders be forwarded a copy of the Inaugural Meeting

Minutes.

MEMBERSHIP:

It was resolved that membership of the Association be confined to English Leicester breeders for the present time, but that the Committee should discuss the introduction of Associate membership at its next meeting.

CORRESPONDENCE:

The secretary tabled a letter (14/9/82) from IBM regarding a 'Stud Management System' to be demonstrated at the 1982 Royal Melbourne Show. Received.

FINANCE:

It was decided on the motion of Mr Gadd, seconded by Mr M. MacFarlane that a bank account should be opened by the Honorary Treasurer at the ANZ Bank, Warragul and that cheques drawn on behalf of the Association should be signed by the Honorary Treasurer and Honorary Secretary.

NAME OF ASSOCIATION:

After discussion, it was decided on the motion of Mr Gadd, seconded by Mr MacFarlane that the name of the Association should be – 'The English Leicester Association of Australia'.

NEXT MEETING:

It was resolved that a Committee meeting should be held at 1.00 pm on Monday, 11th October, 1982 in the Administrative Building, Royal Melbourne Show Grounds to formulate the Association's Constitution.

The secretary was directed to collect copies of Constitutions from other Sheep Breed Societies and distribute same to Committee members for their perusal.

The Chairman thanked members for attending and the meeting closed at 5.00 pm.

Since its formation the English Leicester Association of Australia (ELAA) members have worked tirelessly for the breed. As well as the usual printing of promotional brochures and banners, the association has promoted the unique qualities of the English Leicester's lustrous wool by supplying wool for the wool craft section of shows and from 1992 to 1999 sponsoring knitting workshops as a promotion of the wool. From 1987 until about 1994 special 'lustre wool' sales were promoted and supported by members. For about six years starting in 2007, some members pooled their wool and Woolgrowers Independent Selling Services sold it as Lustre Wool. For a short period, wool was also sold into the doll's hair trade. In 1994, along with the Lincoln breeders, a Lustre Wool feature show was held at the Royal Melbourne Show.

English Leicester knitting workshop "Derby Hill", Maldon, Vic 1995. Ethel Stephenson.

English Leicester knitting workshop. Ethel Stephenson.

Over the years the association has been able to arrange many English Leicester feature shows at several Victorian country shows, displayed sheep at various field days and had a display at the Geelong Wool Museum.

In 1986 the association participated in South Australia's 150 year celebration, 'On the Sheep's Back', and in 1989 represented the breed at The World Sheep and Wool Congress which was held in Launceston. Over 1,000 delegates representing 17 countries attended the congress. Thirty one different breeds of sheep were represented.

Ram class at the World Sheep and Wool Congress, Launceston 1989. Left to Right: Eric Gray, Garry Webb, Brenton Heazlewood, Colin Taylor, unknown. Brenton Heazlewood.

Norm Badcock judging at the World Sheep and Wool Congress, Launceston 1989. Brenton Heazlewood.

Ivan Heazlewood and Anne Heazlewood manning the English Leicester Association of Australia exhibit at the World Sheep and Wool Congress, Launceston 1989. Brenton Heazlewood.

Anne Heazlewood manning the English Leicester Association of Australia exhibit at the World Sheep and Wool Congress, Launceston 1989. Brenton Heazlewood.

British breeds represented at the congress included Southdown, Suffolk, Cheviot, Lincoln, English Leicester, Border Leicester, Romney, Wiltshire Horn, South Dorset Down, Dorset Horn, Ryeland, Dorset Down, Hampshire Down and South Suffolk.

The ELAA was represented at the Congress with a promotion in the sheep pavilion. The display included fleece and crossbred fleece, skin and wool products and commercial breeding facts.

The results of the English Leicester judging were, Champion ram Colin Taylor, Reserve Champion ram I.C. Heazlewood, Champion ewe and Reserve Champion ewe E.B. Gray.

In 1996, Dr Rob Banks, Lambplan Co-ordinator from the University of New England was guest speaker at the September meeting. He spoke regarding English Leicester breeders entering their sheep data into

Lambplan. All breeders sent their flock pedigrees to Margaret Kingman who entered the details into a spreadsheet which then detailed one or a choice of suitable matings. The University was trailing a new program and used the data as a prototype and did not charge the association for the work. The result was that there were not enough animals and diversity of blood lines to get a meaningful result and the project was stopped.

Ten years later in 2006 the association undertook another project to collate the genetics of the breed, 'as a tool to enable the most productive matings and improvement across the national flock'.

This was led through the Department of Primary Industries, Orange, NSW, with assistance from a company called XPrime who were developing a computer programme called Xaim. Again nothing really came of this project mainly due to the failure of the computer programme.

In 2021, when breeders were expressing concern regarding the small genetic pool available to them, particularly due to quarantine restrictions banning any genetic imports, history repeated itself in that the association again had discussions with Meat & Livestock Australia through Lambplan. This time emphasis was on any potential inbreeding problems.

With the help of breeders who have recorded and submitted pedigrees, Lambplan are assisting us by analysing as many pedigrees as possible in this pilot programme to look at the inbreeding, or lack of it, for the breed as a whole.

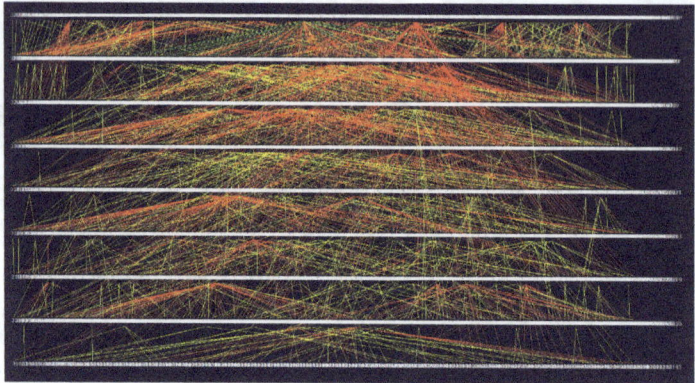

Lambplan English Leicester pedigrees. Meat and Livestock Australia.

One of the more contentious subjects that the association has dealt with was the issue of a potential breed name change to Leicester Longwool. In 1990 there was discussion regarding this possible change of name. The reasoning being that those that market their wool in yarn form do not use the name ENGLISH. It was decided to send out a Referendum Form to all English Leicester breeders (including NZ) for options on the matter.

At the following meeting in February 1991 this motion was rescinded, and it was moved that it be further discussed at the next meeting.

The July 1991 meeting resolved to send a questionnaire to all members to seek their opinion. The results of this were presented to the AGM in September, being 17 against a name change and 1 for. It was moved by G. Henry 'that ELAA not proceed with a change of name'. It was however noted that there was a valid point in the name change when it comes to marketing of wool and sale of products. This subject was again on the table in 2009, with another vote being taken by members who again voted to retain the breed name as English Leicester. Only Australia and New

Zealand use the name English Leicester, all other countries refer to the breed as the Leicester Longwool.

Over the years the association has hosted several overseas people who have been associated with the breed. Some of these have judged our sheep at shows. These include Harold Bennett (NZ), Bob Todhunter (NZ), Betty Garbutt (UK), Harold Nobes (UK), Ivor Robinson (NZ), Elaine Shirley (USA) and Lisa Westervelt (USA).

In 1988 the breeders celebrated the 150th year of the breed being in Australia. This was based on information contained in the book 'Sheep Breeding in Australia' by G.A. Brown. We now know that Leicesters were here earlier than 1838, as is stated by Brown. In conjunction with the July ASBA show, which at that stage was held in Melbourne, a 150th year dinner was held at the Union Hotel, Ascot Vale. It was during this dinner that Life Membership was bestowed on Ian Morrish for his support and dedication to the breed over 40 years.

ASBA Sheep Show Melbourne 1988 (commemorative show for 150 years of English Leicesters in Australia). Holding left sheep is Ian Morrish, President ELAA. Holding right sheep is Ethel Stephenson, Secretary ELAA. On the left is the steward, Joan Sloan.

South Australian Society

For some time South Australia had its own breed promotional society. At the July, 1986 meeting of the English Leicester Association of Australia (ELAA) the possibility of the formation of a branch in South Australia was discussed, noting that SA needed to report further at the next AGM due to be held in September of that year.

At the committee meeting, November 1986 the South Australian Group was again on the agenda. It was decided, after considerable discussion, to contact South Australia allowing them to carry on with their meetings with ELAA approval, giving support and agreeing to the principle of branches under the auspices of the ELAA.

At the following committee meeting held in March 1987, David Parker from SA gave a brief outline of the circumstances leading to the need for a South Australian Society. These included distance to Victoria, young group, hobby farms, the need for mutual help, discussions on breeding needs and ongoing plans to suit local needs. After a long discussion it was generally accepted that there was a real need for a branch (Society) to be formed in South Australia. All present were in verbal favour and no motion was moved.

The Muster, No 4, March 1987 reported the following:

> A group of enthusiastic breeders in SA have recently formed a body for the promotion of the English Leicester Breed.
>
> ELSSA, the English Leicester Society of South Australia, is gearing up for big events in 1987-1988. Since the 1930's English Leicesters have diminished in numbers in SA, for no apparent reason. In most areas of the south-east of the state, Adelaide Hills and south coast, conditions are ideal for the breed.
>
> From one stud in the state in 1983 there are now eight registered breeders with three more pending. But with a lack of large studs from which to draw stock it will take time to build up numbers.

However, new faith in the breed, patience and enthusiasm will prevail.

The English Leicester recently has had many successes in interbred sections at shows, suggesting that an appreciation of the breed is returning.

In 1985 English Leicesters won the All-breeds trophy at the Millicent Show and in 1986 was runner-up in the All-breeds at the Hamilton Sheep Show, Mt. Gambier, Gawler, Coonalpyn and a winner at Murray Bridge and Strathalbyn.

The Southern Districts Show Society, which encompasses four shows, gave a trophy at the end of the season for British Breeds Longwools, counting aggregate points from the four events. For the first time last year the English Leicesters were successful.

Easy care and docility, along with good lambing percentages and improved wool return are some of the real assets of the breed. Also the growing handspinning area holds the English Leicester wool as some of the most valuable and easy to work with.

Some three-month-old English Leicester lamb's wool

microned at 18m, and mature sheep at 30m, indicating the visual assessment is harsher than the microntest.

Plans are underway for more advertising of the breed in 1987 and a delegation of SA breeders will discuss with the English Leicester Society of Australia better joint promotion leading to a healthy interest in and acceptance of the English Leicester across the country.

Kelvin Rorke, President ELSSA.

SA breeders at this time were:

- K.T. & L.A. Rorke
- S. & P. Di Bona
- I. Parker
- P. & J. Foureur
- M. Chigros
- D. Johnson
- J. Seater
- C.F. Spencer

On 17 May 1987, the Secretary of the Association, Ian Morrish, attended a field day at Balhannah SA, conducted by the newly formed English Leicester Association of SA.

I do not know exactly how long this Society lasted for. It was active in 1994 but I suspect that it folded in 1996.

30 Year Meeting

On 24 September 2012, a special meeting of the association was held to celebrate 30 years since the re-establishment of the ELAA. Foundation, current, past and new members attended the meeting which was held at the Royal Melbourne Showgrounds, where the first meeting was held and chaired by Mr Duncan McGregor, Dalmore English Leicester stud, No 115, Maldon, Vic.

President Kirsty Harker opened the meeting and welcomed all, particularly the three members who were present on 19 September 1982 – Colin Taylor, the first secretary; Dianne McFarlane, the first treasurer; and Carol Stuart, past member.

Also in attendance were three life members – Mr Ivan Heazlewood, Mr Colin Taylor and Mrs Ethel Stephenson.

Congratulations went to Colin Taylor as he was the only one of the 17 people present at the first meeting who still attends all meetings and is showing his sheep successfully at all major and regional shows.

A minutes silence was observed for members present at the first meeting and who are now deceased: Ian and Marion Morrish, first president; Alex Habel; Alastair Williamson; Jim Heard and Neville Stuart.

The English Leicester Feast

In April 2019, the English Leicester breeders in Tasmania held an open day to showcase their sheep, followed by an 'English Leicester Feast' to demonstrate the eating qualities of the meat.

The event was organised by Paul Willows of the 'Nant' stud at Evandale. Four English Leicester breeders from Victoria attended the day as well as Tasmanian breeders and those interested in the English Leicester wool for craft work.

The day commenced with morning tea and an inspection of George Willow's sheep at Evandale before travelling to Brenton and Anne Heazlewoods at Whitemore. After a light lunch a tour of Heazlewood Seeds cleaning plant was undertaken before inspecting sheep from the Melton Park stud.

The highlight of the day was the dinner at night with over 35 people in attendance. This was held at The Sebel Hotel's, Bluestone Bar & Kitchen in Launceston. The menu consisted of English Leicester lamb presented in three different ways by chef, Michael. The evening was a great success with everyone enjoying lamb being presented in not the usual way. A very talented group of ladies put on a significant display of cleverly designed crafted garments made from English Leicester wool which very much added to the theme of the evening.

English Leicester fashions displayed at the English Leicester Feast, Launceston 2019. Brenton Heazlewood.

Tasmanian Tour 2023

The Tasmanian breeders hosted their mainland counterparts again in April 2023. Fifteen breeders from New South Wales and Victoria attended. A welcome dinner was held on the Friday night at which the changes that breeders have made to the breed over the past 200 years was discussed.

Visits were to the Melton Park, Malton and Nant studs to inspect and discuss sheep.

A highlight of the weekend was a visit to the Winton Saxon Merino Stud of John and Vera Taylor. This stud was founded in 1835 and has basically been run as a closed flock all its life.

English Leicester gathering Tasmania April 1023. Back Row, L to R: Alison Stewart, Melissa Mamers, Margaret Kingman, John Kingman, Lachlan Stewart, Anne Heazlewood, Wendy Beer (obscured), Ian Morris, Maureen Morris, Ros Cohen, Colleen Southwell, Andrew Colvin, Fiona Hume, Nick Wootton. Front, L to R: Poppy Mamers, Jason Southwell, "Spitfire", Brenton Heazlewood, George Willows, Edward Southwell, Vanessa Wootton. Brenton Heazlewood.

Life Membership of the ELAA has been presented to the following:

- Ian Morrish

- Ethel Stephenson

- Colin Taylor

- Ivan Heazlewood

- Margaret Kingman

Margaret Kingman receiving her English Leicester Association of Australia Life Membership from Vanessa Wootton, April 2023. Brenton Heazlewood.

ELAA positions

Year	President	Secretary	Treasurer
1982	I. Morrish	C. Taylor	D. MacFarlane
1983	I. Morrish	C. Taylor	D. MacFarlane
1984	I. Morrish	C. Taylor	D. MacFarlane
1985	C. Taylor	K. Rorke	D. MacFarlane
1986	C. Taylor	E. Stephenson	D. MacFarlane
1987	I. Morrish	E. Stephenson	E. Stephenson
1988	I. Morrish	E. Stephenson	E. Stephenson
1989	I. Morrish	E. Stephenson	E. Stephenson
1990	B. Sullivan/G. Henry	E. Stephenson	E. Stephenson
1991	I. Morrish	E. Stephenson	E. Stephenson
1992	I. Morrish	E. Stephenson	E. Stephenson
1993	I. Morrish	E. Stephenson	E. Stephenson
1994	N. Cole	E. Stephenson	E. Stephenson
1995	N. Cole	E. Stephenson	E. Stephenson
1996	N. Cole	E. Stephenson	E. Stephenson
1997	N. Cole	E. Stephenson	E. Stephenson
1998	N. Cole	E. Stephenson	E. Stephenson
1999	N. Cole	E. Stephenson	E. Stephenson
2000	N. Cole	E. Stephenson	E. Stephenson
2001	N. Cole		E. Stephenson
2002	N. Cole	M. Kingman	E. Stephenson
2003	N. Cole	M. Kingman	E. Stephenson
2004	N. Cole	M. Kingman	E. Stephenson
2005	N. Cole	M. Kingman	E. Stephenson
2006	N. Cole	M. Kingman	E. Stephenson
2007	N. Cole	M. Kingman	E. Stephenson
2008	K. Harker	M. Kingman	E. Stephenson
2009	K. Harker	M. Kingman	E. Stephenson
2010	K. Harker	M. Kingman	E. Stephenson
2011	K. Harker	M. Kingman	E. Stephenson
2012	K. Harker	M. Kingman	L. Docherty
2013	B. Heazlewood	M. Kingman	L. Docherty
2014	B. Heazlewood	M. Kingman	L. Docherty
2015	B. Heazlewood	M. Kingman	L. Docherty
2016	B. Heazlewood	M. Kingman	L. Docherty
2017	B. Heazlewood	M. Kingman	L. Docherty
2018	B. Heazlewood	V. Wootton	L. Docherty
2019	B. Heazlewood	V. Wootton	L. Docherty
2020	B. Heazlewood	V. Wootton	L. Docherty
2021	B. Heazlewood	V. Wootton	L. Docherty
2022	B. Heazlewood	V. Wootton	L. Docherty
2023	B. Heazlewood	V. Wootton	L. Docherty

Breeders at the Australian Sheep & Wool Show, Bendigo, 2018. Back row L to R: Jason Southwell, Pam Tait (NZ judge), George Willows, Edward Southwell, Vanessa and Nick Wootton, Brenton Heazlewood. Front row: Colin Taylor, Paul Willows.

19
Lustre Wool

'We are well aware that long wool in its most perfect state cannot be expected from sheep destitute of the quantity of food, which nature requires for her support.'[289]

'At the genial season when flocks are disburthened of their coat, and pay the annual tribute due for the protection and sustenance which they have received...'[290]

In 1987 the English Leicester Association received an offer from Elders Pastoral to hold a special English Leicester wool sale. It was believed that there may be a better market for this high lustre wool and breeders were asked to support a lustre wool sale. In New Zealand a similar proposal promoted by Laycock Pty Ltd had met with enthusiastic support.

Long Lustrous Wool. Brenton Heazlewood.

Lustre is an intrinsic and inherent silvery brightness of the fibre which is not lost in the manufacturing process, and which cannot be given to goods not made of pure lustre wool. Lustre is only seen in longwool breeds and should not be confused with colour. In Australia there are only two breeds that produce lustrous wool, those being the English Leicester and Lincoln. The Romney and Border Leicester produce demi-lustrous wool.

It is recorded by using a light recording instrument, which in simple terms records the ability of the wool to reflect light, similar to a light metre in a camera. It compares the result of recording reflected light from a white ceramic tile to the light reflected off the cuticle scales of a wool fibre. The recording measures depth, intensity, purity and hue.

The early wool merchants were well aware of lustre wool and its properties. John Luccock, writing in 1805 states that:

THE HISTORY OF ENGLISH LEICESTER SHEEP IN AUSTRALIA

'There are some other breeds of sheep which yield a wool remarkable for its brilliancy; although the pile be not perfectly opaque, yet the surface of it seems to possess very fine polish, like that of a metallic needle; and the lustre with which it reflects the rays of light has given it among workmen the appropriate appellation of silvery haired wool. This is most frequently found upon the backs of sheep whose pile is remarkably long and hairy.[291]

Elders held the initial 'Lustre' wool auction on 6 May 1987. The top price received for fleece wool was 471 c/kg greasy.

In May 1988, the second lustre wool sale was held with the top price being 418 c/kg greasy.

The 1989 sale consisted of 15 bales of AAA Lustre wool and the top price was 350 c/kg. All the wool was purchased by Modiano Morris Australia.

The 1990 sale consisted of 30 bales of lustre fleece and oddments with the top fleece price being 331 c/kg and was principally bought by Laycock Pty Ltd of Melbourne.

In 1991, 19 bales of lustre fleece wool and oddments were sent to the Gordon Technical College at Geelong. The wool was classed by the Wool and Rural Studies Department of the college and offered for sale by private tender. A three bale lot of AA Lustre was purchased by the Melbourne College of Textiles at 203 c/kg with the remainder going to Carpet Wool Marketers Ltd in Geelong. The top price for fleece lines was 201 c/kg.

Mr. Ian Henry classing S. and G. Oliver's English Leicester wool at the Gordon College, Geelong

Classing wool Geelong. The Muster, August 1992.

Nine breeders from Victoria and NSW supported the 1992 Lustre Wool Collection. A total 23 bales of fleece and oddments were classed by the Wool and Rural Studies Department of the Gordon Technical College, Geelong. All wool was offered for sale by Private Tender.

Five bales were bought privately. The remaining 18 bales were sold to Carpet Wool Marketers, Geelong, where the top price paid for fleece wool was 221 c/kg.

In March 2008, Woolgrowers Independent Selling Services held their annual lustre sale on behalf of the English Leicester Association. That years sale was a combined lustre sale, with the Lincoln Association offering wool for the first time.

The wool bales were displayed traditionally and generated a great deal of attention from several exporting companies, with one exporter being a repeat buyer from the previous years sale. Although prices were down on the previous year due to the increase in the value of the Australian dollar, the sale was still a success and all wool was sold to the trade.

In 1994, in conjunction with the Lincoln breeders, a 'Lustre Wool' feature show was held at the Royal Melbourne Show. Nine English Leicester breeders exhibited sheep, these being I.C. Heazlewood & Co (Melton Vale), P. & E. Stephenson (Ostlers Hill), Colin Taylor (Koenarl), I.N. & M.E. Hay (Karrama), Mrs M. Kingman (Yarn Time), I.J. Morrish (Bardia), Bellbrook Partnership (Marengo), S.L. & C.D. Lyons (Lonarch) and N.J. & N.T. Cole (West Cloven Hills).

Lustre Feature Show 1995 exhibitors.

Ivor Robinson from New Zealand was the judge. The results were as follows:

- Ram 2 ½ years old and over: P. & E. Stephenson

- Ram over 1 ½ and under 2 ½: I.C. Heazlewood
- Ram under 1 ½: I.C. Heazlewood
- Pen of 2 rams under 1 ½ woolly: I.C. Heazlewood
- Ram shorn under 1 ½: I.C. Heazlewood
- Ewe 1 ½ and over with lamb at foot: Colin Taylor
- Ewe 2 ½ and over with lamb at foot: Colin Taylor
- Ewe over 1 ½ and under 2 ½: Colin Taylor
- Ewe under 1 ½: I.C. Heazlewood
- Pen of 2 ewes under 1 ½: Colin Taylor
- Ewe under 1 ½ shorn: P. & E. Stephenson
- Group, 1 ram 2 ewes under 1 ½: I.C. Heazlewood
- Group, 1 ram 2 ewes any age: Colin Taylor
- Champion ram: I.C. Heazlewood
- Reserve Champion ram: I.C. Heazlewood
- Champion ewe: Colin Taylor
- Reserve Champion ewe: I.C. Heazlewood
- Supreme Champion: I.C. Heazlewood

- Most successful exhibitor: Colin Taylor

- Duncan MacGregor Trophy: I.C. Heazlewood

1994 Royal Melbourne Lustre Wool Feature Breed Show Medal. Brenton Heazlewood.

It is a little ironic that today the breed is mainly known for its long, lustrous wool, a feature which Bakewell paid little attention to. However, it is an asset that has helped keep the breed alive, and to some degree relevant, for the past fifty years.

20

Melton Vale / Melton Park English Leicester Stud History

"Moreton" May 15th 1879
I have this day Sold to Robert Heazlewood my Farm at
Melton Vale for the sum of Twelve Hundred
pounds and
24 Leicester ewes and one ram
To be delivered to me after shearing...

Joseph Heazlewood

'40 Leicester ewes – Fathers gift'
Extract from Roy Heazlewood's notebook on the occasion of his marriage, 1911

My great, great, grandfather Henry Heazlewood arrived in the colony of Van Diemen's Land in 1834. He was a carpenter from the village of Asfordby, near Melton Mowbray, Leicestershire. When he became a land owner in 1854 he named his property 'Melton Vale', in memory of his homeland. The name 'Melton' is still used today by Heazlewood farming families. I have named the property that we have purchased recently, 'Melton Park'.

I write this section on my own stud, not because it is different from other studs that have been in existence for a long time, but because I have material that previous generations have kept which has enabled me to record its history. They have kept ribbons, rosettes, certificates, locks of wool, flock books, correspondence with ram buyers and photographs which help document the stud's history. I am privileged to be the fourth generation shepherd to care for this lovely breed of sheep on our property.

It is ironic that even though Henry was not a farmer at Asfordby, however, when he started farming in Van Diemen's Land he most certainly would have been running English Leicester sheep, which were originally bred by Robert Bakewell at 'Dishley Grange', which is not far from Asfordby.

When Henry died in 1861, each of his six sons inherited some land. As these areas were small (Robert's was 70 acres), three of the brothers eventually sold their portions to Robert and moved to larger properties. One of these transactions, in 1879, involved payment partly in cash and the balance in Leicester sheep and grain.

Even though my great grandfather, Robert, did not establish his English Leicester stud 'Melton Vale' until 1871, there can be no doubt that both he and his father would have been running this breed since they started farming as they were the dominant British breed at that time in Van Diemen's Land. Indeed in 1930 they were still the most numerous registered breed in Tasmania. Robert's foundation sheep included 20 ewes from C.W. Allen of Westbury. A further 15 ewes from the celebrated flock of Thomas Gibson, 'Esk Vale' were added in the next year with rams from William Field's 'Enfield' stud being used.

R.G. Heazlewood's handwritten draft stud history, written in 1907 for the flock history which appeared in the Australian Longwool Flock Book, Volume I, 1909, is still in our possession. This was edited for publication with many of the superlatives omitted. This stud was registered with the Australian Longwool Sheepbreeders' Association as Flock No. 6.

R.G. Heazlewood's Melton Vale English Leicester Stud Flock

The Melton Vale, English Leicester Stud Flock was established in 1872 by the purchase of 20 pure ewes from C.W. Allen Esq.

These were the direct descendants of pure English Leicesters, imported by Mr Jones of Jericho, Tasmania, and in lamb to a high class English Leicester Stud Ram.

A few months later 15/fifteen special stud ewes were selected from the celebrated English Leicester stud flock of T. Gibson Esq., Esk Vale who at that time was the premier breeder of Tasmania. By careful selection and good judgment the Esk

Vale Leicesters had attained a very high standard of excellence and was practically invincible in the show yard. Their lineage was exactly similar to the ewes bred by C.W. Allen Esq., all tracing back to the famous Jones and Stokell importations, which were responsible for the early English Leicester flocks in Tasmania.

A ram from the Enfield stud flock was then introduced, a remarkably fine sheep, by an imported English Leicester Stud Ram, selected in England at a high figure for New Zealand but on arrival there some restrictions would not allow him to be landed, consequently he was reshipped to Tasmania and purchased by W. Field Esq. of Enfield and mated within his flock with such good results that his stock became famous. The introduction of the Enfield ram to the Melton Vale Stud Flock was an unqualified success. In addition to an increase in the size, an improvement in the symmetry was also effected, and the constitutional vigour and robustness considerably enhanced, and also an increase in the weight, length of staple and quality of the fleece with a corresponding rise in monetary value per lb. In succeeding years choice stud rams, from C.W. Allen Esq. and T. Gibson Esq. were used with excellent results and by a gradual process and a strict adherence to following on the one line of blood the Melton Vale English Leicester Stud Flock was eventually brought into prominence and for a lengthy period has more than held its own against all comers in the sales and Show Ring in this and neighbouring States.

THE HISTORY OF ENGLISH LEICESTER SHEEP IN AUSTRALIA

The flock and stud rams are famed for their uniformity and even character combined with size, symmetry, general robustness, and high quality of wool.

For many years no outside rams were used the resources of the flock being sufficient to supply the stud rams necessary for home replacements, only the very best being reserved for this purpose which included such notable sires as Sentenial and Marvel, two of the largest prize winners ever bred in Australiasia, the later was considered by competent judges an ideal Leicester Stud ram, these and many other high class sires have been sold at high figures (ranging up to fifty guineas) to flock masters on the mainland who all speak in the highest terms of their value as a means of effecting a general improvement whenever they may be used.

With a view of still further increasing the size, retaining the symmetry and building up the constitution, during the last two years stud rams have been imported from the very best English flocks, notably the ram Lord Sheffield, who after being used for a couple of seasons, secured the Championship of Australia at the Royal Agricultural Show held in Melbourne in 1906 and when submitted for sale he realised the record price of 100 guineas. This innovation has been a pronounced success, the first of Lord Sheffield stock was exhibited at the recent Sheep breeders show in Sydney, competing against Australia and

New Zealand, they succeeded in winning first prize in their class and champions for ewe and ram. Three other English Leicester Stud rams and several ewes have been added this year 1907 and judging from the success achieved by Lord Sheffield's progeny the effect may be readily anticipated being a more even distribution of flesh, a corresponding increase in size, with the symmetry and constitution fully maintained, and the beautiful well known lustrous, weighty high quality fleece so peculiar to Tasmania's environment, will still remain and render material assistance in establishing their superiority.

Stud ewes and rams imported from England between 1905 and 1946 have been:

- 1905 from George Harrison (F42) Ram No.18 Lord Sheffield

- 1906 from E.F. Jordan (F34) 3 ewes

- 1906 from E.F. Jordan (F34) Ram Eastbourne 183

- 1907 from J.E. & C.H. Simpson (F2) Ram Pilmoor Champion No.856. This ram was bred by Simpson but owned by Thomas Nightingale (F39) at the time of purchase.

- 1907 from John Cranswick (F3) Ram Hunmanby Compact No.833

- 1946 from C.E. Simpson (F2) 1 ewe

- 1946 from W.A. Coleman & Sons (F97) Ram Speeton Sample

No.289

English Leicester Ram, Champion of Australasia 1906. The property of R.G. Heazlewood, Glenore, Tasmania. Brenton Heazlewood.

Mr Coleman, Speeton Stud UK, with the author at the Great Yorkshire Show 2013. Brenton Heazlewood.

Two years after founding the stud in 1871, my great grandfather, Robert Heazlewood commenced showing sheep. Between 1896 and 1920 he regularly won Championships at Melbourne and Sydney. My grandfather Roy, my father Ivan, showed sheep at local shows as well as on the mainland and now I am doing this, competing at local shows as well as taking a team to the Australian Sheep and Wool Show in Bendigo each year.

Some of the championship ribbons won by Robert Heazlewood at Sydney and Melbourne shows:

- Sydney 1896: Champion Ram

- Sydney 1896: Champion Ewe

- Melbourne 1905: Champion Ram

- Melbourne 1906: Champion Ram

THE HISTORY OF ENGLISH LEICESTER SHEEP IN AUSTRALIA

- Sydney 1907: Champion Ram
- Sydney 1910: Champion Ewe
- Melbourne 1910: Champion Ewe
- Melbourne 1911: Champion Ewe
- Melbourne 1912: Champion Ewe
- Melbourne 1914: Champion Ewe
- Sydney 1914: Champion Ewe
- Sydney 1914: Champion Ram
- Sydney 1915: Champion Ewe
- Sydney 1916: Champion Ram
- Sydney 1918: Champion Ram
- Sydney 1918: Champion Ewe

R.K. Heazlewood 1918 Sydney Show Certificate. Brenton Heazlewood.

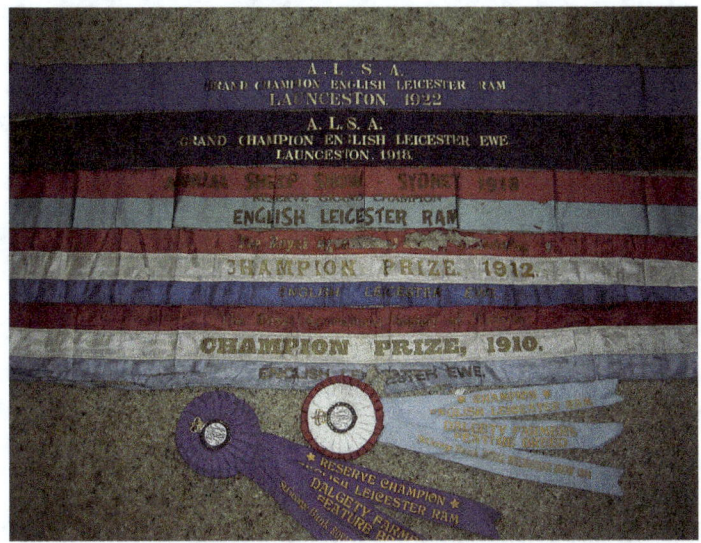

Show ribbons. Brenton Heazlewood.

F3104 Wool specimen, pure Leicester stud ram, bred by R.G. Heazlewood, 'Melton Vale', Whitemore, Scone, Tasmania, Australia, 1892. The ram the sample was taken from was Champion at Hobart, Longford and Albury shows in 1892. The sample tested at 42.3 microns with a staple length of 160mm. Collection: Museum of Applied Arts and Sciences. Purchased 2003 (originally gift of S Bradbury, 1893). Photo: Chris Brothers.

1897 lock of Melton Vale wool. Brenton Heazlewood.

When the Tasmanian Longwool Sheepbreeders Association was formed in Launceston in 1892, Robert and his two brothers, Charles and Edwin, were members. Robert was a regular attendant and on occasion chaired committee meetings.

Around 1905 Robert and several other Tasmanian breeders purchased both ewes and rams from leading Yorkshire breeders. Forty years later in 1947 my grandfather imported a ram and ewe from the Coleman family 'Speeton' flock. These were some of the last live imports of English Leicesters into Australia before live sheep imports were banned.

5 Ewes. Brenton Heazlewood.

When Robert's youngest son, Roy (my grandfather), married in 1911 his wedding present was 40 English Leicester ewes. These ewes formed the foundation of the flock which Roy registered with ALSA in 1912, Flock No. 36. This is the flock which has been handed down to me and now carries the prefix 'Melton Park', having been changed from 'Melton Vale' in 2006. Roberts flock, No.6, was registered up until about 1926 as it is in Vol. 16 (1925), but not Vol. 17 (1927). Robert died in 1927.

Not only did my grandfather and great grandfather take sheep to the mainland shows, but they also sold considerable numbers of flock rams to Victoria, and New South Wales in particular. These sales were either through agents or direct to repeat purchasers.

The Examiner, Thursday 14 July, 1910, reported on the sale and shipment of sheep from my great grandfather to NSW.

THE HISTORY OF ENGLISH LEICESTER SHEEP IN AUSTRALIA

STUD SHEEP

Messers. Robert Gardner, Limited, report having shipped by s. Wakatipu yesterday to Sydney on account Mr R.G. Heazlewood, of Glenore, 74 pure English Leicesters. They comprised – two stud English Leicester rams, sired by his imported ram from England, "Eastbourne", dams prize ewes; 50 stud ewes, 25 being two-tooths, in lamb to the champion ram at the recent Campbell Town Show; and 25 flock rams. All the above were sent to private orders in New South Wales.

The same firm also report having sold privately, on account ... of Mr R.G. Heazlewood, 50 flock Leicester rams. These are also for private orders in New South Wales, and will be shipped at an early date.

The series of letters to my grandfather, Roy K. Heazlewood, from F.W. Peterson enquiring about and ordering rams gives a good insight into the large scale use of English Leicester's on Merino properties. It is interesting to note in the third letter that Border Leicester's are being considered.

18th July 1913

Mr Roy K. Heazlewood
Melton Vale
Whitemore
Tasmania

Dear Sir

Please let me know as soon as possible about how many pure English Leicester Ram lambs you expect to have next lambing and also what months you will lamb during this year. Also if you will give me the offer <u>definitely</u> of first pick of the whole drop for delivery Feb or Mch/14. I am <u>almost</u> <u>certain</u> to buy <u>100</u> of these young rams – I must buy at <u>least 50</u> for extra ewes now on hand & the other 50 would be to enable me to sell some of the older ones now on hand. We had a particularly bad year last year owing to the drought or I should have purchased a consignment out of your last drop. Please let me know if price will be the same 3 ½ guineas delivered Sydney for any number from 50 to 100.

Yours Faithfully

F. W. Peterson
Manager

PS I am now using the X Bred ewes the progeny of the mixed lot of EL Rams I purchased the first lot 1910 – & next year will be using X Bred ewes the progeny of the 60 EL Rams your own breed that I purchased the following year. Do you think the Rams I would get from you next year would be too closely related to the maiden ewes by your Rams to use with them. Of course the older X Bred ewes by the mixed lot of Rams would be alright to mate with your rams – & also all the Merino ewes I have. I merely ask this by the way – it will not make any difference to me about buying your Rams next year.

***

14th May 1914

Roy K. Heazlewood Esq
Melton Vale
Whitemore
Tasmania

Dear Sir

The 95 young EL Rams have just arrived at Oban.

I am pleased with them, they are a very nice lot. Unfortunately about half the number are more or less blind with eye disease – but the weather now being Cool they should with attention soon get over this trouble. I note you will on receipt of this letter send the a/c direct to Mr Morton.

With kind regards

Yours faithfully

F. W. Peterson
Manager
Rawdon Estate
Rylstone

***

26th December 1917

Mr Roy K. Heazlewood
Melton Vale
Whitemore Tas

Dear Sir

<u>*English Leicester Rams*</u>

You will remember a few years back supplying me with some E/L Rams for Oban.

Oban has been sold & I am now Managing 'Rawdon' near Rylstone NSW for Mr Morton. Latterly we have been using

Merino Rams on the first cross ewes to produce Comeback Merinos & now are beginning to want E/L Rams again to mate with Comebacks.

For next year (to work May/June/18) I require <u>50 or 60 Rams</u> – not sure of number until I go through the ones on hand. Can you supply or could I get by going over to Tasmania that number (or even 40) <u>good</u> English Leicester Rams 2Th & 4Th or either age.

The first year I bought in Tasmania (1910) Rams of those ages – the lot was more or less uneven in size & quality – subsequently you gave me first pick of all your drop & I then got a much evener line of rams. I find however that Tasmanian bred E/L Rams dropped say Sept will not work over here the following Autumn, so for next years requirements I must buy at least "2Th" Rams now & if we do business I would like at some time to have offer of <u>first pick 50 or more</u> of your next seasons drop of English Leicester Ram Lambs for delivery FOB Sydney March 1919. Please state price if you grant this offer & also you might give me an idea as to price for the 2Th or 4Th Rams if these can be got <u>now</u> & also as regards quality & eveness of same.

I am getting quotations from other breeders in this state including 'Border Leicesters' but if I can get my requirements from you @ prices that compare reasonably with local prices & quality ditto. I prefer to do business with you. The results so far *from using <u>Your</u> rams have been quite satisfactory in every way.*

Yours ffully

F.W. Peterson mgr
Rawdon Estate
Rylstone

PS Will probably require a draft of <u>50 or more</u> Ram Lambs (E/L) each year for some years ahead & if I now commence using your Rams again would ask for first offer of pick of Ram lambs from your to year. FWP

***

6-2-18

Mr Roy K. Heazlewood
Melton Vale
Whitemore
Tas

Dear Sir

About 10 days ago I wired you as follows

"Please select 60 pick of 100 rams at price quoted. Delivery end February" & now confirm same.

Please consign as before to Mr P H Morton, Sydney advising him direct (his address 43 Hunter St Sydney) when to expect them. Also please render the acct for them to him.

Have I the first offer of pick of 50 or more of your next drop of EL Ram lambs same as in past & if so please state if price is same as before 3 ½ gns FOB Sydney. Will probably want more than 50.

Yours faithfully

F. W. Peterson
Rawdon
Rylstone

***

22-2-18

Mr Roy K. Heazlewood
Melton Vale

Whitemore Tas

Dear Sir

I am in receipt of you letter of 6th inst for which I thank you & have noted Contents.

Rams should now be in Sydney but I have not yet received advice from the Sydney Agents.

I note that I have offer of your next seasons drop of E/L Ram Lambs the pick any number @ 3 ½ g. FOB Sydney delivery March / 19.

Will advise you in due course say end of Sept or early Oct next about my requirements in EL Ram Lambs.

Yours ffully

F.W. Peterson

Our flock has a proud history of exporting both live sheep, semen and embryos to Canada, USA, New Zealand and Sweden.

- 1981: 1 ram to New Zealand
- 1986: 1 ram to Canada as part of the World Sheep and Wool Congress.

- 1987: 2 rams to New Zealand

- 1989: 1 ram and 5 ewes to Colonial Williamsburg Foundation as part of a consignment from Tasmania to re-introduce the breed back into the USA.

- 2008: Semen to the USA

- 2014: Semen to the USA

- 2015: Semen to the USA

- 2018: Embryos to Sweden along with 'Langs Crossing'

When my father, Ivan, exhibited at the 1991 Royal Hobart Show, the Melton Vale English Leicester stud celebrated the centenary of its first showing in Hobart.

It achieved a carbon copy of its 1891 success, again winning the one-year-old ram class. But this time the Melton Vale rams went further, not only securing the championship but also taking out the interbreed award for champion longwool ram of the show.

In 2021, 130 years after my great grandfather first exhibited in Hobart, I again exhibited at the 200th Royal Hobart Show. Again, our stud won the one-year-old ram class as well as the Supreme Longwool exhibit of the show.

Robert Heazlewood's 1891 presence in Hobart was noticed by the local press:

'The yearling Leicester ram bred by Mr R.G. Heazlewood of Glenore which obtained the first prize, was an excellent animal and bore beautiful quality of wool throughout.

Mr Heazlewood has not exhibited at Hobart Show before but has been a prize-taker in the north for many years past. He has also received a second prize for a pair of yearling ewes and highly commended for a pair of Leicester ewes 4-tooth and upwards.'[1]

R.G. Heazlewood's Champion Leicester Ram Royal Hobart Show 1904. The Tasmanian Mail, October 29, 1904.

*R.G. Heazlewood with Leicester Ram, Hobart Show 1910.
The Weekly Courier, October 10, 1910.*

The Melton Vale English Leicesters were shown at Hobart without a break from 1891 through to about 1965.

After a long absence from exhibiting at mainland shows, Ivan took a team to the 1994 Lustre Wool feature held at the Royal Melbourne Show. This was held in conjunction with the Lincolns. Nine breeders exhibited English Leicesters.

The Melton Vale stud won:

- Ram 1½ to 2½
- Ram under 1½
- Pen of 2 rams under 1½ woolly
- Shorn ram under 1½
- Ewe under 1½

- Group 1 ram 2 ewes under 1½

- Champion Ram

- Reserve Champion Ram

- Reserve Champion Ewe

- Supreme Champion

- Duncan MacGregor Trophy

MV Melb show line up. Brenton Heazlewood.

While we have imported genetics from both England and New Zealand for use within the stud, we have also exported genetics to Canada, United States and Sweden. In 1986 one ram was taken to Canada as part of a consignment of sheep sent from Tasmania to the World Sheep and Wool Congress which was held in Edmonton. In 1989 five ewes and one ram were sent to the Colonial Williamsburg Foundation in the United States as part of a consignment of English Leicesters from Tasmania to re-introduce the breed back into that country. Semen was also sent to the United States in 2008, 2014 and 2015. In 2018, 100 embryos were sent to Sweden in conjunction with the Langs Crossing Stud, NSW.

Today the stud consists of 50 ewes. Sheep are shown at several local shows as well as the Australian Sheep and Wool Show in Bendigo, Victoria, each year.

R.G. Heazlewood Champion Ram Launceston Show 1918. The Weekly Courier, October, 1918.

R.K. Heazlewood's Champion Ram and Ewe, Launceston Show 1932. Weekly Courier, October, 1932.

THE HISTORY OF ENGLISH LEICESTER SHEEP IN AUSTRALIA

Three Melton Vale Rams. Brenton Heazlewood.

Melton Vale ewe. Brenton Heazlewood.

One year old Melton Vale ewes. Brenton Heazlewood.

Melton Vale ewe with lambs. Brenton Heazlewood.

THE HISTORY OF ENGLISH LEICESTER SHEEP IN AUSTRALIA

Melton Vale "MV" c1905. Brenton Heazlewood.

Melton Vale ewes. Australian Farm & Home, April 1949.

Melton Vale ewes sent to Colonial Williamsburg, USA. Held by Venetia, Brenton and Duncan Heazlewood. Brenton Heazlewood.

Melton Vale Show Rams 1993. From left Champion, Reserve Champion and first Pairs Launceston Show 1993. Brenton Heazlewood.

Seventh generation Alexander and Celia with Wellington and Spitfire. Photo courtesy Tasmanian Country.

Celia with Spitfire. Photo courtesy Tasmanian Country.

Celia feeding the rams. Jane Heazlewood.

21

How Has the English Leicester Changed, and Should it Have Modernised?

> *'The superior qualities of the Leicestershire breed are, that they will feed quickly fat at almost any age, even on indifferent pastures, and carry the greatest quantity of mutton upon the smallest bone.'* [293]

Should all sheep breeds have kept pace with modern trends? That is, kept 'modern'?

This is a question that stud stock breeders must have been asking themselves ever since Robert Bakewell did the first major sheep breeding improvement in the second half of the 1700's by breeding the New Leicester.

It is certainly true that present day breeds that have continued to develop, such as the Poll Dorset (developed from the Dorset Horn by removing the horns), White Suffolk (developed from the Suffolk by removing the black), Southdown and Border Leicester, are popular and widely used in their sector of the sheep industry. They have kept pace with today's requirements for quicker maturity, more muscle and less fat etc. It is relatively easy for today's stud and commercial breeders to make changes in their breeding programs to keep pace with consumer requirements through the use of performance recording and now DNA testing.

If all breeds kept changing to stay modern, the sheep industry would soon lack the genetic diversity that it has today. I think it is important that the so-called heritage breeds stay as they are, even though they may not be an economical enterprise for their keepers.

The English Leicester has perhaps the greatest role in 'staying as it is'. It has undoubtedly contributed more to other breeds of sheep than any other sheep breed.

How do we tell if the English Leicester has changed significantly in the last 225 years? One way is to look at the early descriptions given to the breed by the breeders and authors of the time. There are many descriptions which give us some idea as to what the Dishley, or New Leicester was like.

Perhaps one of the earliest descriptions of the then prototype New Leicester is found in John Dyer's poem of 1757, THE FLEECE. In this he describes the type of sheep best suited to the fertile grazing ground around Melton Mowbray, Leicestershire.

> 'That larger sort, of head defenceless, seek,
> Whose fleece is deep and clammy, close and plain:
> The ram short-limb'd, whose form compact describes
> One level line along its spacious back;
> Of full and ruddy eye, large ears, stretch'd head,
> Nostrils dilated, breast and shoulders broad,
> And spacious haunches, and a lofty dock. [294]

Even though this was only written a few years after Bakewell took over the management of the farm from his father, it is describing a Leicester as we mostly know it today. Dyer describes a large, hornless (defenceless) sheep with long (deep), dense (close) and plain wool. The ram is short legged but heavy and dense (compact) in structure, with the level and broad (spacious) back that we know today. Dyer describes the eye as being large (full) and having a healthy red colour (ruddy) whereas we today describe it as being large and bright. While I would not describe the ears of today's sheep as large, I would say they are of good size and the head of good length. The nostrils are described as wide, large and open (dilated), which we like today, along with the good breast and shoulders. To finish, Dyer's large hindquarters (spacious haunches) and high tail setting (lofty dock) are again part of today's breed standard.

Culley, writing in 1807, makes the following description:

> 'They are peculiarly distinguished from other long-woolled breeds, by their fine lively eyes, clean heads, straight, broad, flat backs, round (barrel-like) bodies, very fine small bones, thin pelts, and inclination to make fat at an early age.'

He also states that the wool, upon an average, is from 6 to 14 inches long and weighs 8 lb per fleece. Lowe, on the other hand, writing in 1845, describes the wool of a 15–16-month-old sheep as being medium length, 6 to 8 inches long and weighing about 7 ½ lbs.

In comparison our Leicesters today still have that lovely, bright, lively eye. The heads today tend to have the top-knot which I think gives them a lovely look. Looking at photos of my grandfathers and great-grandfathers sheep, they were bare headed with no top-knot. This feature is obviously relatively easy to breed on or off. We still like the straight, broad, flat backs but have tended to lift the head and neck up from the horizontal, perhaps not good that we are tending to do this, and I do not use rams if they have their head too high. The round bodies we have kept. The bones we have made bigger. I think this has been a wise move to thicken and strengthen the bones particularly for rams as they need strong rear legs in particular. I would not describe the pelt of the modern day Leicester as being thin as they have a good reputation for being able to stand cold weather. The inclination to put on fat is still with us today, it is part of the breed, and I do not think that we should try and get rid of this trait, it is just something that we have to watch in our management of the breed.

THE HISTORY OF ENGLISH LEICESTER SHEEP IN AUSTRALIA

We know that Bakewell paid no or little attention to the wool, but if it was from 6 to 14 inches long his disregard for it really did not shorten its length very much. We do not know the age of the sheep that these measurements refer to, but this length would still be a fair average today. If it was only the 6-8 inches as Lowe states, then we have increased this length considerably. The weight of 8 lb is very low for 12 months of growth. We are not given any description of the style, character or lustre of the wool in Bakewell's time. My guess is that all these characteristics have been greatly improved over the years as today the breed is named and known for its wool.

The early breeders had some difference of opinion regarding wool. Both John and Thomas Stone (whose sheep came directly from Bakewell) liked a larger sheep that was *not so heavy in the coat*, whereas Mr Farrow of Loughborough who also liked big sheep *made such an especial point of wool that the weight of each fleece was always printed on the sale-card.*[295]

Mr Grey of Dilston noted that in the early 1800's there were two distinct Dishley families. The 'bluecaps' and 'red legs.' The 'bluecaps' or blue headed Leicesters were described as tender when lambed and soft-woolled on the scalp, of greater length than the 'red legs', good feeders but rather delicate and light in the wool. This type came from Mr Stone. The 'red legs' were closer to the ground, very compact, with less fat and more fibre, were generally hardier and had a more closely planted fleece.[296] This demonstrates that very early on there was change within the breed. Whether this indicates early modernisation is perhaps debatable, but it does indicate that change was possible.

The following description of the New Leicester is taken from the book 'Bischoff on Wool Woollens and Sheep' Vol ll., published in 1842. If this description is compared with the one in today's flock book we can see that the basis of the breed is still the same.

As a lowland sheep, and destined to live on good pasture, the New Leicester is without a rival: in fact, he has improved, if he has not given the principal value to all the other long-woolled sheep.

The head should be hornless, long, small, tapering towards the muzzle, and projecting horizontally forwards. The eyes prominent, but with quiet expression. The ears thin, rather long, and directed backwards. The neck full and broad at its base, where it proceeds from the chest, but gradually tapering towards the head, and being particularly fine at the junction of the head and neck; the neck seeming to project straight from the chest, so that there is, with the slightest deviation, one continued horizontal line from the rump to the poll. The breast broad and full, the shoulders also broad and round, and no uneven or angular formation where the shoulders join either the neck or the back, particularly no rising of the withers, or hollow behind the situation of these bones. The arm fleshy through its whole extent, and even down to the knee. The bones of the legs small, standing wide apart, no looseness of skin about them, and comparatively bare of wool. The chest and barrel at once deep and round, the ribs forming a considerable arch from the spine, so as in some cases, and especially when the animal is in good condition, to make the apparent width of the chest even greater than the depth. The barrel ribbed well home, no irregularity of line on the back or the belly, but on the side the carcase very gradually diminishing in width towards the rump. The quarters long and full, and, as with the fore legs, the muscles extending down to the hock; the thighs also wide and full. The legs of a moderate length, the pelt also moderately thin, but soft and elastic, and covered with a good quality of white wool, not so long as in some breeds, but considerably finer.

The principal recommendations of this breed are its beauty and its fullness of form, comprising, in the same apparent dimensions, greater weight than any other sheep, an early maturity, and a prosperity to fatten, equalled by no other breed; a diminution in the proportion of offal, and the return of most money for the quantity of food consumed.

Another way is to compare today's sheep with early paintings or drawings of the breed. These comparisons can sometimes be not as reliable as we know that there was a considerable amount of 'artistic licence' used by the artists. There are some well known paintings of early Leicester sheep such as 'Five Leicester Sheep', by William Henry Davis with which we can do a comparison. The sheep in these early paintings are certainly very recognisable as Leicesters and comparable to today's sheep, even though they are depicted with a very fine and short leg structure supporting an over fat carcass. We know that this is the way Bakewell bred them because he was more interested in feeding the nation than clothing it, but, I am guessing that they are still depicted slightly out of proportion to what they actually were at the time.

There has been bigger bone bred into the legs since Bakewell's time, but today's Leicesters can still pile on the fat if they are fed too well. While this is one reason why they lost popularity in the latter part of the 1930's and 1940's with the trend to leaner meat, it is still a characteristic that the breed maintains today.

When Bakewell bred the Leicester, it was the quickest maturing sheep breed of its day, but it is now out performed by the modern breeds in regard to growth rate. This does not mean that the Leicester has got slower, just that the other breed has become faster. Again, this 'slower' growth rate is a good characteristic to maintain as it produces *the mutton fine-grained and of superior flavour*', as George Culley described it in 1786. The increasing trend for 'slow food' with good flavour has also seen an increasing demand for these slower maturing breeds.

The worldwide trend to finer wool is perhaps the influence that is putting the most pressure on the English Leicester to change today, but perhaps this very trend is also the strongest reason for the breed to not change. The long, lustrous strong wool with its defined crimp and style makes the breed what it is. I know that the wool craft industry is putting pressure on breeders to produce sheep with finer, more pencilly wool as this is more saleable, but there are other breeds that produce this type of wool. With the Lincoln being the only other really strong woolled breed, it is more important than ever to not, 'go with the flow', but to keep the wool type that has been valued so much in the past and is one of the English Leicesters most recognisable traits.

Should the English Leicester have modernised? In my opinion – no.

I think that we have to accept that the English Leicesters are not now a mainstream commercial breed and therefore there is no real economic reason to change them. We have them because we like them for what they are, not for what we may want to change them to be.

The English Leicester is the breed it is because it has not significantly changed since 1760, when Bakewell 'let' his first ram. The characteristics which Bakewell so strongly fixed in the breed have served it, and the wider sheep industry well for the last two and a half centuries.

Looking down the road to the future. Brenton Heazlewood.

22

Young Breeders

Today the breed is blessed to have several young breeders interested in its future.

Poppy Mamers lives on a small farm in Main Ridge on the Mornington Peninsula, Victoria. She recently started breeding English Leicester sheep and has a small flock of 24. She commenced with 8 young ewe hoggets from Roslyn Cohen and added another 6 ewes from Nick and Vanessa Wootton.

She loves English Leicester sheep because they have real personalities and are very docile. Because she has a small flock, she spends a lot of time halter training them and walking them around her farm. Poppy calls them her Flower Flock and the ewes are all named after flowers; her favourite ewe is Marigold.

Having her own flock has cemented Poppy's dreams of doing agricultural science when she finishes school and owning her own farm. Poppy's goal is to breed her own ram in the next couple of years and take him to shows.

Poppy Mamers. Melissa Mamers.

Poppy Mamers. Melissa Mamers.

George Willows, Tasmania, initially inherited his English Leicester sheep from his grandfather Ian Campbell. Ian established the Nant stud on his 'Nant' property near Bothwell in 1945. After George's family returned from Singapore to Tasmania George had the opportunity to purchase the stud from the Hogarth family, thus returning its care to the grandson of its founder. Over the years George has increased the ewe numbers and taken on showing his sheep with considerable success.

Nant (Tas) Royal Hobart Show Champion 1951. Australian Farm & Home, July 1952.

In 2020 George and his father purchased the sheep from the stud of the late Victor Goring. These sheep made the long trip from Toowoomba, Queensland to Tasmania to be integrated into George's stud. These ewes demonstrated just how tough the English Leicester can be as they came out of drought conditions with lambs at foot and made the long six day trip to Tasmania.

George and Paul Willows inspecting their Queensland sheep after their arrival at Nant, Tasmania, 2020. Brenton Heazlewood.

George regularly exhibits his sheep at Tasmanian local shows as well as the annual Bendigo Sheep Show where in 2022, he won the coveted Duncan McGregor trophy for the best exhibit under one and a half years.

Edward Southwell (left) and George Willows discussing the finer points, 2023.

THE HISTORY OF ENGLISH LEICESTER SHEEP IN AUSTRALIA

Edward Southwell, Orange, New South Wales, became interested in English Leicester sheep back in 2013, when his father bought a small flock of about 30 ewes from down in Tasmania. Being a merino farmer and classer his whole life, his father was fascinated by them, especially how they managed to grow such a long staple on a flat skin, a trait that all merino farmers would love for their flocks to inherit. They first discovered these sheep at the Bendigo Sheep and Wool show, when Edward's dad ventured down to the shed below the Merino Shed and was convinced into buying a flock by Ethel Stephenson.

'Since that moment, we have grown to love the English Leicester, not only because of their unique physical traits, but also because of their distinct personalities; made up of an unlikely mix of cheekiness and laziness. Many a time have we driven through the paddocks and walked up to an English Leicester we believed to have died, just to find it basking in the sun, completely content and still. This is the paradox, compared to trying to get them into the yards, an activity which is always difficult due to 3 or 4 rogues in the flock.'

'Since we got the flock, we have travelled around to many shows, both in NSW and Victoria. It is always a breath of fresh air being a part of the English Leicester group at these shows, providing a break from the relentless competitive nature that comes with showing a lot of the major breeds. Not only have I loved being able to show these unique sheep to people, but also the opportunity to meet and get to know the other breeders. They are always the warmest, kindest, and most considerate group of people at any sheep event we go to, and it has been a privilege getting to know everyone. In the future, I look forward to keeping this rare breed going in Australia.'

Junior handlers Bendigo Show 2023. L to R: Alexander Heazlewood (7), Sophie Stewart (10), Georgina Stewart (6). Brenton Heazlewood.

Left: Willow Drive Champion English Leicester Ewe and Supreme English Leicester Exhibit at the 2023 Australian Sheep & Wool Show, Bendigo. Held by Tahlia Holmes. Right: Melton Park ewe, first under one and a half and Reserve Champion Ewe. Held by Georgina Stewart. Brenton Heazlewood.

23

Bakewell's Legacy in Australia

'The breeder as well as the rope dancer, to preserve his equilibrium, must correct his balance before it is gone too far, and then not by such a motion as will incline it too much in the opposite side.'[297]

Less than a decade before Bakewell's death, a convoy of 11 ships left Britain to found a colony on the further shore of a largely uncharted continent on the far side of the globe.

If the event passed without Bakewell's notice he could well be excused, for the cargo was the, then perceived, human refusal from the overcrowded prisons of the period. The permanent and distant removal of this unfortunate sector of society was a major reason why the British Government chose to occupy the coast which Captain James Cook had discovered only 20 years earlier.

It would be frivolous to suggest that, when in 1787 they prepared for the dispatch of what Australians now call the First Fleet, the authorities may have sought the advice on the production of food, or the choice of animals for the new colony from Robert Bakewell, a tenant farmer whose career of innovative, experimentation and improvement of matters agricultural in general, and animal production in particular, was then at its zenith.

Historians agree that the only positive consideration of the agricultural capability of the site chosen, or indeed of the continent, was based on an uncharacteristically inaccurate assessment by the famous botanist Sir Joseph Banks, who had accompanied Cook to the South Seas and had collected a number of plant specimens from Botany Bay. The Government of the day was not particularly interested in what happened to the unfortunate emigres once they landed.

It is likely then that all the activities surrounding the landing on 26 January 1788 of Captain Arthur Phillip, a detachment of marines and 736 convicts which a nation of now 25 million people (and 65 million sheep) now celebrates as Australia Day, was unknown to Bakewell.

Up until the time of Bakewell's death in 1795 almost no information, except for a few terse official dispatches, would have flowed back to Britain and unless you happened to be an acquaintance of the military personnel or a relative of the involuntary immigrants involved there was little reason to be interested in or even know about Australia.

While we can safely assume that Bakewell was not consulted, his influence, however, on what was to be for 100 years Australia's most important industry was immeasurable.

THE HISTORY OF ENGLISH LEICESTER SHEEP IN AUSTRALIA

The domestic animals for the new colony were embarked at the last, nearest port of call, Cape Town, but the 100 or so sheep which landed had either died or been eaten within a year. Some replacements came from the Cape, but most of Australia's early sheep were hardy, primitive, but prolific little sheep from India with hairy coats but no wool. That was of no consequence to the often starving settlement; there was a dire need for meat and little or no requirement for wool.

Merinos have been so dominant in Australia since about 1840 that most Australians assume that they must have come with the First Fleet. In fact, there were already 2,500 prolific hardy sheep of mainly Indian descent scrounging the unsuitable (for sheep) lands around Sydney when, in 1797, 13 Merinos arrived from the Cape. No more than 30 Merinos arrived in the first 30 years.

Bakewell's influence had arrived on these shores before any Merinos had set foot here. This was in the form of the Irish strong woolled sheep which were infused with New Leicester. These were here in 1796, and perhaps earlier. Eight years later in 1804 Bakewell's influence again arrived in the Teeswater. The arrival of the Teeswaters was announced in the colony's fledgling newspaper, *The Sydney Gazette*, with a flourish.

> *'As a present from His Grace the Duke of Northumberland, Major Johnstone has received three Capital Teeswater Ewes and a fine Ram, a breed that must eventually be advantageous to the general interests of the Colony in propagating one of the most valuable species in the whole Animal Creation.'*

The influence of Bakewell was beginning to settle on Australia. After 1825 it grew dramatically when in addition to Merinos, Cotswolds, further Southdowns, Teeswaters and Bakewell's own breed, the Leicester, arrived in significant numbers. In the 1830's the establishment of new settlements in Victoria and South Australia led to the opening of vast grazing lands and sheep numbers exploded. The relative value of meat and wool was now reversed. Wool could be sold to Europe, meat could not, and the Merino assumed the numerically dominant role in Australia's sheep industry which has never come under challenge since.

There has, however, always been a place in Australia for British sheep, and by the 1830's our graziers had almost universally elected the Leicester as the companion, crossing breed for the Merino. By 1830 the Teeswater, having done sterling service, was dropping from sight, the Cotswold would persist till the end of the century, but only in insignificant numbers, and the Southdown could wait on the sideline, also in small numbers till 1880 when it became possible to successfully ship frozen meat to Europe.

It is obvious that Bakewell's own breed, the New Leicester, came to Australia in numbers because it was then the focus of attention in England and was successful there in its own right. In Australia, however, it became the leading British breed and maintained that position for the rest of the 19th century because it complemented and crossed well with the Merino and delivered satisfactory results to a country whose economy depended on sheep. In the period 1830 to 1910 it was crossed, as the sheep population grew, with hundreds of thousands, probably millions, of Merinos.

Up until at least 1880 Europe was not particularly discriminating about wool quality. Leicester-Merino cross sheep produced more length of staple and more weight resulting in bigger wool cheques for Australian graziers. The only exportable carcass product was tallow, for soap and candles, and the Leicester cross also provided more of this.

William Clarke was one early migrant who transported Leicesters from England to the Colony. On arrival in Van Diemen's Land he quickly amassed large areas of grazing land either by purchase, lease or squatting. The hundreds of thousands of sheep which he owned were either half bred Leicester-Merino, or Merino heavily infused with Leicester. He sometimes came in for criticism for debasing Australia's reputation for fine wool. In a newspaper defence of his breeding strategies he wrote in 1855:

> *'I can say that I have crossed more extensively with Leicesters than any man in the colonies (he may well have claimed – in the world) and have never had occasion to regret it, my fleeces being increased by 30 per cent in weight, the price per pound was not reduced more than 10 per cent, in many cases my wool sold at prices higher than the average Port Phillip wools ... in boiling down they produce on average 40 pounds of tallow of superior quality to that produced by fine wool sheep. ... The cross bred sheep thrive well on poor land ... are hardier ... and also much less liable to foot rot.'*

A Melbourne auctioneer wrote:

'For many years the wethers and cast-for-age ewes of English Leicester-Merino crosses, from the Clarke stations, travelled in on the hoof, in mobs of 500 to 1000, week after week, right through the winters and topped Newmarket for quality and price. Never since have we seen the like in such numbers for evenness of type and quality combined.'

Clarke's activities illustrate the visible role of Bakewell's breed as a half-breed in the land of the Golden Fleece. There is, however, a further hidden aspect which merits comment.

There are two elements of our Australian past which till recently have been suppressed. One is our convict ancestry and the other is the mixed genetic background of our superb Merino sheep.

We now find that our libraries and archives are daily full of Australians busily researching the official lists and proudly proclaiming direct descent from some hapless starving individual who was transported for stealing a loaf of bread. From around 1840 when the Merino began to take its rightful place as the most suitable breed for the interior of Australia, its flock master's have vehemently and vigorously claimed that their stud is exclusively descended from pure Spanish or Saxon sheep. The last 25 years, however, have seen researchers prove with documented authenticity that our Merino is advantageously and gloriously impure and that it has long carried a judicious blending of British Longwool blood.

If Bakewell had been asked to advise on Australia's sheep industry, it is likely he would have first asked 'what are the grazing lands like?' Nobody would have been able to answer him, but had they been able to describe the dry sunny plains on which most of Australia's sheep would eventually graze he would have said that any European sheep would need modification, redirection, adaption; for that is what he had so recently done – to several species of domestic livestock.

That is, in fact, what happened to the ovine emigrants from Spain and Saxony when they came to Australia. It is now being recognised that there were infusions, first of the hardy, primitive Bengals and then British Longwool from 1800 to at least 1860, all designed to achieve increased wool production of a type which the world's manufacturers required from a body which had the size and constitution suitable to thrive in a country which from time immemorial had sustained a sparse population of most un-sheep-like animals. The desire was to put Merino wool on a Leicester carcass.

The Australian Agricultural Company (AA Co.) was perhaps the first Merino based grazing operation to import Leicesters and use its genetics on their sheep to aid in the creation of an Australian Merino.

Many Merino stud properties of the 19th century also ran flocks of pure Leicesters or Lincolns to facilitate the half-breed program which they devoted their aged Merino ewes to, and evidence of accidental or intentional blending of Longwools abounds.

One of the last and most significant dilutions may have taken place on the Peppin Brothers', Wanganella Station, in southern New South Wales, in the 1860's when they turned their attention from dealing in sheep and founded a Merino Stud which has been the keystone of the breed ever since. The 200 ewes on which the stud was founded were already on the property and were 'selections from dealers' mobs', which meant that their genetic background included our original African and Indian sheep. Bakewell's influence was also evident. 'Their size and the importance of some Leicester sheep in some of the flocks from which they were drawn point to a substantial contribution from that breed', wrote J.C. Garron, a recent researcher. The Peppins then purchased Rambouillet rams which had come direct from the flock of the Emperor of France for auction in Melbourne. Rambouillet's had 30 years before being influenced by 'Dishleys'.

The Leicester infusion and transforming influence over the early small, Spanish Rambouillet sheep expressed itself in one of the early and influential sires used by the Peppins. 'Emperor', who Massy describes as *'undoubtedly one of the most influential rams ever used in Australia,'*[298] was a ram with strong bone, large size and barrel-like body which displayed the Leicester influence. His wool weights were almost double the then best in Australia, again coming from the Leicester influence. Finally, the presence of a Lincoln stud of formidable numbers on Wanganella has led to speculation ever since that they too were included in the early breeding strategies of the Peppins.

Australian Merino breeders were slow, compared with many other groups of livestock breeders in establishing a stud register and publishing a flock book. (Perhaps Bakewell would have approved the delay as he was more concerned with the progeny than the pedigree). This did not appear until 1923.

Bakewell would not have objected to the use of the Leicester and a kindred breed in the evolution of two Australasian breeds. He was used to many local breeds in Britain, some the consequence of isolation, but most of them adapted to specific areas and having specific uses. The Leicester had been widely used in Australia for 40 years when around 1870 the Lincoln appeared on the scene. It came with a surge and was the first of numerous British breeds to enjoy an often-temporary boom.

The Lincoln, which itself had been improved by the Dishley Leicester, challenged and temporarily supplanted the English Leicester as the crossing breed with the Merino. By the 1890's several breeders, in both Australia and New Zealand, began to stabilise and fix the Merino-Longwool half-bred, promote it as a new breed and give it the name Corriedale. Most of them used the Lincoln in this exercise, but the Leicester was used in some foundation flocks. A decade later the Polwarth, which is basically ¾ Merino and ¼ Lincoln, was established rather strongly in Victoria and Tasmania. While the Corriedale is still here in large numbers and occupies the grazing areas of South America in the millions, the Polwarth has virtually died out.

Both Robert Bakewell and George Culley would have been pleased with the performance of the Border Leicester in Australia. Culley who farmed in Northumberland, was one of Bakewell's admirers, using his rams and corresponding with him regularly. It was from Culley's farm that a Leicester stronghold was established in the North of England. Although a few examples of the deviant breed reached Australia in the closing decades of the 19th century, it was not till about 1930 that the Border Leicester began to replace the English Leicester as the major half-bred sire.

Just one year after Bakewell's death, sheep which had come under his influence were to be found grazing in the vicinity of Sydney, New South

Wales. Now some 230 years after his death it may now be appropriate to contemplate where his theories and practices have been most beneficial to Australia.

Bakewell's greatest bequest to Australia may not be in the few hundred flocks (now only 30) of pure English Leicester sheep which have existed since 1825, or in the acknowledged, visible contribution to the Leicester-Merino cross which held sway till the early 20th century, or in two dual purpose breeds of Australian evolution, or even in the now dominant Border Cross ewe which so admirably bridges the gap between wool and mutton, but in the clandestine blending of British Longwool genes to evolve the superb Merinos which still contribute so much to our national prosperity.

There is no edifice raised to Robert Bakewell in Australia, however, to our fellow Australian graziers we can say 'if you seek a monument to the great pioneer of animal husbandry, look around you'. From the lush Leicester-like fields of Tasmania, to the sheep-wheat areas of Western Australia and on the arid salt-bush plains of New South Wales you will find millions of living memorials to Robert Bakewell.[299]

Sue Hatton and Brenton Heazlewood at "Dishley Grange". Brenton Heazlewood.

24
Robert Bakewell (1725-1795)

> *'Prices began to rise and therefore cereal growing became more profitable, there was not a strong swing away from livestock farming as it had become part of the mixed farming enterprises that operated over most of the country.'* [300]

Agriculture had gradually changed from the common system, where no one person owned land and everyone's livestock grazed together. There was no, or very limited control of livestock breeding and the land that was cropped was continually cropped until it was worn out and then left to recover as best it could by itself. When Bakewell was born this open field system had changed to more fenced fields and compact farm units, each under the control of an individual, either owner or tenant. The enclosure system concentrated the ownership of land into great estates, and the creation of both large and smaller leased farms. The gradual introduction of enclosures has been described as the first grand improvement which took place in the fields of agriculture as it helped contribute to the fertility of the soil. The conversion of arable fields into good pastures helped prepare the way for the improvement of domestic animals.[301] Leicestershire, the county where Bakewell lived is a good example of this change from unenclosed arable agriculture to enclosed pasture-based agriculture. Marshall notes in 1786 that Leicestershire had not long ago been an open arable county but *now is a continuous sheet of greensward.*[302]

Bakewell was not the first or last of the agricultural revolutionists. The work of the first two, Tull and Townsend, laid some of the foundation work that enabled Bakewell to be the successful farmer he was. Jethro Tull (1674-1741) pioneered the use of a mechanical drill and the practice of horse-hoeing between rows. Charles Townsend (1674-1738) is famed for advocating the inclusion of turnips in the four year rotation previously practised. This rotation of roots, barley, clover and wheat, replacing the custom of two years cereal followed by fallow, increased crop production and for the first time provided, by the use of turnips, winter feed for fattening livestock. Previously, most of the livestock was slaughtered and salted before winter due to the lack of winter feed. After Bakewell, Thomas Coke (1754-1842) was the fourth of the important agricultural figures. He was not so much an innovator as a demonstrator of the techniques developed by Tull, Townsend and Bakewell. He was a pioneer in the dissemination of agricultural knowledge and is described as the father of the agricultural discussion meeting.[303]

Who was Robert Bakewell and what made him such a great livestock breeder?

Robert Bakewell was born at Dishley near Loughborough on 23 May 1725. His grandfather became tenant at 'Dishley Grange', which was owned by Sir Ambrose Phillipps, in 1709.[304] Our Robert Bakewell was the third Bakewell to become a tenant of this farm. This fact alone must say something about the way the Bakewell's farmed and the good relationship that they must have had with their landlords. Bakewell never married.

He was not a landowner, but a tenant farmer at 'Dishley Grange', *a yeoman of considerable property*,[305] as his father had been. Bakewell took over the management and tenancy of the farm in 1760, at the age of 35 from his father from whom he had obviously learnt a lot about livestock breeding. *The instructive conversation of his father* had helped lay the foundation for what Bakewell saw as *but a stupid and indolent adherence to old customs*, and *the land foul and starved for want of stock, or stocked with shabby and ill-sorted animals.*[306] The heritable traits that Bakewell so successfully used with his livestock were perhaps also a factor in his own life. His philosophy that *'like will produce it's like'* is perhaps also true for his own story. In an anonymous necrology published in 1800, the author connects Bakewell's success to his father's qualities and to his social position. Bakewell is described as having had only the minimal education *'generally bestowed on people of his rank in life, and extended no further than to writing and arithmetic'*,[307] and his success is attributed to *'early professional initiation in husbandry'* by an *'orderly'* father who *'was a man of a strong and inquisitive mind'*.[308] Bakewell took this further than his father had by experimenting with his animal breeding and also in his farming practices. While Bakewell obviously had a most ingenious and able farmer in his father to learn from, it was perhaps the fact that his father allowed him to make a series of tours to improve his knowledge of agriculture at that time that was to be of great value to Bakewell throughout his life. He was to continue this practice of making tours for the rest of his life.[309]

Arthur Yong, writing in 1771 about Bakewell's farming practices echoes this view of heritable traits when he states: '*This improvement was begun by his father, now living, and carried on and finished by himself.*[310] The fact that Bakewell inherited these traits from his father should not be surprising as it is a part of evolution for all species. Perhaps what enhanced these traits that Bakewell was given is the fact that he was born in the right place at the right time to make the most use of these traits.

Enclosure

Bakewell's position as a tenant farmer is also important. The progress of the enclosure movement, with the subsequent development of 'commercial' farming, had been very slow. Between 1455 and 1607 the area of land enclosed amounted to only 516,673 acres. In Bakewell's century, however, during the reign of George III, 3,500 Acts of Parliament (each enclosure had to be confirmed in this manner) were passed to enable the enclosure of 5-6 million acres.[311] With more enclosure came more direct control over the land, more and better pastures and better control over the breeding of the livestock, on the part of the occupier of the farm, whether it be the landlord or the tenant and this resulted in more incentive to improve the land.

Enclosure meant that the larger open fields were enclosed (fenced), larger farms were formed by the amalgamation of the smaller peasant holdings and these were let to tenants who cultivated them with wage labour. By the nineteenth century this new society had created great inequality in rural England. Property ownership became very concentrated, rents had risen and wages stagnated. This meant that the landlord's mansion was lavish, the farmer's house modest and the labourer's cottage a hovel.[312] The change to enclosures was particularly rapid in Leicestershire, where by the beginning of the seventeenth century, at least a quarter of all parishes in the county were already enclosed, with about another third at least partially enclosed. Dishley was enclosed by the early sixteenth century.[313] The fact that Bakewell was a tenant farmer on an enclosed farm is important in the fact that the fenced fields gave him greater control over the livestock he carried, especially in relation to the breeding of this sheep and cattle.

Bakewell was naturally in favour of the enclosure system as opposed to the older commons system, and gave the following example of its advantage. He asserted that if two poor men were to each buy a cow in the spring, and one ran his on the common and the other paid a farmer one shilling and sixpence a week to run his cow, and at Michaelmas (29th September) they were both sold, the price difference would more than repay the weekly expense of running the cow.[314]

Could Bakewell have succeeded if he did not have the advantage of an enclosed farm? I guess not.

Bakewell took the subdivision of his pastures and meadows to perhaps a greater extent than most farmers of his time. He contended that all the grass should be eaten off the ground by August to make way for the fresh growth that comes afterwards. If the fields are too big he said this cannot be done and the stock would only eat the sweetest grass and leave great quantities which would go sour. He also believed that in a larger field *they trample and spoil much more in beating about for the sweet spots* than they do in a smaller field.[315]

The more enclosures meant that there were more tenant farmers, the people that Bakewell had to 'sell' his ideas for better animal breeding to. The fact that Bakewell was 'one of them', and not from the land of gentry meant that he could talk to them at their level. *He dressed as a yeoman of his day, in a loose brown coat, scarlet waistcoat, leather breeches and top boots.*[316] The other important factor is that he was a practical farmer and therefore his advice was more likely to be listened to and followed within the farming community. Bakewell did not always receive this admiration and following, especially in later life, as he did encounter some spiteful opposition from particular individuals within the farming community, with for instance the forming of the Dishley Society.

Tenant

Bakewell was evidently a good tenant as Arthur Young concludes his description of 'Dishley Grange' by saying:

> *'No where have I seen works that do their author greater honour: they are not the effect of a rich landlord's determining to be a good farmer on his own land, but the honest, and truly meritorious endeavours of a tenant, performing great and expensive works on the property of another. It is true, he is fortunate in a generous and confederate landlord; and much do I wish, that such excellent farmers may always meet with*

the same encouragement. A truly good farmer cannot be too much favoured, a bad one cannot his rent raised too high. Let me exhort the farmers of this kingdom in general, to take Mr Bakewell as a pattern in many points of great importance; they will find their account in it, and the kingdom in general be benefited not a little.'[317]

It is important to note that Young sees that part of Bakewell's excellence is due to his tenant status and the relationship he has with his landlord. He also sees that Bakewell's practice should be copied by others in his class. The fact that Bakewell's 'generous and considerate landlord' is included, establishes an agricultural ideal in which Young desires the reader to take 'Mr Bakewell as a pattern'.[318]

The news of Bakewell's work spread very quickly considering that at the time there were no means of mass propaganda. While a few newspapers did exist, it was mainly in personal recommendations by word of mouth that his work became widely known. Letter writing would also have played a large part in the spreading of the news of his work. Other prominent farmers of the time, such as his 'disciple', and friend George Cully who was a prolific writer of travel journals and letters would have most certainly helped Bakewell's fame spread. It could perhaps be said that Bakewell 'bred' these pupils such as the Culley brothers to help spread and recommend the use of his livestock. Both Matthew and George Culley stayed at Dishley before farming in their own right in Northumberland where they began to hire Dishley rams for their own use, resulting in them developing the Border Leicester.

Entertainer

Bakewell was also a good salesman, host and entertainer. He ran Dishley as a model farm, hosting visitors from *Russian Princes, French and German Royal Dukes, British Peers and sightseers of every degree*.[319] Dishley became a centre for higher learning with many young men coming to stay and work just for the experience gained. He also received many written requests for advice, such as from Count de Bruhl Envoy Extraordinary from the Elector of Saxony to the Court of Great Britain, requesting to know the best methods of breeding, rearing and feeding the best kind of stock. Bakewell was usually happy to help by answering, as there was *not any knowledge I should not with great readiness communicate.*[320]

Bakewell's goods and farming practices were also their own advertisement, but it was his Dishley, or New Leicester sheep that ultimately carried his fame to most parts of the world.

Irrigation & Farm

In 1771 Dishley Grange was about 440 acres, 110 which were arable and the rest grass. It was running 60 horses, 400 large sheep and 150 beasts of all sorts. The cropping consisted of about 15 acres of wheat, 25 acres of spring corn and about 30 acres of turnips. Young stated that he did not buy in hay or straw, yet *it must at once appear, that he keeps a larger flock on a given number of acres, than most men in England: the strongest proof of all others, of the excellence of his husbandry.*[321] He did not feed the turnips in the paddock, but carted them all in to be fed in the stall to the cattle.

He also fed the stalled cattle only the amount of straw that they would eat, thus reducing wastage.

He was also an early user of flood irrigation over his pastures to increase hay production by up to three times.[322] In 1790 Marshall wrote that *Mr Bakewell is, in truth, a master in the art; and Dishley, at present, a school in which it might be studied with singular advantage.*[323] It must be remembered that Bakewell was not the first to practise water meadows (flood irrigation). He learnt this system from George Boswell (1735-1815) of Puddletown, Dorset, who was considered the expert in watering meadows. Bakewell ploughed the fields, cropped them for two years with turnips to remove the traditional ridge and furrow. He then planted barley undersown with 10 lb of common broad clover and half a bushel of ryegrass. He cleaned out the local brook and dug ditches to run the water to the paddocks where he had sluices installed in such a manner that he could run the water evenly over the fields. With this system he could irrigate 60 to 80 acres of pasture. To prove the value of the irrigation he would create *proof pieces* in most of his meadows. These were small squares, over which the water was prevented from running by having ditches dug around them. Arthur Young described the difference as *so great, as to bring complete conviction to the mind of every person who views them.*[324] He also grew a small area of potatoes and was growing cabbage for the first time.[325]

Bankruptcy

While Bakewell was obviously a very good manager of his livestock, the same could not perhaps be said of his management of his financial affairs.

BRENTON HEAZLEWOOD

In November of 1776 The London Gazette published the following Commission of Bankruptcy against Bakewell.

> *Whereas a Commission of Bankrupt is awarded and issued forth against Robert Bakewell, of Dishley in the County of Leicester, Dealer and Chapman, and he being declared a Bankrupt, is hereby required to surrender himself to the Commissioners in the said Commission named, or the major Part of them, on the 6th Day of December next at Two in the Afternoon, on the 10th Day of December next at Ten in the Forenoon, and on the 4th Day of January following at Three in the Afternoon, at the House of Thomas Barry, called the Lion and Lamb, in Leicester, and make a full Discovery and Disclosure of his Estate and Effects; when and where the Creditors are to come prepared to prove their Debts, and at the Second Sitting to chuse Assignees, and at the last Sitting the said Bankrupt is required to finish his Examination, and the Creditors are to assent to or dissent from the Allowance of his Certificate. All persons indebted to the said Bankrupt, or that have any of his Effects, are not to pay or deliver the same but to whom the Commissioners shall appoint, but give Notice to Mr Thomas Pares, jun, Attorney, in Leicester; or Mr Coore, Attorney, Broad street, London.*[326]

What had led Bakewell to become insolvent? At the time of his bankruptcy he was not receiving the high prices for the 'letting' of his stock that were to come later, and the terms of payment were such that he did not receive the income until the rams had 'proved their worth'. This was often a lengthy delay and combined with the report that a number of Scottish and Irish customers had failed to pay for animals hired must have put a strain on his cash flow.

> *Sometime ago the clients who had hired animals of the breeds perfected by Mr Bakewell failed to pay properly; none of the Scotsmen whom he had supplied had paid him and so forth. The result was that the continual and heavy expense which he necessarily incurred reduced him to a condition of bankruptcy, although he was not in the least to blame himself.*[327]

Some have suggested that his very generous hospitality was to blame, for as there was no inn at Dishley visitors would stay at Dishley Grange where Bakewell's maiden sister, Hannah, would receive the visitors and look after them, sometimes for weeks.

Bakewell had been in a poor financial situation for some years before he was declared bankrupt. In 1770 he mortgaged the paternal estate at Swepstone for five hundred pounds.[328] This must have kept him from bankruptcy until 1776.

Perhaps he could even be described as extravagant in his management of the farm.

> *The expanded heart of this man demanded more capacious means for the gratification of its generous desires; and it is*

evident, from his conduct, that he was ambitious rather of the honour than the profit of his calling.[329]

Culley commenting on Bakewell's farming practices as early as 1765 and 1771, noted that:

Mr Bakewell employs more than he has any occasion for, he is without doubt at a most consumed expense, I suppose pays more money in wages than rent. No man says Mr Huskisson has the spirit to make things good that Mr Bakewell has but I apprehend he carries this and some other things into the extreme...[330]

His whole business is carried on still with that surprising order and regularity, which I never saw anywhere else, and which is rather to be admired than imitated.[331]

The settlement of Bakewell's bankruptcy was a long and protracted affair, much to the annoyance of his creditors. In 1778 the Commissioners published the following notice.

The Commissioners in a Commission of Bankrupt awarded and issued forth against Robert Bakewell, late of Dishley in the County of Leicester, Dealer and Chapman, intend to meet on the 2nd Day of June next, at Three of the Clock in the Afternoon, at the Lion and Lamb in Leicester, in order to make *a Dividend of the said Bankrupt's Estate and Effects; when and where the Creditors, who have not already proved their Debts, are to come prepared to prove the same, or they will be excluded the Benefit of the said Dividend. And all Claims not then proved will be disallowed.*[332]

Some of his stock was originally offered for sale in March 1777, but nothing was apparently sold. John Smellie of Nottingham, one of his creditors, in April 1778 complained that although a dividend of 10s. 0d. had been promised in September 1777 and a further 6s. 0d. in May 1778, nothing in fact had been paid.[333] In July of 1778 part of Bakewell's livestock was advertised for sale.

To be SOLD by Mr CART, (without reserve) at Nottingham, on Wednesday the 5th of August, in the Race Week, in 20 Lots or upwards, agreeable to such Conditions as will then be produced. The entire STOCK of BROOD MARES and FOALS, two and Three-year-old Fillies, of Mr Robert Bakewell, of Dishley, being of the Coach and Cart Kind. The above Stock were bred by Mr Bakewell, and chiefly got by Blake and George, which Horses covered several Seasons at Five Guineas a Mare.

N.B. The Stock at Dishley, consisting of Bulls, Cows, Rams and Ewes, are selling off by private Contract.[334]

The livestock listed in this sale notice were not sold until March 1779 and then at Dishley where a ten-year-old stallion was sold for 140 Guineas and a bull for 130. While these prices seem high for the time 'many capital judges are of the opinion they were remarkably cheap'.

On 5 December 1778, a further notice appeared in The London Gazette stating that his Certificate of Bankruptcy would be allowed and confirmed.

> *Whereas the acting Commissioners in the Commission of Bankrupt awarded against Robert Bakewell, of Dishley in the County of Leicester, Dealer and Chapman, have certified to the Right Hon. Edward Lord Thurlow, Lord High Chancellor of Great Britain, that the said Robert Bakewell hathconformed himself according to the Directions of the several Acts of Parliament made concerning Bankrupts. This is to give Notice, that, by virtue of an Act passed in the Fifth Year of His late Majesty's Reign, his certificate will be allowed and confirmed, as the said Act directs, unless Cause be shewn to the contrary on or before the 26th of December instant.*[335]

In August 1779 all his rams, over 100 in total, were advertised for sale.

To Be Sold

At Dishley, near Loughbrough, Leicestershire On Monday, Sept the 27th, and the four following days. All the RAMS (consisting of more than 100 in number) bred by Mr Robert Bakewell. Proper persons will be appointed to treat with customers for the same, or any other part of the stock; as it is the determination of the Assignees, that the whole shall be disposed of by Lady-day next.

Dishley, 12 August, 1779[336]

Nearly two years later the remainder of the stock and lease of Dishley Grange were offered for sale. Again, little apparently happened as in June 1783 the long-suffering Smellie placed an advertisement in the Leicester Journal complaining that the assignees had decided to dispose of all of Bakewell's stock by Lady Day 1780 but although the creditors had been waiting seven years they had not received eight shillings in the pound towards the payment of their principal.[337]

In April 1780 the Commissioners advertised that they were going to make another dividend to Bakewell's creditors.

The Commissioners in a Commission of Bankrupt awarded and issued against Robert Bakewell, of Dishley in the County of Leicester, Dealer and Chapman, intend to meet on the 12th Day of May next, at Four o'Clock in the Afternoon, at the House of Thomas Barry, known by the Sign of the Lion and Lamb, in Leicester, in order to make a Further Dividend of the Estate and Effects of the said Bankrupt; when and where the Creditors, who have not already proved their Debts, are to come prepared to prove the same, or they will be excluded the Benefit of the said Dividend, And all Claims not then proved will be disallowed.[338]

For the years after 1776 when he was declared bankrupt, Bakewell was not allowed to manage his own affairs at Dishley Grange. This must have been a strain on his confidence and probably his health. During the period from 1776 to 1778-9 Dishley was run by the assignees to the bankrupt estate. There were five of them with John Ashworth of Daventry, a former pupil of Bakewell having the principal direction of the business. Bakewell was fortunate to also have been previously associated with three other of the assignees, Samuel Huskinson of Croxton who was Bakewell's companion on his early Midland tours, William Bently, the first regular banker in Leicester and a fellow Presbyterian and treasurer of the Great Meeting congregation in Leicester, and Henry Walker of Thurmaston who was also associated with Bakewell as a breeder and grazier. The fifth was another banker of Leicester, William Hodges. Their appointment as assignees is perhaps one reason for the inordinate delay in settling Bakewell's bankruptcy. Bently was too old to travel far from home, Huskinson was said to be rather infirm and not competent to undertake the administration of the estate on his own. Ashworth ran a busy inn at Daventry, nearly 40 miles from Dishley. Both Walker and Hodges were themselves subsequently declared bankrupt.[339]

While Ashworth must have done a good job of keeping the business viable, he must have relied heavily on the head stockman John Breedon to keep it running in his absence. One can also presume that Bakewell would have relied on receiving information from Breedon as to what was happening with the stock and no doubt Bakewell would have been making his opinions known to Breedon as to what he thought should be happening. Breedon's good stock sense came to the fore on at least one occasion.[340]

> *Young A was the sheep which Mr. Ashworth would have sold to Mr. Roby if Jno.Breedon had not convinced him of his folly. At that time they had two favorite Tups at Dishley – Vis – Charles and Young A. Mr Ashworth who had the principal direction of the business during the Bankruptcy, had used Charles two Seasons, contrary to Jno. Breedons Superior Judgment and at the last would have sold A to Mr. Roby if Breedon had not fortunately prevented him.*[341]

Sometime before 1784 Bakewell appealed for financial help to enable him to carry on his work.

> TO THE NOBILITY GENTRY AND OTHERS.
>
> *The Humble Petition of Robert Bakewell of Dishley in the County of LEICESTER*
>
> *SHEWETH*
>
> *That your Petitioner has for a Series of Years employed his Attention on a Plan for improving the Breed of Horses for*

Cavalry, Harness and Draught, as also of Meat Cattle and Sheep.

That your Petitioner in Pursuit of this Plan had many Difficulties to surmount having the Prejudices of other Breeders to combat and various Experiments to make in Order to ascertain which were the best kinds to breed from; and that such Experiments were attended with considerable Expense, and more trouble than he can well convey a Sense of

That your Petitioner apprehends he has brought all the Different Kinds of Stock abovementioned to a greater Degree of Perfection than has been done by any other person and thereby rendered important Services to this Country, and in this Opinion he hopes he is justified, by the best Judges having purchased from his Stock at higher Prices than from any other, and having sent them into the Counties of Bedford, Bucks ,Cambridge, Chester, Cumberland, Derby, Devon, Dorset, Durham, Essex, Gloucester, Hereford, Herts, Huntingdon, Kent, Lancaster, Leicester, Lincoln, Norfolk, Northampton, Northumberland, Nottingham, Oxon, Rutland, Salop, Somerset, Southampton, Stafford, Suffolk, Sussex, Warwick, Westmoreland, Wilts, Worcester, Yorks, into North-Britain, Wales, Ireland, Germany and Jamaica.

That your Petitioner has made considerable Improvements in Agriculture, Division of Lands, Watering of Meadows, &c.

That your Petitioner in Consequence of the aforesaid Difficulties and Expenses, as well as by many great and unavoidable Losses to the Amount of many Thousand Pounds, is rendered incapable of pursuing his Plan; and as considerable Part of the Stock is soon to be sold, and probably will fall into the Hands of those who for want of Experience or other Causes cannot be supposed to manage it to the same Advantage, consequently little if any further Improvement can be expected therefrom.

But if the Publick should take his Case into their Consideration and grant him such Assistance as would enable him to purchase the whole or the best part of his Stock, he is fully persuaded he could be highly instrumental to the general Good of this Nation, by continuing in his late Line of the Breeding Business, and carrying it forward in such a Manner as will be most conductive to the Public Service, and he apprehends he could make a great Improvement from the State the Stock is now in, as he has done from the State of Stock in general at the time he began this Business, an Object he thinks of great Importance to the Honour and Interest of the British empire; for if it be allowed that the Increase of Herbage by Improvement in agriculture is a real Advantage to the Public in general, he conceives the Improvement of Stock, so as to gain a greater Quantity and better Quality of Flesh from such Herbage to be of equal if not of greater Importance.

Your Petitioner therefore most humbly Solicits &c.[342]

This petition had the desired effect as over one thousand pounds was raised. The list of subscribers was long and included some very influential people of the day. The Duke of Richmond was said to have given 500 guineas and The Duke of Rutland 200 guineas. Lord Sheffield and the Earls of Hoptoun and Middleton were contributors as well as some of his close associates such as Thomas Paget and Matthew Culley.[343] Obviously Bakewell's influence on the agriculture of the day, as is stated in the petition, was such that his friends and acquaintances felt that they could not stand by and see him fail.

We are not sure when Bakewell was able to clear his bankruptcy, but it was still in effect in 1789 as Bakewell himself states in this letter to Arthur Young in August of that year.

> *'Your favour of the 18th instant I rec'd but of late have been so much engaged on the Ram business and other matters relative to my Bankruptcy which I hope will soon be settled that I have not had time to attend to anything else, and on account of the state my affairs have lately been in I should think any opinion as coming from me will rather injure than benefit any publication, and probably may have a tendency to make the whole disregarded, as most people think well or ill of any scheme by the success with which it is attended but this I leave to you.*[344]

Bakewell had been officially bankrupt for 13 years when he wrote this letter and it, along with other letters he wrote while bankrupt, do raise the question of how much say or do in his business he actually had while bankrupt. In this letter he talks about being '*so much engaged on the Ram business*', and in a letter of March 10, 1788 while in London to show the King a Black Stallion he states: '*I have sold a Bull and Two Heifers which are to put on board this week to go to Mary land N.America and if these please I hope to have further orders*'.[345]

Although we do get conflicting information, it would appear that Bakewell must have still been able to have some influence on how the business was run, especially the breeding side of it. I cannot imagine how his business would have progressed during this time if he had been totally shut out from it. In November 1789 in a letter he wrote to Culley he states '*I never made near as much as this season or with so little trouble. I have not fixed a price on anything for more than twelve months past....*'[346] Indicating that for some time he has had the say in price setting. Perhaps after his successful petition for financial help and the eventual payment of his overdue accounts which enabled him to clear his debts, he was given back the control of the business even though he was still officially bankrupt.

Would Bakewell have been able to achieve even greater success with his livestock breeding if he had not been burdened with the worry and control of bankruptcy over such a long period of time. While he presents a picture that it did not greatly affect him, it must have been a constant worry to him and thus perhaps an influence in some of his decision making.

Breeding Principles

Environment

The effect that the environment and feed had on animals was often thought to strongly influence the characteristics expressed by livestock, and therefore it was supposed that stock type was entirely determined by the climate and nutritional features of the area where they lived. It was also presumed that if the animals were moved, that they or their offspring would change permanently to match the changed conditions that they now experienced. Therefore, the place was presumed to have developed the breed, and if it was moved it would change to match the new conditions. While we now know that this is not the case it was a strongly held view at the time of Bakewell, and even Darwin would argue well into the nineteenth century that environmental change could alter the form of the animal or plant by changing the nature of the seed which it transmitted.[347]

Bakewell did not mind using animals from different regions as he believed that the nature of the animal functioned independently from its external world and therefore a change in location did not necessarily change the animal. This was very radical thinking in a time when most people believed that the environment alone produced the different types of animals.

Principles

Bakewell bred the Dishley or New Leicester to feed the nation, not clothe it. He was not at all interested in wool. The fact that coarse wool was not worth very much at the time probably helped him come to this conclusion.

This is perhaps ironic for those of us who are still breeding his sheep, as one of their great attributes today is their long lustrous wool. In fact it has been the quality of their wool, its length, lustre and weight, and their ability to so strongly transfer these characteristics to other breeds that has saved the New Leicester from becoming extinct. Today the Leicester has been superseded in regards to producing meat by the more modern, quicker maturing breeds such as the Poll Dorset and Suffolk. Perhaps it is not possible to have a true 'dual purpose' breed, for as we concentrate on breeding for one trait, we naturally tend to lose other traits. This is certainly true in most cases if we consider today's breeds where the Merino is the prime wool producer, but a poor mutton producer, and at the opposite end of the scale the Poll Dorset is the prime meat producer with little wool of low value.

The greatest population of mutton eaters during Bakewell's time were in the fast-expanding manufacturing towns, and they would always choose the fattest meat and pay the greatest price for it.[348]

Arthur Young has summarised the breeding principles that Bakewell used in breeding both his sheep and cattle as *'the leading idea, then, which has governed all his exertions, is to procure that breed which in a given food will give the most profitable meat-that in which the proportion of useful meat to the quantity of offal is the greatest: - also in which the proportion of the best to the inferior joints is likewise the greatest.'*[349]

Bakewell was not the first person to try and improve the 'Old Leicestershire' or 'Warwickshire' breed. Marshall states that there appeared to have been only one breed of long-woolled sheep in the Midland district. In Warwickshire and Staffordshire it was known by the name

'Warwickshire' and in Leicestershire, Rutlandshire, Northamptonshire and Nottinghamshire by the name 'Old Leicestershire'.[350]

These sheep were poor types, with Marshall describing a *true old Warwickshire ram* as the worst sheep he had ever seen.

> *'His large frame, and remarkably loose. His bone, throughout, heavy. His legs long and thick, terminating in large splaw feet. His chine, as well as his rump, as sharp as a hatchet. As to fat, he has none; nor flesh enough to ascertain its quality; though his pasture was good: his skin might be said to rattle upon his ribs, and his handle be conceived to resemble that of a skeleton wrapped in parchment.*[351]

While not all sheep were this bad, it did represent the poor state of the stock before serious improvement had started on them.

The Sheep He Used

The first person who had a serious attempt to improve the 'old' breed was a Joseph Allom of Clifton. He *had raised himself, by dint of industry, from a plowboy, seems to be acknowledged, on all hands, as the first who distinguished himself, in the Midland District, for a superior breed of sheep.*[352] It was the fashion of the superior farmers to purchase their ram lambs from Allom in the summer. He appeared to be the only person who became a distinguished breeder of sheep before Bakewell.[353] Wykes states that Bakewell obtained the origins of his improved stock from Allom.[354]

While it might be true that he obtained some stock from Allom, we know that he bought sheep from whoever he thought that their sheep had something to offer him. Hunt relates that when Bakewell was searching the country for suitable sheep, he purchased two from Mr Wall, the predecessor of Mr Chaplin, one for five guineas, and one for fifty guineas. Mr Wall obviously thought that the one that he charged fifty guineas for was the best, but Bakewell believed the five-guinea sheep to be the best and subsequently gave preference to it which *much improved his flock,* while Mr Wall continued to breed from the relations of the fifty-guinea sheep, *in consequence of which his flock degenerated.*[355]

Bakewell was not the first to use selective breeding or 'in and in' breeding, but because of his ability to see the wider picture and the end result he was aiming for, he was the first to successfully practise these breeding principles and achieve the result he was desiring. He would select one characteristic that he wanted to change, for example bone, and breed from those animals which possessed that feature. *You can't eat bone, therefore, give the public something to eat*[356] Bakewell said, so he set about reducing it. Perhaps the reason Bakewell succeeded where some others had not, was his ability to see form and its relationship to certain characteristics, for example its fattening ability. He could perhaps also see, and used, the heritability of his desired characteristics for quicker gain even if it meant ignoring another factor such as the wool. He was also prepared to ignore some 'fancy points', such as colour which often distinguished local breeds from one another, but which he considered to be of no account, instead viewing the animal as a marketable commodity.[357]

We will never know what breeds Bakewell used, as he never left any record of this part of his breeding programme. We do know that he ran many different breeds of sheep for comparison purposes, but we do not know if he used any of these breeds to produce the Dishley Leicester. In a 1792 report to Sir John Sinclair, Chairman of the Society for the Improvement of British Wool, Bakewell showed the authors of the report the following breed that he had at Dishley Grange.

> Dorsetshire [Dorset Horn] – a horned breed with a good fleece but not well shaped.
> Iceland – with four horns, a spotted, long-carcassed animal, fleece and shape both bad.
> Cape of Good Hope – more like a goat than a sheep, fleece very hairy and ill shaped.
> Norfolk – a horned long-legged sheep; fleece and shape both bad.
> Cheshire – neither fleece nor shape good.
> Sussex – fleece good but shape bad.
> Hereford [Ryeland] – superior fleece.[358]

One feature of the breed that we look for and cherish today is the dark pigmentation on the face and ears. Those sheep that exhibit this well, tend to have a blueish pigmentation. How this came about we do not know. Valentine Barford, in 1828 referred to Bakewell using a black ram of his neighbour Richard Astley. Barford, whose flock had an unbroken descent from Bakewell's, mentions his own flock producing black lambs, though none of them had been retained for breeding.[359] A black ewe from Leicester Forest has also been mentioned, but there is little evidence to substantiate this. He may also have made a cross with some of the dark-faced Down or Forest breeds, or it just may have been a characteristic of the larger breeds like the Southam Notts and Old Leicester which Bakewell most likely used.

In Breeding

> *Mr Bakewell had certainly the merit of destroying the absurd prejudice which formerly prevailed against breeding from animals between whom there was any degree of relationship.*[360]

The use of inbreeding in controlled livestock production is today still as controversial as it was when Bakewell brought this practice to greater public attention. In nature there is no such thing as incest, but in the non-controlled populations the laws of nature, that is the survival of the fittest, weed out the weakest animals that may have resulted from the closely related natural matings that have taken place. It is when livestock breeders impose their control over who gets mated to who that careful selection and particularly careful culling has to take place.

While other breeders were not willing to use inbreeding, Bakewell considered it essential. Bakewell probably learnt about inbreeding from pigeon fanciers and race horse breeders, and that it could be used to fix characteristics. If his best animals were related to each other, he still bred them to each other regardless. In-and-in breeding, as it was called, was used to speed up the results that could be achieved in comparison with cross-breeding. While it will strengthen the good characteristics, it will also strengthen the undesirable ones. Some believed the practice would abate their vigour, weaken their constitution, and shorten their natural lives. An inbred animal was more likely to be prepotent, which meant that it would strongly stamp its characteristics on its progeny.

Before enclosures it was very difficult to control the breeding of livestock, with rams running with the ewes all the time as separating them was difficult. John Hunt described their *propagation* as *promiscuous, without the least attention being paid to the improvement of the breed*.[361] Ernle said that prior to Bakewell, *stockbreeding as applied to both cattle and sheep was the haphazard union of nobody's son with everybody's daughter*.[362]

Bakewell used normal breeding practices to establish the foundation of the Dishley breed before making use of in-breeding to complete and fix the characteristics he desired. Many authors give the impression that he only used in-breeding to achieve his aims, but Hunt, who knew Bakewell well, clearly states that this is not the case.

> '... yet it must be evident that after his own breed of sheep had obtained a decided superiority over all others, it was only by family connection or what has since been called breeding in-and-in, that any additional improvement could be obtained; and though this was not the basis on which the first improvements were established, yet it must be looked up to as the grand principle which gave a complete finish to the whole.'

He goes on to say that:

> *'His success did not depend either upon crossing or of breeding in-and-in, but on his accurate knowledge of the nature of the animal.'*[363]

Bakewell may well have learnt about in-breeding from his shepherd, John Breedon, who he employed in 1760. Breedon was a breeder of game-cocks and evidently excelled in the breeding of these birds. Whenever he exhibited his poultry, he *would boastingly exclaim, these Sir, I tell you have been bred in-and-in for more than twenty years.*[364]

Hunt does not deny that degeneration can occur through in-breeding, but reinforces that good management is required for it to be successful.

> *'It is not by promiscuous intercourse, though limited to the same family, that improvement is to be insured; but by judicious selection alone that success is to be obtained.'*[365]

Bakewell's success in using inbreeding was due to his very careful selection and culling, as he obviously knew the pitfalls that could occur through this practice. Sir J.S. Sebright described the fine line in selection that the breeder must keep as, he *must observe the smallest tendency to imperfection in his stock, the moment it appears, so as to be able to counteract it, before it becomes a defect; as a rope-dancer, to preserve his equilibrium, must correct the balance, before it is gone too far, and then not by such a motion, as will incline it too much to the opposite side.*[366] The breeder's success will depend on his ability to be able to do this.

Bakewell was not the only breeder of the New Leicester sheep to practise inbreeding. In 1776, Mr Paget, then of Ibstock, used the ram of Bakewell's known as 'P'. Because he paid a high price for the ram, he selected the *most perfect and beautiful females for the seraglio of this grand plenipotentiary of the Dishley cause.* Two years later he joined the same ram to thirty of his own daughters with evidently such great success that he repeated the experiment.[367]

The Influence of the Church

The role that the church played in Bakewell's life, and particularly its influence in his use of in-breeding, at a time when the Church of England had much influence, has to be considered. The Bakewell's were members of the Unitarian congregation, not the dominant Church of England. The Unitarians are a nonconformist church group dating from the mid-seventeenth century in England.[368] They are historically a Christian theological movement named for the affirmation that God is one entity, in direct contrast to Trinitarianism, which defines God as three persons in one being, i.e. The Father, Son and Holy Ghost. While it was technically illegal to be a Unitarian until the 20th century, they gradually gained acceptability.

Both the Loughborough and Mountsorrel chapels were described as Presbyterian, which at that date meant they were liberal nonconformists.

While we do not know what Bakewell believed, we do know that the Unitarian tended to be rational freethinkers, and were men with a wider variety of social and intellectual interests than other nonconformist sects.

They had a preference for freedom of conscience and to interpret the scriptures for themselves.[369]

Because Bakewell thought for himself, he fitted well into the 18th century Unitarian mould. His use of in-breeding, which was then seen as incest, and went against everything that the Church of England and the Establishment stood for. This was a time when religious beliefs were sacrosanct, and his going against the then widely held views of correct livestock breeding created him enemies. It was perhaps one of the reasons why Bakewell was not so forth-coming about his breeding methods. Bakewell's status as a Yeoman, tenant farmer, may also have made it easier for him to go against the grain, as it were, in regards to the way animals should be bred. Had he been a landlord, it may well have been more difficult for him to so openly practise inbreeding.

Erasmus Darwin, the grandfather of Charles Darwin and the author of Zoonomia, give his opinion that there was no such thing as incest in nature.[370]

Bakewell attended the Warners Lane Unitarian Chapel in Loughborough. He became a trustee of the chapel in 1774. Even throughout his later financial embarrassment he continued with monetary assistance to the Loughborough Chapel, and also the nearby Mountsorrel Chapel, of which he had been made a trustee in 1761. Religion was a serious part of his life and he would never conduct business on a Sunday, regardless of how important a visitor might be or how far they might have travelled. Later when he formed the Dishley Society, it was written into the rules *that no member shall let a Ram on Sundays*.[371]

While Bakewell had set ideas about the best methods to go about animal breeding, and the results that should be aimed for, he obviously realised that things may change in the future. In a letter to Culley he wrote, *'I would recommend to you and others who have done me the credit of adopting my opinions to pursue it with unremitting zeal as far as shall be consistent with prudence and common sense, always open to conviction when anything better is advanced.*[372]

Bakewell was a great believer in examining his animals by feeling them. He would sooner trust *the hand in the dark* over *the eye in the light*.[373] The carcass of the animal could only be critically assessed by feeling it and he believed that *the disposition to fatten is discovered only by feeling*.[374]

Another factor that had to be considered to a certain degree was the keeping ability of the carcass in hot weather. As there was no refrigeration the butchers would have had a limited time in which to sell the meat. Arthur Young discusses the differences in keeping ability between the Norfolk (a horned black faced breed) and the Southdown. In hot weather the Southdown carcass would keep *sweet* for 24 hours longer than the Norfolk, and thus be *worth a halfpenny a pound more than the Norfolks*, but in cooler weather there would be no difference. There is a reason suggested as to what may cause this difference in keeping ability. A butcher, John Vyse, does suggest *the probability of much more fat within* the Norfolk meat. Bakewell adds that the loose texture of the marbled meat admits more air, *and will consequently be disposed the sooner to ferment and putrify.*[375]

Darwin

Darwin's interest in Bakewell lay in the variation revealed by his technique of selection.[376] The great changes that mankind had achieved in a short period of time encouraged Darwin to speculate on what might be achieved by natural selection over the enormous spread of geological time. He wrote in his essay of 1844, *'let this work of selection, on the one hand, and death on the other, go on for a thousand generations, who would pretend to affirm it would produce no effect, when we remember what in a few years Bakewell effected on cattle and Western on sheep, by this identical principle of selection'.*[377]

Because of the variation between domestic animals, Darwin rejected the popular opinion that members of a species differed only in superficial, 'non-essential' characters. According to this point of view, variation in domestic animals could only be explained on the basis that different varieties were descended from 'primordially distinct' species or from crosses between such species.[378] The new breeds, particularly those which had been bred by selection within a single variety, without crossing, lent support to Darwin's point of view.

Darwin wrote:

> *'Remember how soon Bakewell on the same principle altered cattle and Western sheep, carefully avoiding a cross... with any breed'.*[379]

Darwin even touches on the origins of Bakewell's foundation sheep. In 1868 he wrote:

> '*What methodical selection has effected for our animals is sufficiently proved, ...So greatly were the sheep belonging to some of the earlier breeders such as Bakewell and Lord Western, changed, that many persons could not be persuaded that they had not been crossed.*'

This perhaps suggests that he believed that the New Leicester had been bred directly from a single variety with no outcrossing.[380]

But Darwin is a little ambiguous in relation to this subject of the foundation stock that Bakewell may have used. He also writes '*So with our British sheep: almost all the races, except the Southdown, have been largely crossed*'.[381] Thus Bakewell's sheep provided Darwin with a less certain example of the power of natural selection within a single breed than did his cattle.[382]

Darwin obtained most of his information on farm animals from the authors of books on the subject, such as David Lowe, Marshall, and his main source of reference Youatt. He did state that *the difficulty is to know what to trust* on the subject.

Ram Breeding and the Letting of Rams

The sheep breeding system which operated in Bakewell's time was not too dissimilar to the system that takes place today. The modern system has the dedicated ram producers, mostly 'stud' sheep breeders who concentrate on providing the 'commercial' sheep producer with rams. The major difference between the way it operated in Bakewell's time and today, is that today the rams are sold outright rather than being 'let' for the season.

Bakewell had a group of principal ram breeders in the Midland district who used his rams, including Mr Paget of Ibstock, as well as breeders in Lincolnshire, Yorkshire, Northumberland, Worcestershire and Gloucestershire. Bakewell was not the first or only ram breeder to let his rams. A Mr Palfrey let a considerable number of the Warwickshire breed, and Mr Frizby who let no less than 80 rams per season at five guineas each of the old Leicestershire breed.

The ram letting season started on 8 June and lasted until the Michaelmas (29 September). Each ram breeder kept an open house, with breeders from all over the country going from one ram breeder to another to either hire rams or just pass judgement on the rams on offer. The private showings closed with a public show at Leicester on the 10th of October, where rams of every description, but mostly of the inferior sort of the improved breed, were exhibited, some to be sold, but chiefly to be let. This show had been held from *time immemorial*, not for letting rams, but for selling rams.[383]

While the letting of rams had long been a practice in Lincolnshire, its origin in the Midland district can be traced to a ram let by Bakewell at the Leicester fair in about 1750 for sixteen shillings. Bakewell also let two more rams on the same day for seventeen shillings and sixpence each.[384] The date that Bakewell let this first ram cannot be exactly fixed. John Hunt states that it took place in 1760, and the ram was let to Mr Wilbore, of Illson-on-the-Hill.[385] This is ten years after the date given by Marshall. While all early writers on the subject agree that the cost of the transaction was sixteen shillings, I would tend to believe that Hunt may have recorded the date most accurately as *he was the neighbour and companion of Bakewell.*[386]

The usual practice of ram letting was for the ram breeder to keep the rams available for letting in small enclosures or paddocks, with several rams in each according to their age. A small pen was erected in each paddock into which the rams were put while being inspected.

Bakewell developed a new system of showing his rams, in which each ram was shown individually, they being never seen together. This system drew some criticism as it was harder for the less experienced judge to compare the animals with each other. He also stopped setting the price at which the rams were to be let, instead, letting the customers set the price by placing their own value on the rams. This is in effect what happens today when rams are sold through an auction system, the purchasers, through supply and demand, are setting the price.

There was generally no contract signed between the two parties, it being left to their honour to fulfil their side of the agreement. The agreed price was not usually paid until after *the ewes have brought proofs of the ram's efficiency.*[387]

The prices which were obtained for letting the rams for each season kept gradually rising. In 1780 Bakewell let several rams at ten guineas, while in that same year a Mr Parkinson let a ram of Dishley blood for twenty-five guineas, a price which then astonished the whole country. In 1786 Bakewell let two thirds of a ram for two hundred guineas, valuing the ram at three hundred guineas for the season. In that year he let twenty rams for more than a thousand pounds. In 1789 this had risen to twelve hundred guineas for three rams, and for his whole letting three thousand guineas.

Bakewell was not the only one letting rams, with six or seven other breeders making from five hundred to a thousand guineas each. Marshall estimated that in 1789, the incredible sum of ten thousand pounds was made by ram letting in the Midland Counties that year.[388] These high prices can in part be explained by the fact that they were being paid by the principal breeders who were aiming to produce rams which would be let *to inferior tupmen, as ram getters.*[389]

Rams were often sent two or three hundred miles for a season's work. They were transported in two-wheel carriages, with springs or hung in slings, some of them large enough to hold four rams. They would travel twenty or thirty miles a day.

The method of using the rams also changed, perhaps due to the increased prices being paid and the need for the rams to serve more ewes per season. Bakewell invented the use of teaser or aproned rams to signify when ewes were in season. Instead of the rams just running with the ewes in a paddock, the ram was kept from the ewes, they only being brought to the ram when in season and the ram only being allowed to serve each ewe once. With this system the ram could cover more ewes per season.

Bakewell used the letting of his rams to not only increase the rate at which his breed spread its influence throughout the country, but to also help him increase the speed at which he could do his improvements to the breed. He assessed the breeding ability or value of the sires he used by the quality of the lambs they left, not just by how good the ram itself looked. By letting rams, rather than selling them outright, he retained ownership of them and therefore the ability to use them himself if he wished to. This system was also a good proving ground for his sheep. By having individual rams being used in different parts of the country, under different management systems and mated to different breeds of ewes, it provided him with the ideal 'test bed' to see how they would perform. If he had sold his best rams to individual purchasers, then the spread of their genetics would have been restricted to one breeder alone, rather than having its progeny more widely circulated and improving more stock. It also allowed the other breeders who were hiring the rams to experiment more extensively than they would otherwise have been able to.

Ram letting was not without its risks to Bakewell. He was opening his breed up to criticism if they did not perform well, particularly if a high price had been paid for the ram's use. While we would presume that the breeders who paid high prices for rams would select their best ewes to be mated with the ram, we could also presume that if the resultant off-spring were not considered to be up to standard that it would be the ram that received the blame, not the ewes.

The modern day version of ram letting is the selling of semen. Today ram breeders can retain the ownership of their better rams, but sell semen packages to other breeder's, not only in their own country, but worldwide if restrictions do not prevent it. Modern technology has also allowed this to happen on the female side through the collection of eggs and the sale of the fertilised embryo.

Perhaps the only argument that could be mounted against Bakewell's practice of ram letting and the modern day sale of semen is that it could lead to the genetic concentration of the better animals at the cost of a wider overall genetic pool for the breed.

What would Bakewell think of these modern day techniques to disseminate genetics? My guess is that he would embrace it whole heartily and use it to great advantage. I think he would be pleased to know that the use of semen has helped in the spread of his modern day Leicester breed of sheep. It could perhaps be argued that without this modern day system of genetic transfer the Leicester breed numbers worldwide would be a lot lower than they are, and the breed may well have become a lot more inbred, thus putting its existence in jeopardy.

Quiet stock

Bakewell was described as a kind-hearted man, and that his natural kindliness made him the friend of both man and animals. This kindliness to the farm animals led to the breeding of extremely quiet and docile animals. Housman reports that young boys could handle large bulls with ease and move them from one part of the farm to another with just a switch.

An example of the way Bakewell handled animals is with the quietening of a horse thought to be ungovernable. When going on a long journey he had his own bridle and saddle put on the horse and then led it until out of sight of the farm. When he returned the horse was as gentle as a lamb and would obey his verbal orders. Bakewell never revealed how he broke the horse in, he never rode with a whip or spur, but carried a strong walking stick when on horseback.[390]

This kindness which he showed to his livestock set Bakewell in advance of the other farmers of his time.

Regarding his sheep, the Dishley or New Leicester has always been described as a quiet or docile animal, and this is still the case today. The description of an English Leicester Sheep, taken from the Flock Book of the Australian Stud Sheep Breeders Association describes them as having a 'quiet temperament', the UK breed society describes them as 'docile', while in the United States as 'docile and easy to handle'. This very good attribute is an example of the way Bakewell selected for what he wanted, and then so strongly fixed it in the breed that we are still reaping the benefit of it some 200 years later.

As a breeder of these docile sheep, I appreciate this quietness every time I handle them, and I know that this characteristic is a strong part of what makes the breed what it is. I can remember that as a young boy I would go with my father to the local shows where he would be showing English Leicester, Border Leicester and Dorset Horn sheep. I was always allowed to hold the English Leicesters for judging as they were 'quiet', even though I probably stood no taller than the sheep I was holding.

This docility that the breed exhibits is also displayed in their good mothering ability, a factor which must help in lamb survival rates. Compared to the other breeds of sheep that I run, I say that the English Leicesters are good counters at lambing time; if they have two or three lambs, that is what they will look after.

It is now being recognised by abattoirs that docile animals produce the best quality meat, whether they be sheep or cattle, again an example of Bakewell being ahead of his time.

Bakewell looked after his staff in the same way as he did his stock. He was surrounded by old and faithful servants. He would not engage a farm hand for a period of less than four years, with most staying much longer. In July 1793 John Breedon, the senior shepherd, had been there for thirty-two years and William Peet, superintendent of the horses for nearly forty years.[391]

Bakewell's Success

Some of Bakewell's closest associates were fellow Presbyterians, and in many cases his cousins. They gave Bakewell considerable support in his early days in particular, by hiring his rams which helped to disseminate his progeny. They would have also supplied Dishley blood lines to farmers who could not afford Bakewell's prices. Perhaps the most important help that they gave to Bakewell and the development of the Dishley breed was to help extend the gene pool which Bakewell had to work with by giving him access to further stock which had been bred on his own principles with which to experiment.[392]

It was a slow process for Bakewell to get his stock breeding recognised and accepted. The superiority of the Dishley breed was acknowledged by Arthur Young when he wrote that Bakewell's principal market of later years has been in Lincolnshire, Warwickshire, and other counties, were once those breeders who had been his rivals, *now establish the superiority of his breed by that most incontestable of all methods, becoming purchasers of it.*[393]

A lot of Bakewell's success can be attributed to his ability to identify and fix the characteristics that he was after.

Dishley Society

> *There can be little doubt that the object which he had principally in view in forming this Society was to promote his own interest, and that of the members of the Society, and its establishment has very mainly contributed to preserve the purity of the new Leicester breed, and this has been very beneficial to this country.*[394]

Bakewell formed the Dishley Society in 1783, with the aim of preserving the purity of the breed. No doubt increased profit was also a factor in its formation and Bakewell also made sure that the sheep he bred continued to play the major role in the breed's progress. The members of the Dishley Society used it to tightly control the letting of rams.

The members who were present at that first meeting were Wm. Walker, J.V. Stone, Jno. Bennett, Jno. Manning, Jos. Robinson, Nath. Stubbins, Nich. Buckley, R. Bakewell, F. White, Jno. Breedon, Saml. Knowles. Mr Paget was elected president of the society and Mr Honeybourn treasurer.

The subscription was set at ten guineas, a lot of money at that time (the yearly wage for a farm worker at this time was approximately eighteen pounds, four shillings).

The Society's principal rules are listed below. These rules were decided at meetings of the Dishley Society between 1790 and 1796 and are numbered as they appear in the hand written records of those meetings starting with number 4 in January 1790 to number 17 in May 1796.

> 4. That secrecy be kept by all the members respecting the business of these meetings except to the members absent, and that any member quitting the society keep secret upon his honour the transactions before he left.
>
> 8. That no member shall give his Rams at any season of the year any other kind of food than green vegetables hay and straw.
>
> 42. No member shall sell any ewes or rams of his own breed, to breed from, unless he sells his whole flock of sheep or sells them to members of the Society.
>
> 25. That no member shall take in a less number than 40 Ewes to tup, and those the property of one person.
>
> 23. That every member before he sets price of or lets a Ram, enquire what number of Ewes he is to serve and whose

property they are.

24. That no Ram shall be let to serve more than one hundred Ewes.

13. That no member shall let a Ram share or part of Ram to any Ram breeder residing within thirty miles of Leicester not being a member, who hired a Ram of Mr Bakewell last season 1789.

38. That Mr Bakewell shall not to let a Ram to any person within one hundred miles of Dishley for less than fifty guineas, and that beyond that distance prices and all other circumstances to be left to his own discretion.

18. From the first to the eighth of June in future the members shall not show their Rams except to each other, that they begin their general show on the eighth of June and continue till the eighth of July, from that time to the eighth of September they shall not show them to any one; but shall then open their show again and continue till the business of the season is over.

29. That no one ram shall in future be let to serve (the flocks of) more than two persons.

21. That the number of Rams to be let in each season shall be

thirty (per member).

20. That the members shall not show more than 24 Rams to any person or company at one time.

22. Resolved that no member shall let a Ram on Sundays.

41. Resolved that no member shall sell any ewes in future except to kill, at less than ten guineas each.

17. That any member letting a Ram shall have liberty to send 50 of his own Ewes to him in case of the failure of the Ram he keeps at home for his own use, except the Ram he wishes to send his own Ewes to shall be let to two persons in which case he shall send only 20 Ewes and the above privilege to extend to the failure of Dishley Rams.[395]

It is interesting to note that on 14 January 1793, a committee composed of Messrs. Buckley, Bennett, Bakewell, Paget, Stone and Stubbins was appointed *'with full powers to transact any business which may occur in the interval of general meetings'*.[396] Stubbins and Stone were early sellers of sheep to come to the colony and obviously prominent members of the Dishley Society.

The hierarchy of the ram breeding business was very important to the society, being one of the main reasons for its existence. The Dishley Society members aimed to sell or let their rams to the *ram breeders*, who in turn sold or let the rams they bred to the general sheep grazier of the day, whose aim was to produce meat for the local market. The average price that these graziers gave for hiring a ram was 5 or 6 guineas. The *ram breeders* could thus afford to pay more for good quality rams. They were also concerned about the *ram breeders* selling or letting their rams at fairs or markets and did what they could to prevent this.

This hierarchy system is reflected in some of the rules of the Dishley Society.

- That any person shewing ten Rams in one season shall be considered as a Ram Breeder.

- That every member in future letting a Ram share or part of a Ram to any Ram Breeder residing within thirty miles of Leicester who lets Rams at Fairs or Markets shall forfeit to the society the sum for which he shall so let any ran or share of a Ram.

- That no member let a Ram share or part of a Ram to Messrs. Barkby of Stanton Grange near Newark, Hobb of Claypole near Newark and Cooke of Claypole near Newark until they agree not to sell Rams or let them at Fairs or Markets.[397]

The Dishley Society was obviously a great success as prices paid by buyers and leasers of rams rose sharply. In 1786 Bakewell made 1,000 guineas by letting 20 rams, and in 1789 1,200 guineas for 3 rams and 2,000 guineas for another 7 rams. After 1788 Bakewell no longer set prices for the rams but left his customers to set a price before deciding whether it was acceptable.

The Perfect Form

Selection for type, in some form or another, has been practised by sheep owners since time immoral. The Old Testament tell us that Jacob became exceedingly rich through his sheep selection and breeding practices.[398] While Jacob may have been the exception, rather than the rule in his sheep management at that time, it appears that his practices were not really followed. Up until Bakewell's time selection had mostly been done on a subconscious basis, rather than as an aid to serious improvement in the livestock. What was there and available was kept and used to produce the next generation regardless of how ideal or good it was.

What was the perfect form for a sheep in Bakewell's time? Probably the same answer could have been given then as now, that is, that each person has their own idea of the perfect sheep. As a breeder of sheep, I am like every other breeder in striving to achieve this goal, but what I think is ideal or getting near to perfection may not be what the other breeders are looking for in an animal. Did Bakewell ever stand back in admiration of one of his creations and say, 'this is it, the perfect sheep.' My guess is no. He may have been justly proud of what he had achieved in bringing the unimproved ovines of the day into the modern world of his time, but I am sure that he was always looking for more improvement.

Being the first real improver of sheep, Bakewell achieved more than perhaps any breeder since his time has accomplished. While he did start from a low base, his achievements were remarkable. Starting with poor base stock meant that any small improvement would have a major beneficial effect on the sheep industry of the time, but it was also a disadvantage to him in that he did not have anything of great benefit to start with. If we were wanting to breed a 'new' breed of sheep today, as has happened with breeds such as the Poll Dorset and White Suffolk, we would have a lot of very good genetic material from which to pick, choose and blend to achieve the results we are after. Bakewell did not have this choice, just a lot of poor localised breeds to pick from.

The qualities he was looking for in this perfect form were:

1. The ability to reach a marketable state in the shortest possible time, *'that is, a natural propensity to acquire a state of fatness, at an early age, and when at full keep, in a short space of time'.*

The primary objective in the development of the Dishley Leicester was to have a sheep that was marketable in the shortest possible time with the least consumption of food. While the sheep of the day were not ready for market until they were two and a half or three years old, the Dishley Leicester was marketable from fifteen months old.

2. Eating quality.

Bakewell did not mind how much fat he bred into his sheep, in fact the more the better as he believed that you could not feed the hard-working labourer too much fat. Today as a breeder of these sheep I have to make sure that they do not put on too much fat. Another example of the degree to which Bakewell implanted these traits into the breed that we have to watch for it today. The early writers on the Leicester have mixed opinions regarding the eating quality of the Leicester meat. It was stated that the flesh, as with all longwool breeds, was more lax in the fibre and therefore less delicate in taste. Some described it as *coarse in grain* and *somewhat insipid to the taste*, though whether it was inferior to the other longwool breeds opinions varied.[399] Today the slow maturing (which the Leicester is, compared with the modern breeds) meat with a nice covering of fat is considered very tasty.

3. Proportion of parts.

Bakewell had an aversion to any part of the animal that did not contribute to its profitability. He was against large bone and a carcass full of offal. Henry Dixon stated that *'the hogshead* [a large cask] *or truly circular firkin* [small cask] *shape' with 'short, light boned legs, not much exceeding six inches in length' was his Improved Leicester sheep mould, 'on the plain principle that the value lies in the barrel and not the legs'.*

4. Conversion efficiency.

A sheep that was the most efficient convertor of food to meat was Bakewell's aim and he was perhaps the first breeder to consider this when selecting stock.

5. Beauty of form.

This of course is in the eye of the beholder, but as a breeder of Leicesters I think that Bakewell got it pretty right. While I know that the look of an animal does not necessarily relate to its efficiency or profitability, I do think that the Leicester of today has a beautiful form and is a pleasant animal to look at.

We know from early writings some of the things that he was after and some of the traits he did not want. He wanted *a barrel or an egg form* in his sheep, particularly the rib cage being like a barrel, but *ridgy backs and big bellies were his aversion.* The *scrag* or collar of a ram should be *thick and bowed like a swan*, and that he should have *an eye like a hawk, the head long and thin between the ears, and the ears thin and free from wool.* He did not like *ewe-headed* rams as he considered them to be invariably light in their lean flesh and delicate in constitution, *nothing but first-rate loins, thighs and scrags can support in-and-in-breeding.*

6. Docility.

Bakewell selected for docility in all his stock as he rightly believed that docile or quiet animals do better.

Today, the qualities we emphasise and aim to achieve when breeding the modern day Leicester have changed slightly from Bakewell's time. As the Leicester is now classed as a heritage breed, we are not so concerned about growth rate or conversion efficiency, but we are looking for:

1. Beauty of form.

We are looking for a well balanced animal that has its head, shoulders, barrel and hind-quarters in the correct proportion. We want a good looking animal that can walk well. The rams must be masculine in looks and the ewes feminine. We have really not changed this beauty of form from Bakewell's time. A few small cosmetic changes such as adding the top-knot and increasing bone size have been made, but in essence we are still looking at his sheep.

2. Long lustrous wool.

While Bakewell paid little, if any attention to wool, today we place a lot of emphasis on the Leicester's long, lustrous wool. In fact, it is the Leicester's wool that has played such an important part in keeping the breed alive today. The wools use in the craft industry has particularly helped the breeds success in both the USA and Sweden.

3. Docility.

We as breeders today are so fortunate that Bakewell so strongly implanted this characteristic into the breed because it is one of the breeds strong points today. How lovely it is to handle sheep that are naturally quiet and docile.

Criticism

> *No one can deny* (says Sir John Sebright) *the abilities of Mr Bakewell in the art of which he may fairly be said to be the inventor. But the mystery with which he is well known to have carried on every part of the business, and the various means which he employed to mislead the public, induce me not to give that weight to his assertions, which I should do to his real opinion, could it have been ascertained.*[400]

Both Bakewell and his sheep received much criticism for many years, perhaps a case of cutting the tall poppy down as we would say today. We know that Bakewell concentrated his breeding objectives on early maturity and feed conversion efficiency. His use of in-breeding to obtain and fix these traits meant that some not so desirable traits also surfaced.

As previously mentioned, the eating quality of the meat was questioned. How bad this was relative to other breeds of the day is hard to qualify, but we can assume that if it had been too bad the butchers and customers would not have purchased them, but, how much choice did the general public have as to the meat they purchased? The amount of fat produced by the Leicester also came in for criticism, but Bakewell stated that *I do not breed mutton for gentlemen, but for the public.* It must also be remembered that the labourers of the day would have burnt off, due to their hard physical work, a lot of the fat that they consumed with their meals. Fat was also an important by-product at that time for candle making and lubricant for industry.

There were differences of opinion regarding the meat quality. Culley said the Dishley breed *is as remarkable for the fineness of its grain, as the Lincolnshire sheep are for coarse grain; the former is also as fine-flavored and sweet as a mountain sheep.*[401]

It is also stated that the New Leicester suffered from a low fecundity rate and was delicate. This, if it was the case, would most likely have been as a result of in-breeding.

Being the leader of the Dishley Society also meant that Bakewell took the brunt of criticism levelled at the members of this society. The exclusiveness, secrecy and high prices were all areas that were readily criticised by some members of the farming fraternity.

THE HISTORY OF ENGLISH LEICESTER SHEEP IN AUSTRALIA

Summary

The Dishley Leicester was hailed a great success during Bakewell's lifetime, a significant factor as most success stories are not realised as such until well after the death of its inventor, or in this case, the breeder. The Leicesters use in many countries of the world, and the significant role that it played in improving many other breeds of sheep confirmed this high status. There has not been any other single breed that has been used so widely to improve other breeds, from the many well-known crosses, to the how many unacknowledged crosses.

In summing up Robert Bakewell there are many adjectives that we can use to describe him.

He was a pioneer animal breeder who had a clear objective of what he wanted to achieve and how he was going to do it. He did not mind experimenting in all his farming aspects to achieve the result he was after. He was obviously a lateral thinker who did not mind going against the practices of the day. Obviously, a good observer of livestock who could not only see the form of the animal in front of him, but how it could help him achieve his aim. He developed selective breeding principles and performance testing some 200 years before these became common practice. He was a traveller thirsty for knowledge in a time when most farmers never left the district they were born in. He was a religious man, but at the same time did not mind going against the church's views regarding breeding principles. He was kind to his fellow mankind as well as the animals in his care. He, on the one hand, freely gave information to all his visitors, but also was very secretive about how he actually bred the

Dishley Leicester. He was a good publicist and promoter of his livestock in an age when publicity was not easy to do.

Robert Bakewell knew what type of livestock had to be bred by the farmers of the day to feed the growing population and he spent his life achieving this for the betterment of the country. But it is perhaps, not so much in the actual livestock that Bakewell bred that we should remember him, but more in the principles and methods that he used to achieve this result. He set the rules and principles that we still use and benefit from today with our sheep breeding.

References

1. The Complete Grazier 1805, p46.

2. Garran & White. Merinos, Myths and Macarthurs, p107. The Wool Act (28 Geo. III, c.38).

3. British Parliament Hansard, 21 May 1824. Bischoff. Woollen and Worsted Manufactures, Vol I, p243, 244.

4. The Sydney Gazette & New South Wales Advertiser. Sunday 9 September 1804.

5. Historical Records of New South Wales, vi, p179.

6. Ian Parsonson. The Australian Ark, p32.
   Garran & White. Merinos, Myths and Macarthurs, p107.
   Carter HB. His Majesty's Spanish Flock, p293.

7. Carter HB. His Majesty's Spanish Flock, p433.

8. The Sydney Gazette & New South Wales Advertiser. Sunday 9 September 1804.

9. Historical Records of Australia, series 3, vol. 3, p251.

10. Carter editor. The Sheep and Wool Correspondence of Sir Joseph Banks 1781-1820, p431.

11. The Sydney Gazette & New South Wales Advertiser. Friday 28 June 1822, p3.

12. Skamp RB. The Wool Trade of Australia and Tasmania in The Year Book of Australia, 1889, p149.

13. On Wool Growing in Australia. The Australian and New Zealand Monthly Magazine, Vol 1 No. 5 (1842).

14. Ellis MH. John Macarthur, p221.

15. Ellis MH. John Macarthur, p221.

16. George Culley. Observations on Livestock. 1807, p166.

17. Henry Parker. The Rise, Progress and Present State of Van Dienam's Land. 1833. p159.

18. David Low. On the Domesticated Animals of the British Islands. 1845. p180, 181.

19. Garran & White. Merinos, Myths and Macarthurs quoting GW Evans. P61.

20. Hobart Town Gazette and Van Diemen's Land Advertiser, Saturday 14 January 1823.

21. Widdowson Henry. The Present State of Van Diemen's Land. 1828.

22. Hobart Town Gazette and Van Diemen's Land Advertiser, Friday 6 February 1824, p2.

23. Sydney Gazette and New South Wales Advertiser, Monday 21 November 1825, p3.

24. Launceston Examiner, 17 May 1843, p5.

25. Henry Widowson. Present state of Van Diemen's Land. 1829.

26. Colonial Times (Hbt), 2 April 1833 p2.

27. Margaret Mason-Cox. Lifeblood of a Colony. P6.

28. Van Diemen's Land Chronicle, 19 November 1841, p1.

29. Van Diemen's Land Chronicle, 19 November 1841, p1.

30. The Courier. Friday 10 December 1841, p2.

31. Hobart Town Courier, 12 November 1841, p3.

32. Historical Records of Australia. Series 1, Vol. 12.

33. Heazlewood Ivan. Old Sheep for New Pastures, p23-33.

34. Adrian Collins. George Stokell Pioneer Settler at Clarence Plains. From The Knopwood Historical Lectures 1988. P40.

35. Adrian Collins. Private correspondence September 1989.

36. Ivan Heazlewood. Old Sheep for New Pastures, p41, 42, 43.

37. Michael Clarke. 'Big' Clarke. P76.

38. Brown GA. Sheep Breeding in Australia, p360.

39. Flock Book for British Breeds of Sheep in Australia. Vol.7, p87.

40. Flock Book for British Breeds of Sheep in Australia. Vol.6, p90.

41. Weekly Times, Saturday 24 March 1928, p32.

42. Weekly Times, Saturday 10 March 1928, p2.

43. VDL Co Annual Report, 7 March 1826.

44. VDL Co Annual Report, 15 March 1831.

45. Launceston Advertiser, 20 October 1836.

46. VDL Co Annual Report, 20 March 1841.

47. Examiner, 28 November 1851.

48. Examiner, 6th March 1852.

49. Heazlewood Ivan. Old Sheep for New Pastures. p40.

50. Brown GA. Sheep Breeding in Australia. p395.

51. GA Brown. Sheep Breeding in Australia. p 394-398.

52. Bethell L. S., The Story of Port Dalrymple, Blubber Head Press, Hobart, p20.

53. Meredith, L. & Griffin, D. 1894, Early Deloraine, Regal Press, Launceston, p52.

54. Heazlewood Ivan. Old Sheep for New Pastures, p65-67.

55. Heazlewood Ivan. Old Sheep for New Pastures, p67, 68.

56. Flock Book for The Longwool Breeds of Sheep in Australia, Vol. 1.

57. Cornwall Chronicle, 1867.

58. The Muster No.13, June 1989.

59. Daily Telegraph, Launceston, August 19, 1913, p3.

60. The Examiner, November 13, 1913, p2.

61. Weekly Times, August 2, 1913, p49.

62. Australian Star, Sydney, July 6, 1893, p3.

63. Daily Telegraph (Launceston), Monday 2 May 1892, p2.

64. The Australasian Review. September 15 1893, p336.

65. Launceston Examiner, Friday 22 June, 1894, p7.

66. Heazlewood Ivan, Old Sheep for New Pastures, p103, 104.

67. The Sydney Herald, Monday 21 May 1832.

68. Port Phillip Gazette, 12 July 1843.

69. Pemberton P. Private correspondence.

70. The Monitor, Sydney. Friday 27 April 1927.

71. Pemberton PA. Pure Merinos and Others.

72. Colonial Office to AACo, 12 Oct 1835, AACo 1/17 DD22.

73. London Minutes, 27 Oct 1835, AACo, 160/92.

74. London Minutes, 22 Dec 1835, AACo 1960/92.

75. H.T. Ebsworth to Dumaresq, 5 February 1836, AACo 1/17 DD27.

76. Despatch 27, 6 Jun 1836, AACo 78/1/15, f 475.

77. Philip Oakden arrived in Launceston in 1833. He had gone to Hamburg in 1816 after suffering a business failure in London. When he returned to London in 1827 he was experienced in all aspects of wool trading. In Launceston he was an active merchant buying wool and other merchandise for shipment to England. He probably imported the first Lincoln sheep into Australia. In Tasmania he had several rural properties including 'Dunedin' at St Leonards and 'Bentley' at Chudleigh. It was during his temporary return to England in 1836 that he would have written the letter referred to.

78. Committee of Management, 6 Jan 1837, AACo 160/114.

79. Committee of Management, 31 January 1837, AACo 160/114.

80. Committee of Management, 14 March 1837, AACo 160/114.

81. Despatch No. 41, 29 Jul 1847, AACo, 78/1/15, f 609.

82. Despatch No. 42, 24 Aug 1837, AACo, 78/1/15, f 618.

83. Massy Charles. The Australian Merino, the story of a nation. p295.

84. Massy Charles. The Australian Merino, the story of a nation. p295 quoting Gregson (1907) p107.

85. Massy Charles. The Australian Merino, the story of a nation. p295.

86. J. Wilkinson to Lesley Melville, 14 Feb 1838, forwarded to Dumaresq by Hart Davis. AACo, 1/17 DD54.

87. London Despatch 33, 16 March 1838.

88. H.T. Ebsworth to Dumaresq, 11 May 1838. AACo 1/17 DD 59a enclosing Report 1/17 DD59c.

89. Despatch 11, 26 Oct 1838, AACo 78/1/16, f 83.

90. Despatch 10, 22 July 1839, 78/1/16, f 191.

91. Despatch No. 2, 1 Jan 1840 NBAC AACo, 78/1/16, f 131, with attached memo on weight of Leicester fleeces.

92. Despatch No. 31, 28 July 1840, AACo 78/1/16, f 422.

93. Despatch No. 38, 13 Nov 1840, AACo 78/1/16, f 609.

94. Despatch No. 46, 19 Mar 1841, AACo 78/1/16, f 691.

95. Despatch No. 48, 1 May 1841, AACo, 78/1/16, f 710.

96. Despatch No. 49, 24 June 1841, AACo, 78/1/16, f 741.

97. P.P. King to H.T. Ebsworth, London, Report of Proceedings, AACo 78/1/17, f 17.

98. Despatch No. 57, 26 Jan 1842, AACo, 78/1/17, f 89, with full Reports as attachment A.

99. London Despatch No. 47, 19 Ag 1842, AACo 78/3/1.

100. Despatch No 62, 30 April 1842, AACo, 78/1/17, f 161.

101. Despatch No 63, 6 Jun 1842, NBAC AACo, 78/1/17, f 193.

102. Despatch 73, 26 Dec 1842, AACo, 78/1/17, f 311.

103. Despatch No. 107, 25 Oct 1844, AACo, 78/1/17, f 991.

104. London Despatch No. 69, 19 Jul 1844, AACo 78/3/1.

105. Despatch 110, 26 Dec 1844, AACo, 78/1/17, f 1070.

106. P.P. King, to H.T. Ebsworth, 19 Dec 1844, AACo, 78/1/17, f 1174.

107. Despatch No 122, 29 Sep 1845, AACo, 78/1/17, f 373.

108. Despatch No. 142, 15 Aug 1846, 78/1/17, f 730.

109. Information received from P.A. Pemberton. AACo records held by ANU Archives.

110. Coughlan Timothy, Sheep and Wool in New South Wales, 1893.

111. Parsonson Ian. The Australian Ark, p77-79.

112. Perth Gazette, Saturday 16 September 1843, p3.

113. Flock Book for The Longwool Breeds of Sheep in Australia, Vol. X.

114. Stud and Farm, February 1967, p29.

115. Varro. The Husbandry of Livestock.

116. Bonwick James, Romance of the Wool Trade, 1887, p275-279.

117. South Australian gazette and Colonial Register, Saturday 18 June 1836, page 6.

118. Bishop Collection located in the SLSA.

119. Massy Charles. The Australian Merino, the story of a nation. p247.

120. Australasian (Melbourne) Saturday 26 March 1881, p23.

121. Massy Charles. The Australian Merino, the story of a nation, p460.

122. Flock Book of The Australian Longwool Sheepbreeders' Association, Vol 1, p43.

123. The Australasian, Saturday 16 October 1909, p7-8.

124. Flock Book for British Breeds of Sheep in Australia, Vol. 34, p32.

125. Rees Abraham.

126. Brown GA. Sheep Breeding in Australia, 1880, p188.

127. Brown PL editor. Clyde Company Papers Vol. II, p389.

128. Brown PL editor. The Narrative of George Russell of Golf Hill, p99.

129. Clarke Michael. 'Big' Clarke, p63-65.

130. The Australian Pastoralists' Review, October 15, 1898, p108.

131. Geelong Advertiser & Intelligencer. Saturday 22 May 1852. p3.

132. GA Brown. Sheep Breeding in Australia. p360.

133. The North Eastern Ensign, Tuesday 14 October 1873, p2.

134. Lee Timothy. Wanganella and the Merino Aristocrats, p103.

135. The Argus (Melbourne), Thursday 11 January 1877, p3.

136. The Argus (Melbourne), Friday 11 October 1861, p5.

137. The Argus (Melbourne), Friday 13 November 1863, p7.

138. The Argus (Melbourne), Saturday 25 November 1876, p10.

139. The Argus (Melbourne), Thursday 11 January 1877, p3.

140. The Argus (Melbourne), Thursday 13 September 1877, p7.

141. Argus, Thursday 1 June 1865, p3.

142. Australasian, Saturday 30 September 1865, p5.

143. Brown GA. Sheep Breeding in Australia. 1880 p360.

144. Hay Standard and Advertiser, Wednesday 9 October 1872, p3.

145. Brown GA. Sheep Breeding in Australia, p392.

146. James Grant, 'Sladen, Sir Charles (1816-1884)', Australian Dictionary of Biography, National Centre of Biography, Australian National University, https://adb.anu.edu.au/biography/sladen-sir-charles-4589/text7541, published first in hardcopy 1976, accessed online 23 September 2021.

147. Fay Woodhouse. The Enterprising Mr MacGregor, pvii.

148. Flock Book for British Breeds of Sheep in Victoria. Vol 1, p58.

149. Flock Book for British Breeds of Sheep in Victoria. Vol 1, p62.

150. J. Ann Hone, 'MacGregor, Duncan (1835–1916)', Australian Dictionary of Biography, National Centre of Biography, Australian National University, http://adb.anu.edu.au/biography/macgregor-duncan-4096/text6509, published first in hardcopy 1974, accessed online 15 December 2018.

151. Fay Woodhouse. The Enterprising Mr MacGregor, p135, 138.

152. Harry H. Peck. Memoirs of a Stockman, p114.

153. The Live Stock Annual of Australia. 1903, p159.

154. Flock Book for British Breeds of Sheep in Australia. Vol. 28, 1936, p47.

155. Flock Book for British Breeds of Sheep in Australia. Vol. 35, 1943, p24.

156. Flock Book for British Breeds of Sheep in Australia. Vol. 80, 1987, p27.

157. Flock Book for British Breeds of Sheep in Australia, Vol. 39, 1947.

158. The Muster, August 2014.

159. The Royal College of Surgeons of England, Plarr's Lives of the Fellows.

160. Flock Book for British Breeds of Sheep in Australia. Vol. 9.

161. Commonwealth of Australia Gazette. Periodic Friday 15 May 1987.

162. The Muster, No.64, August 2005.

163. The Muster, No. 79, August 2010.

164. Professor Owen, lecture at the Great Exhibition of 1851.

165. Roberts Stephen H, The Squatting Age in Australia 1835-1847, p175.

166. Queensland Sheep and Wool Industry, p75, 77.

167. Queensland Sheep and Wool Industry, p77. Daily Standard (Brisbane), August 8, 1919, p8.

168. Queensland Times, Ipswich Herald and General Advertiser, August 2, 1873, p3.

169. The Brisbane Courier, September 16, 1876, p6.

170. The Queenslander, June 30, 1906, p39.

171. The Queenslander, July 25, 1908, p38.

172. Queenslander, February 3, 1906, p35.

173. Queenslander, August 10, 1907, p35.

174. The Queenslander, March 2, 1912, p35.

175. The Sydney Stock and Station Journal, February 8, 1918, p4.

176. The Queenslander, July 26, 1919, p34.

177. Queensland Sheep and Wool Industry, p97.

178. Youatt W. Sheep, their Breeds, Management and Diseases. 1862, p322.

179. Woollen and Worsted Manufactures, Vol I, p390/391. Bischoff.

180. The Australian Merino, Charles Massy, p209.

181. Woollen and Worsted Manufactures, Vol I, p390, Bischoff.

182. The Australian Merino, Charles Massy, p207.

183. Ditto. P209 Taintor, cited in Randall, Fine Wool Sheep Husbandry p22.

184. Bischoff on Wool Woollens and Sheep, Vol II, p262. Bischoff.

185. Ditto, Carman et al, Special Report p269.

186. The Australian Merino, Charles Massy, p210, Bernardin, op cit.

187. The Australian Merino, p210. Charles Massy.

188. The Australian Merino, quoting Bernardin, p212. Massy.

189. The Australian Merino, p213, 214. Charles Massy.

190. Sydney Monitor and Commercial Advertiser, Monday 8 April 1839, p6.

191. Charles Massy. The Australian Merino, 1990, p145 quoting Gregson, The Australian Agricultural Company, 1824-1875.

192. Charles Massy. The Australian Merino, 1990, p145.

193. Charles Massy. The Australian Merino, 1990, p175, 176.

194. Charles Massy. The Australian Merino, 1990, p217.

195. Charles Massy. The Australian Merino, 1990, p279-282.

196. Charles Massy. The Australian Merino, 1990, p285, 286.

197. Charles Massy. The Australian Merino, 1990, p296.

198. Charles Massy. The Australian Merino, 1990, p917.

199. Selbright John. The Art of Improving the Breeds of Domestic Animals. 1809.

200. Michael Clarke. 'Big' Clarke p160.

201. Michael Clarke. 'Big' Clarke p160, 161.

202. Michael Clarke. 'Big' Clarke p161.

203. Harry H. Peck. Memoirs of a Stockman p98.

204. John Graham. A treatise on The Australian Merino.

205. Carnaby. Sydney Morning Herald 27 January 1843 p3.

206. The Pastoral Review, August 15, 1913 p777.

207. Nicolson Andrew. A Sheep for all seasons. A history of the Australian Corriedale, pp11, 13.
Nicolson Andrew. Australian Corriedale Century 1882-1982, pp9, 22.

208. Butler Alan. Sheep.

209. Garran & White. Merinos, Myths and Macarthurs. p218.

210. Garran & White. Merinos, Myths and Macarthurs. p219.

211. Massy Charles. The Australian Merino. The story of a Nation. p267.

212. Sydney Morning Herald. Tuesday 5 September 1843, p2.

213. Bonwick J. Romance of the Wool Trade, p132.

214. Clarke Michael. 'Big' Clarke. p88.

215. Bonwick James. Romance of the Wool Trade, 1887. p133.

216. Clyde Company Papers, Vol IV, p50.

217. Clyde Company Papers, Vol IV, p588.

218. Clyde Company Papers, Vol IV, p580.

219. Clyde Company Papers, Vol V, p156.

220. Bonwick James. Romance of the Wool Trade, 1887. p132.

221. Southey Thomas. The Rise, Progress and Present State of Colonial Sheep & Wools, Continued from 1846. p40.

222. Holy Bible. Revised Standard Version.

223. The Farmers Magazine. Vol 4, p310. 1803.

224. The Farmers Magazine. Vol 4, p166. 1803.

225. Marshall. The Rural Economy of the Midland Counties, 2nd edition. 1796. P355.

226. Russell. Like engend'ring like. P200.
Copus. Changing markets and the development of sheep breeds in Southern England 1750-1900.

227. Yorkshire Gazette. Saturday Jan 15, 1820. (British Library Newspapers).

228. Marshall. The Rural Economy of the Midland Counties, 2nd edition. 1796. P376.

229. Marshall. The Rural Economy of the Midland Counties, 2nd edition. 1796. P355.

230. Animals of Agriculture. V19 P537, 1793.

231. Marshall. The Rural Economy of the Midland Counties, 2nd edition. 1796. P357, 358.

232. Anne Orde. Matthew and George Culley Travel Journals and Letters 1765-1798. P203.

233. Marshall. The Rural Economy of the Midland Counties, 2nd edition. 1796. P358.

234. Sydney Morning Herald, Saturday 17 June 1843, p3.

235. Argus, Tuesday 1 May 1849, p2.

236. Clyde Company Papers Vol III, p511.

237. The Australasian Pastoralists' Review. October 16, 1893, p379.

238. Pastures New. Billis RV & Kenyon AS, p150.

239. Blainey Geoffrey. The Tyrany of Distance, p271.

240. Parsonson Ian. The Australian Ark, p 244-247.

241. North British Agriculturist as quoted in the Leader (Melb) Saturday 5 July 1873, p8.

242. Daily Telegraph, Sydney, Tuesday 20 April 1886, p3.

243. Parsonson Ian. The Australian Ark, p 247-249.

244. Reverend Samuel Marsden to Sir Joseph Banks, 27 April 1803, HRH of NSW, vol. V, p413.

245. Archibald David. On the Border Leicester breed of sheep. 1881.

246. Coleman J (ed.). The Cattle, Sheep and Pigs of Great Britain 1877, p286.

247. The Tasmanian Mail, January 12, 1907, p34.

248. Examiner (Launceston) Wednesday 6 February 1907, p2.

249. Mansell Alfred. Pure Bred Flocks, their formation and management. 1902.

250. The Tasmanian Mail, January 12, 1907, p34.

251. The Leicester Flock Book (UK) Vol. XV, 1907, p62.

252. RBA & UK Inflation Calculators have been used.

253. Weaver Lawrence Trevelyan. Painter of Pedigree, Thomas Weaver of Shrewsbury, Animal Artist of the Agricultural Revolution, p221.

254. McIvor. The History and Development of Sheep Farming, p451.

255. Youatt p320.

256. Youatt p315.

257. Archibald David. On the Border Leicester breed of sheep. 1881.

258. Thaer. 1806.

259. McIvor Clarence. The history & Development of Sheep Farming.

260. New Zealand Flock Book, Vol. 106.

261. The diary of John Brown.

262. Blancou Jean & Parsonson Ian. Historical perspectives on long distance transport of animals.

263. Blancou Jean & Parsonson Ian. Historical perspectives on long distance transport of animals. Kennedy Malcolm. Hauling the Loads, p7.

264. Carter HB. His Majesty's Spanish Flock, p362.

265. Toosey James Denton. Dottings of my life.

266. The Pastoralists' Review. June 15, 1907, p354.

267. Clarke M. 'Big' Clarke p14-22.

268. HT Ebsworth (London Secretary) to Robert Williamson, 3 Jul 1841, AACo, 1/19 DK70c.

269. Bishop Collection located in the SLSA.

270. Henry Parker. The Rise, Progress & Present State of Van Dieman's Land (1833), p160.

271. The Imlay Brothers' Account Book, 1837-1840.

272. Harcourt Rex. Southern Invasion Northern Conquest, p73, 74.

273. Harcourt Rex. Southern Invasion Northern Conquest, p84.

274. Perth Gazette, May 9 1840, p2. Clyde Company Papers Vol.II, p411, 447.

275. Australian Stud and Farm Monthly, 1958.

276. The Hobart Town Advertiser, 14 July 1843, p2.

277. Newcastle Chronicle, Saturday 12 February 1870, p2.

278. Maitland Mercury, February 5 1870, p2.

279. McIvor C. The History and Development of Sheep Farming from Antiquity to Modern Times, p97.

280. Russian Proverb.

281. The Australian Pastoralists' Review. August 19 1893, p286, 291.

282. The Improved Leicester Flock Book. Vol I 1893.

283. The Pastoralists' Review. July 15, 1907, p461.

284. Flock Book for British Breeds of Sheep in Australia. Vol 17 1925.

285. J. Ann Hone, 'MacGregor, Duncan (1835–1916)', Australian Dictionary of Biography, National Centre of Biography, Australian National University, http://adb.anu.edu.au/biography/macgregor-duncan-4096/text6509, published first in hardcopy 1974, accessed online 18 November 2018.

286. Matthew G. Hamilton, concluding his speech to the Douglas Agricultural Society 1923.

287. The Examiner (Launceston). October 21, 1905, p2.

288. The Australian Farm and Home. February 2, 1948, p25.

289. Luccock John. An Essay on Wool 1809, p100.

290. Luccock John. An Essay on Wool 1809, p114.

291. Luccock John. The Nature and Properties of Wool 1805, p79.

292. The Tasmanian Mail of October 24, 1891.

293. Berwick Thomas. A General History of Quadrupeds, p64.

294. Dyer John. The Fleece, p37, 38.

295. Journal of RAS, Vol.4 1868, p345.

296. Journal of RAS, Vol.4 1868, p348.

297. Hunt John. The Dishley System. 1812.

298. Charles Massy. The Australian Merino, 1990, p217.

299. Heazlewood Ivan. the Ark, August 1995, p283.

300. Nicholas Russell. Like engend'ring like. P54, 55.

301. John Hunt. Agricultural Memoirs; or History of the Dishley System. 1812. P26.

302. Lord Ernle. English Farming Past and Present. 4th edition. P168.

303. Ivan Heazlewood. Old Sheep for New Pastures. P1, 2.

304. David L. Wykes. Robert Bakewell (1725-1795) of Dishley: farmer and livestock improver. AgHR 52,I, p39.

305. Anonymous. The Annual Necrology 1797-8. P199.

306. Anonymous. The Annual Necrology 1797-8. P203.

307. Anonymous. The Annual Necrology 1797-8. P199-200.

308. Anne Milne. Sentient Genetics: Breeding the Animal Breeder as Fundamental Other. Journal for Eighteenth-Century Studies Vol. 33 No. 4 (2010).

309. David L. Wykes. Robert Bakewell (1725-1795) of Dishley: farmer and livestock improver. AgHR 52,I, p39.

310. Arthur Young. The Farmers Tour through the East of England. 1771, p124.

311. HC Pawson. Robert Bakewell. P5.

312. Robert C. Allen. Enclosure and the Yeoman. P1.

313. David L. Wykes. Robert Bakewell (1725-1795) of Dishley: farmer and livestock improver. AgHR 52,I, p39.

314. Arthur Young. Annals of Agriculture V6, 1786. P497.

315. Arthur Young. Annals of Agriculture V6, 1786. P496.

316. Arnold Caddy. Robert Bakewell and his times. P4.

317. Arthur Young. The Farmers Tour through the East of England. 1771, p134.

318. Anne Milne. Sentient Genetics: Breeding the Animal Breeder as Fundamental Other. Journal for Eighteenth-Century Studies Vol. 33 No. 4 (2010).

319. Pawson. Robert Bakewell. P41.

320. Bakewell letter to G Culley. February 8, 1787. From Pawson p107.

321. Arthur Young. The Farmers Tour through the East of England. 1771, p124.

322. Bakewell letter to G Culley, June 30, 1787. From Pawson p113.

323. Marshall. 1790, p286-287.

324. Arthur Young. Annals of Agriculture V6, 1786. P491.

325. Arthur Young. The Farmers Tour through the East of England. 1771, p126, 127.

326. The London Gazette No 11720, From Nov19-Nov23, 1776, p7.

327. Pawson. Robert Bakewell p40. Extract from a Frenchman in England 1784.

328. David Wykes. Robert Bakewell (1725-1795) of Dishley: farmer and livestock improver, p47.

329. The Annual Necrology 1797-8, p203.

330. Anne Orde. Matthew and George Culley Travel Journals and Letters 1765-1798, p37.

331. Anne Orde. Matthew and George Culley Travel Journals and Letters 1765-1798, p67.

332. The London Gazette No 11873, May 12, 1778.

333. David Wykes. Robert Bakewell (1725-1795) of Dishley: farmer and livestock improver, p47.

334. 'Advertisements and Notices.' St. Jame's Chronicle or the British Evening Post, July 21, 1778 – July 23, 1778. 17th and 18th Century Burney Collection.

335. The London Gazette No 11932, December 5, 1778.

336. 'Advertisments and Notices' General Evening Post. August 24, 1779-August 26, 1779. 17th and 18th Century Burney Collection.

337. David Wykes. Robert Bakewell (1725-1795) of Dishley: farmer and livestock improver, p47, 48.

338. The London Gazette No 12076, April 22, 1780.

339. David Wykes. Robert Bakewell (1725-1795) of Dishley: farmer and livestock improver, p48.

340. Wood & Orel. Genetic Prehistory in Selective Breeding p113, Pawson. Robert Bakewell, p195.

341. Pawson. Robert Bakewell p195.

342. Pawson. Robert Bakewell p181, 182.

343. Pawson. Robert Bakewell p40, 182. David Wykes. Robert Bakewell (1725-1795) of Dishley: farmer and livestock improver p48.

344. Pawson. Robert Bakewell p175.

345. Pawson. Robert Bakewell p172.

346. Pawson. Robert Bakewell p38.

347. Nicholas Russell. Like engend'ring like. P16.

348. Arthur Young. Annals of Agriculture V6 1786 p479.

349. Arthur Young. Annals of Agriculture V6 1786 p466.

350. Marshall. The Rural Economy of the Midland Counties, 2nd edition. 1796. P335.

351. Marshall. The Rural Economy of the Midland Counties, 2nd edition. 1796. P336.

352. Marshall. The Rural Economy of the Midland Counties, 2nd edition. 1796. P337.

353. Marshall. The Rural Economy of the Midland Counties, 2nd edition. 1796. P338.

354. David L. Wykes. Robert Bakewell (1725-1795) of Dishley: farmer and livestock improver. AgHR 52,I, p42.

355. John Hunt. Memoirs 1812. P49.

356. Pawson. Robert Bakewell. P50.

357. Roger Wood. Robert Bakewell, pioneer animal breeder, and his influence on Charles Darwin. Folia mendeliana February 1973. P233.

358. Bischoff. Woollen and Worsted Manufactures, Vol.1, p261.

359. Journal of the Royal Agricultural Society of England 1894, Vol.5, p21.

360. Sir JS Sebright. The Art of Improving the Breeds of Domestic Animals. In a letter addressed to Sir Joseph Banks. 1809. P10.

361. John Hunt. Memoirs 1812. P32.

362. Lord Ernle. English Farming Past and Present. 4th edition. P181.

363. John Hunt. Memoirs 1812. P45,46.

364. John Hunt. Memoirs 1812. P15.

365. John Hunt. Memoirs 1812. P15.

366. Sir JS Sebright. The Art of Improving the Breeds of Domestic Animals. In a letter addressed to Sir Joseph Banks. 1809. P6.

367. John Hunt. Memoirs 1812. P111, 112.

368. Janet Spavold. The Bakewell Family and the Local Unitarian Chapels. P1.

369. Janet Spavold. The Bakewell Family and the Local Unitarian Chapels. P1.

370. John Hunt. Memoirs 1812. P113.

371. Dishley Society Meeting, Rule number 22, June 4, 1795.

372. Pawson. Robert Bakewell. Letter to G Culley. P109.

373. Arthur Young. Annals of Agriculture V6 1786, p469.

374. Arthur Young. Annals of Agriculture V6 1786, p470.

375. Arthur Young. Annals of Agriculture V6 1786, p474.

376. Roger Wood. Robert Bakewell, pioneer animal breeder, and his influence on Charles Darwin. Folia mendeliana February 1973. P236 quoting Darwin 1868, Vol.1, p92, pp99-100.

377. Roger Wood. Robert Bakewell, pioneer animal breeder, and his influence on Charles Darwin. Folia mendeliana February 1973. P236.

378. Roger Wood. Robert Bakewell, pioneer animal breeder, and his influence on Charles Darwin. Folia mendeliana February 1973. P237.

379. Roger Wood. Robert Bakewell, pioneer animal breeder, and his influence on Charles Darwin. Folia mendeliana February 1973. P237 (Darwin 1842, p48).

380. Roger Wood. Robert Bakewell, pioneer animal breeder, and his influence on Charles Darwin. Folia mendeliana February 1973. P237 (Darwin 1868. Vol.2 p198).

381. Roger Wood. Robert Bakewell, pioneer animal breeder, and his influence on Charles Darwin. Folia mendeliana February 1973. P238 (Darwin 1868, Vol. 1 p95).

382. Roger Wood. Robert Bakewell, pioneer animal breeder, and his influence on Charles Darwin. Folia mendeliana February 1973. P238.

383. Marshall. The Rural Economy of the Midland Counties, 2nd edition. 1796. P373.

384. Marshall. The Rural Economy of the Midland Counties, 2nd edition. 1796. P374.

385. John Hunt. Memoirs 1812. P35.

386. Monthly Magazine. 1813. P81.

387. Marshall. The Rural Economy of the Midland Counties, 2nd edition. 1796. P382.

388. Marshall. The Rural Economy of the Midland Counties, 2nd edition. 1796. P384.

389. Marshall. The Rural Economy of the Midland Counties, 2nd edition. 1796. P385.

390. John Hunt. Memoirs 1812. P127.

391. Housman. Journal of the Royal Agricultural Society of England, 1894, V15. P5.

392. David L. Wykes. Robert Bakewell (1725-1795) of Dishley: farmer and livestock improver. AgHR 52,I, p43.

393. Arthur Young. Annals of Agriculture V6, 1786. P484.

394. Youatt William. Sheep, their Breeds, Management, and Diseases. 1862 p317.

395. University of Nottingham Manuscripts and Special Collection, reference number 24A. Papers of the Dishley Sheep Society.

396. University of Nottingham Manuscripts and Special Collection, reference number 24A. Papers of the Dishley Sheep Society.

397. University of Nottingham Manuscripts and Special Collection, reference number 24A. Papers of the Dishley Sheep Society.

398. Genesis 30, verse 37-43.

399. Wykes David. Robert Bakewell (1725-1795) of Dishley: farmer and livestock improver.

400. John Hunt. Memoirs 1812. P51, 52.

401. Culley. Observations on Livestock 1804. P43.

Appendix 1: English Leicester Flock Numbers

Flock No.	Prefix or Property	Owner	Address	Founded	Flock Book	Dispersed
1	Scale Park	L.R. Carter	Clunes, VIC	1883	BBSV, Vol. 1, 1898	1907
1	Oaklands	A.L. Bennett	The Oaks, NSW	1907	ALSA, Vol 1, 1909	
2	Dalmore	D. M'Gregor	Pakenham, VIC	1869	BBSV, Vol. 1, 1898	
2	Hillyfields	A.S. Fotheringham	Dashwoods Gully, SA	1903	ALSA, Vol 1, 1909	
3	Dalveen	W. Richardson	Woodchester, SA	1904	ALSA, Vol 1, 1909	
3	Westbrook	F.G.G. Couper	Darling Downs, QLD	1905	BBSV, Vol. 3, 1907	
4	Booloocooroo	J. Fleming Douglas	Curlewis, NSW	1907	ALSA, Vol 1, 1909	
4	Redcourt	McArthur Bros.	Lindenow, VIC	1904	BBSA, Vol. 3, 1907	1920
5	Malton	Ritchie Bros.	Chudleigh, TAS	1851	ALSA, Vol 1, 1909	
5	Glenvale Park	Fred H. Carter	Whittlesea, VIC	1907	BBSA, Vol. 4, 1909	1930
6	Melton Vale	R.G. Heazlewood	Glenore, TAS	1871	ALSA, Vol 1, 1909	
6	Summerleigh	James McKinnon Carter	Romsey, VIC	1907	BBSA, Vol. 4, 1909	1913
7	Brook Hill	A. Oliver & Sons	Chudleigh, TAS	1883	ALSA, Vol 1, 1909	
7	Scale Park	J.R. & A.W Carter	Clunes, VIC	1907	BBSA, Vol. 4, 1909	1924
8	Leicesterville	C.W. Allen	Westbury, TAS	1864	ALSA, Vol 1, 1909	
8	Fairfield	J.H. Fairchild	Lang Lang, VIC	1907	BBSA, Vol.	1927

8[1]	Viladale	J.G. Habel & Sons	Yulecart, VIC	1907	BBSA, Vol. 27, 1935	
9	Willow Vale	John Badcock & Sons	Glenore, TAS	1872	ALSA, Vol 1, 1909	1970
9[2]	Viladale	John G. Habel	Hamilton, VIC	1907	BBSA, Vol. 4, 1909	
10	Westwood	James Viney	Hagley, TAS	1892	ALSA, Vol 1, 1909	1945
10	Byrneside	Charles Hogan	Byrneside, VIC	1907	BBSA, Vol. 4, 1909	1918
11	Wesley Dale	Henry Reed	Wesley Dale, TAS	1906	ALSA, Vol 1, 1909	
11	Riverlea	D. Porter	Tallarook, VIC	1908	BBSA, Vol. 4, 1909	1918
12	Esk Farm	Boyes Bros.	Longford, TAS	1901	ALSA, Vol 1, 1909	1934
12[3]	Yalart	F.J. Stansmore	Pomborneit, VIC	1907	BBSA, Vol. 4, 1909	1912
12[4]	West Cloven Hills	Nicholas Cole	Camperdown, VIC	1907	BBSA, Vol. 6, 1914	1945
12[5]	East Hill	A.S. Bradshaw	Camperdown, VIC	1945	BBSA, Vol. 38, 1946	1946
12[6]		I.D. Macdonald		1946	BBSA, Vol. 39, 1947	1946
12	Warra Yadin	Nicholas Cole	Camperdown, VIC	1946	BBSA, Vol. 39, 1947	
13	Hill Grove	W.J. Trethewie	Hagley, TAS	1864	ALSA, Vol 1, 1909	
13	Gisborne Park	John A. Beattie	Gisborne, VIC	1908	BBSA, Vol. 5, 1913	1950
14	Leicester	Alfred H. Blake	Lang Lang, VIC	1905	BBSA, Vol. 5, 1913	1916
14	Drumreagh	B. & E. Sadler	Deloraine, TAS	1864	ALSA, Vol 1, 1909	
15		E.H. Heazlewood	Glenore, TAS	1906	ALSA, Vol 1, 1909	
15	Bolinda Vale	Sir R.T.H. Clarke	Lancefield Junction, VIC	1913	BBSA, Vol. 5, 1913	1928
16	Somerset Farm	Sir R.T.H. Clarke	Mooroopna, VIC	1851	BBSA, Vol. 5, 1913	1921
16		R. Norman Smith	Launceston, TAS	1906	ALSA, Vol 1, 1909	
17	Everton	H.R. Trethewie	Evendale, TAS	1907	ALSA, Vol 1, 1909	
17	Baafields	F.J.C. Minchin	Cowwarr, VIC	1910	BBSA, Vol. 5, 1913	1916
18	Brookside	W.R. Jones	Sheffield, TAS	1867	ALSA, Vol 1, 1909	
18		McMillan Bros.	Caldermeade, VIC	1912	BBSA, Vol. 5, 1913	1921
19	Leith	Percy Hart	Westbury, TAS		ALSA, Vol 1, 1909	
19	Milma	H.M. Osborne	Poowong, VIC	1905	BBSA, Vol. 5, 1913	1918
20	Rockland	G. Holmes	White Hills, TAS	1907	ALSA, Vol 1, 1909	

THE HISTORY OF ENGLISH LEICESTER SHEEP IN AUSTRALIA

20		W.K. Stokes	Echuca, VIC	1912	BBSA, Vol. 5, 1913	1914
21	Rathmolyou	Richard Hughes	Breadalbane, TAS		ALSA, Vol 1, 1909	
21	Woodside	John Langham	Woodside, VIC	1905	BBSA, Vol. 5, 1913	1923
22	Dalmore	J. M'Gregor	Pakenham, VIC	1869	ALSA, Vol 1, 1909	
22	Perringrove	J.T. Woods	Almond, VIC	1905	BBSA, Vol. 5, 1913	1948
23	Wickford	Falkiner Bros.	Longford, TAS	1879	ALSA, Vol 1, 1909	
23	Temoin	Robert Howard	Tangie, NSW	1913	BBSA, Vol. 6, 1914	1916
24	Enfield	W. Field	Bishopsbourne, TAS	1847	ALSA, Vol 2, 1910	
24	Woodstock	Oscar Inglis	Avoca, VIC	1912	BBSA, Vol. 6, 1914	1923
25	Coonmoor	McArthur Bros.	Hillside, VIC	1913	BBSA, Vol. 6, 1914	1933
25	Booloocooroo	W. & W.T. Grant	Curlewis, NSW	1907	ALSA, Vol 2, 1910	1946
26	Northumbria	H.P. Harrisson	Jericho, TAS	1900	ALSA, Vol 2, 1910	
26	Liberton	A. Francis	Oakbank, SA	1915	BBSA, Vol. 7, 1915	1923
27	Westbrook	F.G.G. Couper	Westbrook, QLD	1905	ALSA, Vol 2, 1910	
27		F.J. Scott	Kongwak, VIC	1912	BBSA, Vol. 7, 1915	1920
28	Delvine Park	John D. Scott	Bairnsdale, VIC	1913	BBSA, Vol. 7, 1915	
28	Umagarlee	R. & A. Scott	Wellington, NSW	1910	ALSA, Vol 3, 1911	1934
29	Zaretan	Richard J. Ball	Colac, VIC	1915	BBSA, Vol. 8, 1916	1943
29	Merrindie	W.S. Kelly	Giles Corner, SA	1907	ALSA, Vol 4, 1912	1927
30	Emu Park	Albin Clothier	Lindenow South, VIC	1916	BBSA, Vol. 8, 1916	1920
30	Glengrove	H. Morphett	Kangarilla, SA	1911	ALSA, Vol 4, 1912	
31	The Glen	D. Henry	Korumburra, VIC	1914	BBSA, Vol. 8, 1916	1925
31	Mount Ireh	F.J. Dumaresq	Pateena, TAS	1875	ALSA, Vol 4, 1912	
32	Clifton	A.J. Simpson	Hamilton, VIC	1915	BBSA, Vol. 8, 1916	1924
32	Winderradeen	Joseph Gill	Breadalbane, NSW	1910	ALSA, Vol 5, 1913	
33	Wy Yung	Mrs H.T. Taylor	Bairnsdale, VIC	1915	BBSA, Vol. 8, 1916	1916
33	Brymedura	Fraser E. Churchill	Manildra, NSW	1910	ALSA, Vol 5, 1913	
33a	Chandpara	Arnold Caddy	Tylden, VIC	1916	BBSA, Vol. 9, 1917	1945

34	Haye's Park	T.T. Chadwick	Welshpool, VIC	1916	BBSA, Vol. 9, 1917	1918
34		Govt. NSW Exp. Farm	Wagga Wagga, NSW	1910	ALSA, Vol 5, 1913	
35	Rose Hill	W.J. Eades	Colac, VIC	1915	BBSA, Vol. 9, 1917	1925
35	Fair Bank	F.H. Badcock	Hagley, TAS	1903	ALSA, Vol 5, 1913	
36	Riversleigh	E.O. Hobson	Yarram, VIC	1916	BBSA, Vol. 9, 1917	1934
36	Melton Vale	Roy K. Heazlewood	Whitemore, TAS	1912	ALSA, Vol 5, 1913	
36[7]	Melton Park	B.P. Heazlewood	Whitemore, TAS	2006	ASSBA, Vol. 98, 2006	
37	Yarragam	William Johnston	Tallarook, VIC	1917	BBSA, Vol. 9, 1917	1959
37	Glenroy	Norman Heazlewood	Whitemore, TAS	1907	ALSA, Vol 6, 1914	1934
38	Cowes	Clements Langford	Phillip Island, VIC	1916	BBSA, Vol. 9, 1917	1919
38	Alanvale	E.G. Hall	Newnham, TAS	1911	ALSA, Vol 6, 1914	1951
39	Worraine	P.H.V. Le Roux	Lang Lang, VIC	1916	BBSA, Vol. 9, 1917	1922
39	Manaree	John P. Miller	Gunnedah, NSW	1911	ALSA, Vol 6, 1914	
40		C.A. Teague	Wangaratta, VIC	1916	BBSA, Vol. 9, 1917	1918
40	Bettowyun	Hyland Bros.	Araluen, NSW	1912	ALSA, Vol 6, 1914	
41	Navarino	P.D. Ferrier	Echuca, VIC	1914	ALSA, Vol 7, 1915	
41		Harrison Bros.	Sale, VIC	1917	BBSA, Vol. 10, 1918	1920
42	Dalby	R.L. Elliott	Cumnock, NSW	1914	ALSA, Vol 7, 1915	
42		D.A. MacRae	Tourello, VIC	1917	BBSA, Vol. 10, 1918	1923
43	Jessiefield	Ernest L. Cox	Longford, TAS	1905	ALSA, Vol 7, 1915	
43	Brookvale	T.J. McCormack	Balmuttum, VIC	1917	BBSA, Vol. 10, 1918	1924
44	Tiveron	S.E. Roberts	Barwang, NSW	1914	ALSA, Vol 7, 1915	
44	Carinya	James Trembath	Corswold, VIC	1917	BBSA, Vol. 10, 1918	1958
45	Sterling Chase	F.A. Webb	Cudal, NSW	1914	ALSA, Vol 7, 1915	
45	Kewita	James Allen	Toora, VIC	1918	BBSA, Vol. 11, 1919	1925
46	Thelma	A.E. Swan & Son	Taralga, NSW	1914	ALSA, Vol 7, 1915	
46	Springvale	Francis Blennerhassett	Meerlieu, VIC	1918	BBSA, Vol. 11, 1919	1920
47		H. Leslie Heazlewood	Whitemore, TAS	1902	ALSA, Vol 7, 1915	

THE HISTORY OF ENGLISH LEICESTER SHEEP IN AUSTRALIA

47		O.J. Brewer	Drouin South, VIC	1918	BBSA, Vol. 11, 1919	1922
48		S. Tulloch Scott	St. Leonards, TAS	1911	ALSA, Vol 7, 1915	
48		J.S. Burchett	Poowong, VIC	1917	BBSA, Vol. 11, 1919	1923
49	Leven	Humphrey C. Dixon	Gisborne, VIC	1909	ALSA, Vol 7, 1915	1967
49		A.T. Creswick	Melbourne, VIC	1917	BBSA, Vol. 11, 1919	1919
50	Wootoona	H.G.M. Thackeray	Young, NSW	1915	ALSA, Vol 7, 1915	
50		W. Dettmann	South Kyneton, VIC	1918	BBSA, Vol. 11, 1919	1920
51	Lochinver	John williamson	Carisbrook, VIC	1915	ALSA, Vol 7, 1915	1981
51		H.P. Duke	Bairnsdale, VIC	1918	BBSA, Vol. 11, 1919	1919
52		Thomas G. Hudson	Longford, TAS	1911	ALSA, Vol 7, 1915	
52	Cluan	W.G. Gibson	Lismore, VIC	1918	BBSA, Vol. 11, 1919	1928
53	Fairview	G.C. Brunskill	Wagga Wagga, NSW	1915	ALSA, Vol 8, 1916	
53		W. Henley	Bittern, VIC	1918	BBSA, Vol. 11, 1919	1920
54		F.G. Dawson	One Tree Hill, SA	1915	ALSA, Vol 8, 1916	
54	View Bank	N.W. Howard	Meadows South, VIC	1918	BBSA, Vol. 11, 1919	1922
55	Green Hills	David Dunn	Pakenham East, VIC	1917	ALSA, Vol 9, 1917	
55		C.B. Irvine	Nicholson, VIC	1918	BBSA, Vol. 11, 1919	1919
56	Tullochgorum	Gerald P. Potter	Tullochgorum, TAS	1917	ALSA, Vol 9, 1917	
56		J.C. Rockliff	Numurkah, VIC	1918	BBSA, Vol. 11, 1919	
57		Robert A. Morice	Evendale, TAS	1917	ALSA, Vol 9, 1917	
57		J.B. Talbot	Longford, VIC	1916	BBSA, Vol. 11, 1919	1925
58	Glenallen	Moffat Bros.	Duri, NSW	1916	ALSA, Vol 9, 1917	1942
58		S.O. Wood	Bacchus Marsh, VIC	1917	BBSA, Vol. 11, 1919	1924
59		H.K. Nock	Nelungaloo, NSW	1917	ALSA, Vol 10, 1918	1952
59		W.H. Carpenter	Woodside, VIC	1920	BBSA, Vol. 12, 1920	1923
60		William Padbury	Guildford, WA	1913	ALSA, Vol 10, 1918	
60	Bank Glen	R.J. Clement	Springfield, VIC	1919	BBSA, Vol. 12, 1920	1936
61		Ellis Bros.	Campania, TAS	1907	ALSA, Vol 10, 1918	

61	Woodlands	G.F. Hall	Beech Forest, VIC	1919	BBSA, Vol. 12, 1920	1936
62	Waverley	Middleton Bros.	Greensborough, VIC	1919	BBSA, Vol. 12, 1920	1922
63	The Den	George Lee	Mole Creek, TAS	1907	ALSA, Vol 10, 1918	1988
63	Kooringal	A.C. Morrish	Kingston, VIC	1923	BBSA, Vol. 16, 1924	1928
64		J.F. Burn	Richmond, TAS	1889	ALSA, Vol 10, 1918	
64[8]	Excelsis	W.J. Clark	Sunbury, VIC	1924	BBSA, Vol. 17, 1925	1944
65	Strathalbyn	A.W. Edgar & Co	Gingin, WA	1916	ALSA, Vol 12, 1920	
65	Rathlea	James Ford	Hopetoun, VIC	1924	BBSA, Vol. 17, 1925	1936
66	Calcamine	W.G. Spencer	Grass Valley, WA	1921	ALSA, Vol 13, 1921	
66	Cattarno	H.B. Slaney	Moorooduc, VIC	1924	BBSA, Vol. 17, 1925	1934
67		Miss Olive M. Dabb	Baringhup East, VIC	1921	ALSA, Vol 13, 1921	
67	Strathalbyn	A.W. Edgar & Co	Gingin, WA	1916	BBSA, Vol. 18, 1926	1974
68	Highfield	J.R. Simpson	Woodstock, NSW	1924	ALSA, Vol 15, 1924	1934
68	Garden Hill	William Padbury	Guildford, WA	1913	BBSA, Vol. 18, 1926	1926
69		H. Birkett	Windermere, VIC	1924	ALSA, Vol 15, 1924	
69	Calcamine	W.G. Spencer	Grass Valley, WA	1921	BBSA, Vol. 18, 1926	1965
70	Homeden	R.F. Taylor	Gundary, NSW	1926	ALSA, Vol 18, 1928	1943
70[9]	Fern Hollow	W.T. Adams	Barfold, VIC	1908	BBSA, Vol. 19, 1927	1928
71	Little Hampton	J. & G.H. Selby	Bishopsbourne, TAS	1929	ALSA, Vol 18, 1928	1934
71	Coolamon	Mrs M.C. Alt	Avoca, VIC	1914	BBSA, Vol. 19, 1927	1943
72		Argustus W. Scott	Bungendore, NSW	1928	ALSA, Vol 18, 1928	
72	Fersfield	A.S. Bloomfield	Gisborne, VIC	1926	BBSA, Vol. 19, 1927	1929
73		G.A. Boreham	Grenfell, NSW	1929	ALSA, Vol 18, 1928	1949
73	Staughton Vale	Est.R.H. Dugdale	Balliang, VIC	1925	BBSA, Vol. 19, 1927	1945
74	Kyby	Kybybolite Exp. Farm	Kybybolite, SA	1909	BBSA, Vol. 19, 1927	1937
75	Blue Hills	C.H. Neilson	Baringhup East, VIC	1923	BBSA, Vol. 19, 1927	
76	Glenham	Sewage Farm	Islington, SA	1926	BBSA, Vol. 19, 1927	1933
77	Liberton	H.H. Shillabeer	Oakbank, SA	1923	BBSA, Vol. 19, 1927	1951

THE HISTORY OF ENGLISH LEICESTER SHEEP IN AUSTRALIA

78	Hollyburton Park	W.L. Armstrong	Bolinda, VIC	1927	BBSA, Vol. 20, 1928	1929
79	Drungroben	M. Domaschenz	Horsham, VIC	1927	BBSA, Vol. 20, 1928	1929
80	Moonyoonooka	D.E. Grant	Geralton, WA	1927	BBSA, Vol. 20, 1928	1933
81	Gilberside	A.T. Hill	Tarlee, SA	1927	BBSA, Vol. 20, 1928	1937
82	Greendale	J.P. Stokes	Echuca, VIC	1927	BBSA, Vol. 20, 1928	1936
83	Mardenoora	F. Hay	Tocumwal, NSW	1926	BBSA, Vol. 21, 1929	1929
84	Wroxton Grange	James Johnston	Angaston, SA	1927	BBSA, Vol. 21, 1929	1937
85	Clonmel	W.K. Stokes	Echuca, VIC	1928	BBSA, Vol. 21, 1929	1929
86	Burwood Downs	F.B. Woodcock	Bolivar, SA	1927	BBSA, Vol. 21, 1929	1931
87	Hollyburton Park	W.J. Moore	Bolinda, VIC	1928	BBSA, Vol. 21, 1929	1929
88	Glover	Mrs O.M. Glover	Baringhup East, VIC	1928	BBSA, Vol. 21, 1929	1936
89	Bickley Park	G.V. Mitchell	Donnybrook, WA	1928	BBSA, Vol. 21, 1929	1934
90	Euraba	F.E. Clarke	Darraweit Guim, VIC	1929	BBSA, Vol. 22, 1930	1939
91	Cleveland	C.R. Kurll	Narre Warren, VIC	1929	BBSA, Vol. 22, 1930	1932
92	Rubyvale	Burns Bros.	Yarroweyah North, VIC	1929	BBSA, Vol. 22, 1930	1932
93	Meadows	W.S. & S.J. Marwick	York, WA	1928	BBSA, Vol. 22, 1930	1932
94	Fulham	W. & J. Cockbill	Rockbank, VIC	1929	BBSA, Vol. 22, 1930	1950
95	Griffin's	W.J. Griffin	Derrinal, VIC	1930	BBSA, Vol. 23, 1931	1934
96	Paringa	V. Smart	Clarendon, SA	1930	BBSA, Vol. 23, 1931	1965
97	Sunset	H.F. Berwick	Yea, VIC	1930	BBSA, Vol. 23, 1931	1934
98	Longleat	D.H. Wollaston	Mt Barker, SA	1930	BBSA, Vol. 24, 1932	1941
99	Normanton	Harold Norman	Willunga, SA	1931	BBSA, Vol. 24, 1932	1938
100	Cornalla West	W. Hogan	Deniliquin, NSW	1932	BBSA, Vol. 25, 1933	1934
101	Berrangman	J.A. Williams & Son	Brimin, VIC	1932	BBSA, Vol. 25, 1933	1939
102	Stewarton	Miss Nora A. Stewart	Launceston, TAS	1932	BBSA, Vol. 25, 1933	1934
103	Athol	J. & H.B. McCrabb	Wunghnu, VIC	1933	BBSA, Vol. 26, 1934	1935
104	Wattnella	Paterson Bros.	Rochester, VIC	1933	BBSA, Vol. 26, 1934	1944
105	Yurnga	O.R. Talbot	Lancefield, VIC	1934	BBSA, Vol. 27, 1935	1945

106	Connaughtville	N.S. & I.A. Badcock	Westbury, TAS	1934	BBSA, Vol. 27, 1935	
107	Camellup	R.R. Viney	Lowden, WA	1934	BBSA, Vol. 27, 1935	1937
108	Fairvale	M.C. Dobson	Whitemore, TAS	1934	BBSA, Vol. 28, 1936	1954
109	Brookdale	A.F. Peck	Bangor, TAS	1935	BBSA, Vol. 28, 1936	1947
110	Loddon Valley	Baringhup Y.F. Club	Baringhup, VIC	1935	BBSA, Vol. 28, 1936	1938
111	Mysia	Gordon M. Chalmers	Mysia, VIC	1935	BBSA, Vol. 28, 1936	1950
112	Wakool	J.W. & R.H. Clarke	Bendigo, VIC	1935	BBSA, Vol. 28, 1936	1949
113	Smythesdale	R.H. Greaves	Deniliquin, NSW	1935	BBSA, Vol. 28, 1936	1936
114	Nayook	W.J. Jenkins	Yahl, SA	1935	BBSA, Vol. 28, 1936	1952
115[10]	Dalmore	D.S. MacGregor	Baringhup, VIC	1936	BBSA, Vol. 28, 1936	
116	Merkah	V.A. Watson	Tarcutta, NSW	1935	BBSA, Vol. 28, 1936	1938
117	Tipperary	W.G. Burges	Burges Siding, WA	1936	BBSA, Vol. 29, 1937	1964
118	Rosaville	Stanley J. Hingston	Whitemore, TAS	1936	BBSA, Vol. 29, 1937	1968
119	Wynarling	J.B. Stephenson	Sevenhills, SA	1936	BBSA, Vol. 29, 1937	1954
120	Kanimba	R.O. Zander	Angaston, SA	1936	BBSA, Vol. 29, 1937	1948
121	Woodbury	A.A. Armstrong	Deniliquin, NSW	1936	BBSA, Vol. 29, 1937	1950
122	Gladstone Park	James E. Barrow	Broadmeadows, VIC	1936	BBSA, Vol. 29, 1937	1937
123	Buraja	J. Howard Izon	Corowa, NSW	1936	BBSA, Vol. 29, 1937	1937
124	The Gums	R.K., L.N. & R.L. Johnston	Angaston, SA	1936	BBSA, Vol. 29, 1937	1938
125	Moondooma	J. Murray Woods	Tatura, VIC	1936	BBSA, Vol. 30, 1938	1938
126	Aswarby	T.J. Terffry	Exton, TAS	1936	BBSA, Vol. 30, 1938	1944
127	Twin Hills	R. & L. Green	Lancefield, VIC	1937	BBSA, Vol. 30, 1938	1940
128	Munyabala	W.J. Scott	Henty, NSW	1937	BBSA, Vol. 30, 1938	1944
129	Glen Darnel	Sharp & Taylor	Tallarook, VIC	1937	BBSA, Vol. 30, 1938	1945
130	Lesma	Les Wangler	Beaufort, VIC	1937	BBSA, Vol. 30, 1938	1954
131	Glenelg	G.R. Williams	Warral, NSW	1937	BBSA, Vol. 30, 1938	1941
132	Olive Dale	W.A. Oliver	Chudleigh, TAS	1937	BBSA, Vol. 30, 1938	1941
133	Vendale	I.E. Nicholls	Auburn, SA	1937	BBSA, Vol. 30, 1938	1943

134	Glen Hope	J.B. Stuart	Whitemore, TAS	1938	BBSA, Vol. 31, 1939	1948
135	Emu-Plains	Arch Wilson	Mole Creek, TAS	1938	BBSA, Vol. 31, 1939	1953
136	Amesbury	J.W. Anderson	Bamawm, VIC	1938	BBSA, Vol. 32, 1940	1955
137	Golden View	A. & J. Chester	Beverley, WA	1938	BBSA, Vol. 32, 1940	1946
138	Wooregong	Newton P. Burges	Burges Siding, WA	1938	BBSA, Vol. 32, 1940	1945
139	Deepdene	R.W. Clarke	Yea, VIC	1939	BBSA, Vol. 32, 1940	1950
140	Hilldene	F.N. Everitt	Bilbarin, WA	1938	BBSA, Vol. 32, 1940	1965
141	Keddies Hill	Nottinghill YF Club	Nottinghill, VIC	1939	BBSA, Vol. 32, 1940	1940
142	Glen Mona	Mrs A.M. O'Connor	Woodside, VIC	1939	BBSA, Vol. 32, 1940	1941
143	Melton	J. Truman	Calcarra, WA	1939	BBSA, Vol. 32, 1940	1944
144	Gold View	Keith Cumming	Acheron, VIC	1939	BBSA, Vol. 32, 1940	1967
145	Eastlea	R. Watson	Hynam, SA	1939	BBSA, Vol. 32, 1940	1945
146	Campbell's	Keith Campbell	Broadmeadows, VIC	1939	BBSA, Vol. 33, 1941	1949
147	Bonadale	K.B. Cutts	Northdown, TAS	1939	BBSA, Vol. 33, 1941	1941
148	Brierley Grove	Walter Cutts	Sassafras, TAS	1939	BBSA, Vol. 33, 1941	1953
149		E. Gorman & Sons	Berrigan, NSW	1940	BBSA, Vol. 33, 1941	1949
150	Benacre	F.J.C. Layton	Forth, TAS	1940	BBSA, Vol. 33, 1941	1942
151	Bracknell	F.J. Leonard	Bracknell, TAS	1939	BBSA, Vol. 33, 1941	1945
152	Creekside	James A. Leonard	Whitemore, TAS	1940	BBSA, Vol. 33, 1941	1950
153	Loyola	G.F. Maud	Mansfield, VIC	1940	BBSA, Vol. 33, 1941	1951
154	Coolgardie	Percy G. Weaver	Boort, VIC	1939	BBSA, Vol. 33, 1941	1951
155	Brambletye	J.N. Archer	Llewellyn, TAS	1939	BBSA, Vol. 33, 1941	1943
156	Kerry	R.C. Grubb	Launceston, TAS	1940	BBSA, Vol. 33, 1941	1945
157	Glendessary	T.G. Stancombe	Western Junction, TAS	1939	BBSA, Vol. 33, 1941	1944
158		H.F. & K.L. Vincent	Bendigo, VIC	1940	BBSA, Vol. 33, 1941	1941
159	Bonnaville	B.A.R. Bonney	Moriarty, TAS	1940	BBSA, Vol. 33, 1941	1948
160	Karralee	G.W. Down	Tatura, VIC	1940	BBSA, Vol. 33, 1941	1942
161	Delmore	M.C. Heazlewood	Hagley, TAS	1940	BBSA, Vol. 33, 1941	1955

162	Plainsland	F.N. Scott	Kongwak, VIC	1940	BBSA, Vol. 33, 1941	1952
163	Goomalibee	Max T. Armstrong	Benalla, VIC	1940	BBSA, Vol. 34, 1942	1944
164	Pyreness	E.G. Howell	Avoca, VIC	1941	BBSA, Vol. 34, 1942	1945
165	Ohio	J.H.W. Mules	Brighton, SA	1940	BBSA, Vol. 34, 1942	1942
166	Mulavon	J.E. Mullins	Birdwood, SA	1941	BBSA, Vol. 34, 1942	1954
168	Arrawatta	Arrawatta Past. Co.	Deniliquin, NSW	1941	BBSA, Vol. 34, 1942	1971
169	Mayberry	H.R. Byard	Mole Creek, TAS	1940	BBSA, Vol. 34, 1942	1943
170	Fairway	G.R. French	Whitemore, TAS	1941	BBSA, Vol. 34, 1942	1951
171	Beradale	Morse Bros.	Sheffield, TAS	1941	BBSA, Vol. 34, 1942	1949
172	Ardgour	A.L. McLean	Gnarkeet, VIC	1941	BBSA, Vol. 34, 1942	1967
173	Hany	I.L. Reid	Tolberry, TAS	1941	BBSA, Vol. 34, 1942	1944
174	Coolibah	G.R. Hilderbrand	Leitchville, VIC	1941	BBSA, Vol. 34, 1942	1944
175	Moat	Mrs M.I. Lucadon	Carrick, TAS	1941	BBSA, Vol. 34, 1942	1951
176	Mt Violet	Trevor O. Smith	Camperdown, VIC	1941	BBSA, Vol. 34, 1942	1980
177	Glen Villa	N. Gibson	Westbury, TAS	1941	BBSA, Vol. 34, 1942	1950
178	The Valley	J. Watson	Trafalgar South, VIC	1941	BBSA, Vol. 34, 1942	1952
179	Craigie Mains	Thos. Borthwick & Sons	Melbourne, VIC	1941	BBSA, Vol. 34, 1942	1948
180	Scoribreac	C. Macrae	Tourello, VIC	1942	BBSA, Vol. 35, 1943	1953
181	Bardia	I.J. Morrish	Kingston, VIC	1942	BBSA, Vol. 35, 1943	2002
182	Mt Helen	H.J. Pethybridge	Buninyong, VIC	1941	BBSA, Vol. 35, 1943	1949
183	Barwite Park	Barragunda Est. Pty Ltd	Mansfield, VIC	1941	BBSA, Vol. 35, 1943	1950
184	Hallfield	E.W. Blair	Mole Creek, TAS	1941	BBSA, Vol. 35, 1943	1950
185	Haddington	Mikkelson Bros.	Moolort, VIC	1941	BBSA, Vol. 35, 1943	1964
186	Alvira	C. McCulloch	Whitemore, TAS	1941	BBSA, Vol. 35, 1943	1953
187	Glenorn	C.R. Taylor	Howth, TAS	1942	BBSA, Vol. 35, 1943	1943
188	Green Banks	John Badcock	Whitemore, TAS	1910	BBSA, Vol. 35, 1943	1949
189	Pisa	A. Morrison	Oatlands, TAS	1910	BBSA, Vol. 35, 1943	1949
190	Cotswold Rise	A.J. Archie & Sons	Burnie, TAS	1942	BBSA, Vol. 35, 1943	1945

191	Thule	J.R. Hay	Flinders Island, TAS	1942	BBSA, Vol. 35, 1943	1946
192	Macare	W. McRobert	Toolleen, VIC	1942	BBSA, Vol. 35, 1943	1971
193	Ripplee	Mrs H.M. Limbert	Millicent, SA	1942	BBSA, Vol. 35, 1943	1947
194	Barrabogie	H.R. Mitchell	Watchupga, VIC	1942	BBSA, Vol. 35, 1943	1945
195	Tamar	A.G. Berryman	Womboota, NSW	1943	BBSA, Vol. 35, 1943	1963
196	Springhead	F.H. Carter	Newlyn, VIC	1942	BBSA, Vol. 36, 1944	1945
197	Kareu	R.K. Menzel	Armytage, VIC	1943	BBSA, Vol. 36, 1944	1947
198	Fairville	Henry Badcock & Son	Hagley, TAS	1942	BBSA, Vol. 36, 1944	1955
199	Home View	N.B. Flowers	Chudleigh, TAS	1943	BBSA, Vol. 36, 1944	1949
200	South Wonwondah	Heard Bros.	Horsham, VIC	1943	BBSA, Vol. 36, 1944	2008
201	Hilly View	J.O. Hobill	Ballarat, VIC	1943	BBSA, Vol. 36, 1944	1945
202	Hillway	H.H. Lewis	Tatong, VIC	1943	BBSA, Vol. 36, 1944	1968
203	Millson	R.B. Mills & Son	Nullawil, VIC	1943	BBSA, Vol. 36, 1944	1964
204	Dalthill	Mrs C.S. Schultz & Son	Balmoral, VIC	1943	BBSA, Vol. 36, 1944	1956
205	Woodside	Bruce Scott	Hagley, TAS	1942	BBSA, Vol. 36, 1944	1950
206	Toiberry	H.G. Shipp	Little Hampton, TAS	1943	BBSA, Vol. 36, 1944	1948
207	Tyf	Toolleen YF Club	Toolleen, VIC	1943	BBSA, Vol. 36, 1944	1953
208	Laroona	D.R.G. Woodiwiss	Pipers River, TAS	1943	BBSA, Vol. 36, 1944	1950
209	Battery Park	L.J. & A.L. Graves	Mansfield, VIC	1943	BBSA, Vol. 36, 1944	1949
210	Piper	W. Baxter & Son	Pipers River, TAS	1943	BBSA, Vol. 36, 1944	1951
211	Bradford Hills	L.A. Balmer	Maldon, VIC	1943	BBSA, Vol. 36, 1944	1950
212	Walma	F.H. Barber	Horsham, VIC	1943	BBSA, Vol. 36, 1944	1951
213	Coal River	M.J. Burn	Campania, TAS	1943	BBSA, Vol. 36, 1944	1950
214	Glenynille	O.C. Hurst	Pipers River, TAS	1943	BBSA, Vol. 36, 1944	1950
215	Bowness	F.T. Longmire	Werona, VIC	1943	BBSA, Vol. 36, 1944	1961
216	Roland Vale	E.R. Padman	Sheffield, TAS	1943	BBSA, Vol. 36, 1944	1945
217	Kentish	Sheffield School Farm	Sheffield, TAS	1943	BBSA, Vol. 36, 1944	1948
218	Coe Glen	J. & O. Wardlaw	Falmouth, TAS	1943	BBSA, Vol. 36, 1944	1945

219	Woodwestern	W.E. Wuttke & Son	Woodside, SA	1943	BBSA, Vol. 36, 1944	1948
220	Moonaree	C.W. Bird	Bacchus Marsh, VIC	1943	BBSA, Vol. 36, 1944	1950
221	Creek Junction	P. & E. Walter	Creek Junction, VIC	1943	BBSA, Vol. 36, 1944	1949
222	Salt Lake	R.J. Brown	Corop, VIC	1943	BBSA, Vol. 36, 1944	1950
223	Glengowrie	A. Facey & Sons	Wharparilla North, VIC	1942	BBSA, Vol. 37, 1945	1946
224	Hallston	A. Heazlewood	Hagley, TAS	1944	BBSA, Vol. 37, 1945	1950
225	Myrtlebrae	T.R. & Miss J.R. Robinson	Myrtle Bank, TAS	1944	BBSA, Vol. 37, 1945	1945
226	Chateau	Tahbilk Pty Ltd	Tabilk, VIC	1943	BBSA, Vol. 37, 1945	1949
227	Glen-Moidart	J.J.B. Coldwell & Sons	Berrigan, NSW	1944	BBSA, Vol. 37, 1945	1949
228	Bickford	D.W. Parsons	Exton, TAS	1944	BBSA, Vol. 37, 1945	1958
229	Romarnie	E.A. Pearn & Son	Deniliquin, NSW	1944	BBSA, Vol. 37, 1945	1947
230	Red Banks	R.E. Smith	Deloraine, TAS	1944	BBSA, Vol. 37, 1945	1945
231	Birrabee	J. Hamilton-Smith	Tallangatta, VIC	1944	BBSA, Vol. 37, 1945	1974
232	Daisy Lea	R.R. Adams	Clunes, VIC	1944	BBSA, Vol. 37, 1945	1947
233	Cooali	Miss M. McCrabb	Deniliquin, NSW	1944	BBSA, Vol. 37, 1945	1949
234	Landford	F.R. Archer	East Tamar, TAS	1940	BBSA, Vol. 37, 1945	1986
235	Kangaroo	Mrs N.C.M. Smith	Hay, NSW	1944	BBSA, Vol. 37, 1945	1950
236	Mardon Fields	C.W. Donnie	Dromana, VIC	1945	BBSA, Vol. 38, 1946	1951
237	Wyndon	Eric Bramich	Deloraine, TAS	1945	BBSA, Vol. 38, 1946	1950
238	Warilya	D.K. McMillan	Benalla, VIC	1945	BBSA, Vol. 38, 1946	1953
239	Minnow	A.G. & T.S. Oliver	Beulah, TAS	1944	BBSA, Vol. 38, 1946	1946
240	Conaire	H.M. Osborne	Poowong, VIC	1945	BBSA, Vol. 38, 1946	1953
241	Tulliallan	F.A. Palfreyman	Berwick, VIC	1945	BBSA, Vol. 38, 1946	1947
242	Hamley	H.L. Lawrie	Babakin, WA	1945	BBSA, Vol. 38, 1946	1983
243	Bridwood	L. Waters	Bridgetown, WA	1945	BBSA, Vol. 38, 1946	1973
244	Wirruna	H.W. Browne	Young, NSW	1945	BBSA, Vol. 38, 1946	1955
245	Kilderry	His Majesty's Farm Goal	Hayes, TAS	1944	BBSA, Vol. 38, 1946	1951
246	Wimmera Downs	Millar Bros.	Horsham, VIC	1945	BBSA, Vol. 38, 1946	1954

THE HISTORY OF ENGLISH LEICESTER SHEEP IN AUSTRALIA

247	Leicesterfield	A.J. Page	Pingelly, WA	1945	BBSA, Vol. 38, 1946	1953
248	Dugdale	R. Hartley Dugdale	Violet Town, VIC	1945	BBSA, Vol. 38, 1946	1949
249	Glen Allen	F.W. Moffat	Duri, NSW	1945	BBSA, Vol. 38, 1946	1946
250	Gladstone	George Clyde	Rushworth, VIC	1945	BBSA, Vol. 39, 1947	1947
251	Bylands	C.G. Meier	Kilmore, VIC	1946	BBSA, Vol. 39, 1947	1950
252	Early Rise	R.L. Atkins	Penguin, TAS	1946	BBSA, Vol. 39, 1947	1949
253	Bardale	S.P. Barlow	Rushworth, VIC	1946	BBSA, Vol. 39, 1947	1949
254	Comus	James J. McDonald	Nagambie, VIC	1946	BBSA, Vol. 39, 1947	1949
255	Karthina	C.F. Whitton	Winslow, VIC	1946	BBSA, Vol. 39, 1947	1950
256	Denbarrie	G.D. Chapman	Spring Hill, NSW	1946	BBSA, Vol. 39, 1947	1951
257	Warrnambool	John O'Donohue	Panmure, VIC	1946	BBSA, Vol. 39, 1947	1948
258	May Vue	J.A. French	Whitemore, TAS	1946	BBSA, Vol. 39, 1947	1950
259	Somerset Park	R.B. Walton	Buckley, VIC	1946	BBSA, Vol. 39, 1947	1947
260	Craven	A.J. Webb	Albury, NSW	1943	BBSA, Vol. 39, 1947	1948
261	Leicsetley	E.P. Hart	Mole Creek, TAS	1946	BBSA, Vol. 39, 1947	1954
262	Tulliallan	J.M. Elder	Berwick, VIC	1946	BBSA, Vol. 40, 1948	1949
263	Antrim	F.J. Higgenson	Pingelly, WA	1946	BBSA, Vol. 40, 1948	
264	Hill View	P.M. Liston	Lismore, VIC	1946	BBSA, Vol. 40, 1948	1952
265	Keltonlea	W.B. Firth	Hobart, TAS	1947	BBSA, Vol. 40, 1948	1949
266	Dranelg	Edwin A. Gee	White Hills, TAS	1947	BBSA, Vol. 40, 1948	1954
267	Rhyanna	Glenville Lawton	Goulburn, NSW	1946	BBSA, Vol. 40, 1948	1951
268	Kallumally	W. Copland Mackie	Kyneton, VIC	1947	BBSA, Vol. 40, 1948	1950
269	Rockwood	K.A. Francombe	Beulah, TAS	1947	BBSA, Vol. 40, 1948	1950
270	Nant	Ian Campbell	Bothwell, TAS	1946	BBSA, Vol. 40, 1948	1969
271		Bidgemia Past. Co. Ltd.	Spring Hill, WA	1947	BBSA, Vol. 40, 1948	1949
272	Rennat	O.G.A. Tanner	Burges Siding, WA	1947	BBSA, Vol. 40, 1948	1984
273	Guy's	W.C. Gadd	Guy's Forest, VIC	1948	BBSA, Vol. 41, 1949	1959
274	Ovens	S. Hargreaves	Bright, VIC	1948	BBSA, Vol. 41, 1949	1950

275	Gum Tree	N.D. Kingston	White Hills, TAS	1948	BBSA, Vol. 41, 1949	1953
276	Knoxborough	E.G. Knox	Woodend, VIC	1948	BBSA, Vol. 41, 1949	1950
277	Campbelltown	Campbelltown YF Club	Campbelltown, TAS	1948	BBSA, Vol. 41, 1949	1950
278	Oldfold	W.J.A. Higham	Williams, WA	1948	BBSA, Vol. 41, 1949	1960
279	Kempton Park	Thomas Laffey	Carrabubula, NSW	1946	BBSA, Vol. 41, 1949	1952
280	Enderdale	L.M. Margetts	Flowerfield, TAS	1947	BBSA, Vol. 41, 1949	1949
281	Leslie Vale	A.R. Park	Hobart, TAS	1948	BBSA, Vol. 41, 1949	1952
282	Willow Dene	G.G. Curtis	Rushworth, VIC	1948	BBSA, Vol. 41, 1949	1951
283	Hedgeways	Mrs E. Dennis	Nambrok, VIC	1948	BBSA, Vol. 41, 1949	1954
284	Lumeah	F.E. & H.G. Sudholz	Horsham, VIC	1948	BBSA, Vol. 41, 1949	1950
285	Northcote	O.M. Phillips	Bacchus Marsh, VIC	1950	BBSA, Vol. 43, 1951	1951
286	Single Oak	W.T. McAlpine	Hopetoun, VIC	1950	BBSA, Vol. 43, 1951	1954
287	Yahlville	Yahl YF Club	Yahl, SA	1950	BBSA, Vol. 44, 1952	1952
288	Yahl	A.A. Hill	Yahl, SA	1951	BBSA, Vol. 44, 1952	1964
289	Werrook	Rowley & son	Mt Gambier, SA	1951	BBSA, Vol. 44, 1952	1958
290	Burringamyth	G.T. & M.J. Tippett	Inverloch, VIC	1952	BBSA, Vol. 45, 1953	1962
291	Hatchellhill	J. Hatchell-Brown	Broadford, VIC	1952	BBSA, Vol. 46, 1954	1958
292	Rob-Bobbin	Hodge Bros.	West Tree, VIC	1952	BBSA, Vol. 46, 1954	1956
293	Freestone Park	A.J. Caldwell	Morwell, VIC	1953	BBSA, Vol. 47, 1955	1955
294	Glenlothie	D.W. Edgar	Gingin, WA	1954	BBSA, Vol. 47, 1955	1955
295	Talisker	M. Gee	Relbia, TAS	1954	BBSA, Vol. 47, 1955	1955
296	Woodvale	Woodvale Past. Co.	Deniliquin, NSW	1953	BBSA, Vol. 47, 1955	1962
297	Glentromie	Berrinvale Grazing Co.	New Norcia, WA	1954	BBSA, Vol. 48, 1956	1956
298	Barjendy	G.B. Chapman	Spring Hill, NSW	1955	BBSA, Vol. 49, 1957	1957
299	Craggan	Cumming Bros.	Alexandra, VIC	1955	BBSA, Vol. 49, 1957	1973
300	Byfrons	F.B. Morgan-Payler	Phillip Island, VIC	1956	BBSA, Vol. 49, 1957	1960
301	Rusbet	Russell H. Richardson	Dinninup, WA	1956	BBSA, Vol. 50, 1958	1965
302	Cliffdale	Stanley J. Freeman	Briagolong, VIC	1957	BBSA, Vol. 50, 1958	1962

THE HISTORY OF ENGLISH LEICESTER SHEEP IN AUSTRALIA

303	Bungalally Park	James D. McIntyre	Horsham, VIC	1957	BBSA, Vol. 50, 1958	1965
304	Essington	J.H. Pleydell	Briagolong, VIC	1957	BBSA, Vol. 50, 1958	1959
305	Miamup	T. & M.L. Hick	Cowaramup, WA	1957	BBSA, Vol. 50, 1958	1962
306	Marnook	P.S. Marwick	Noggerup, WA	1957	BBSA, Vol. 51, 1959	1969
307	Elswood Lodge	R.E. Sharp	Castlemaine, VIC	1958	BBSA, Vol. 51, 1959	1968
308	Lochanside	H.C. & D.W. McKenzie	Horsham, VIC	1959	BBSA, Vol. 53, 1961	1965
309	Suncrest	J. & J. Sloan	Gingin, WA	1959	BBSA, Vol. 53, 1961	1973
310	Coralling Brook	R.H. Matthews	Williams, WA	1960	BBSA, Vol. 53, 1961	1962
311	Werrook	Mrs D.J. Rowley	Mt Gambier, SA	1959	BBSA, Vol. 53, 1961	1962
312	Preston Valley	O. Jenkins	Lowden, WA	1960	BBSA, Vol. 54, 1962	1962
313	Pieracle	Patrick T. Balkin	Casterton, VIC	1960	BBSA, Vol. 54, 1962	1965
314	Connaughtville	N.S. & S.B. Badcock	Westbury, TAS	1961	BBSA, Vol. 55, 1963	2001
315	Inverell Park	W.C. Langham	Finley, NSW	1962	BBSA, Vol. 56, 1964	1973
316	Narradale	R.M. & M.C. Hart	Naracoorte, SA	1963	BBSA, Vol. 56, 1964	1966
317	Pine Creek	D.P. Humble & Sons	Nannup, WA	1963	BBSA, Vol. 57, 1965	1966
318	Evalyn	John Blakiston	York, WA	1963	BBSA, Vol. 57, 1965	1966
319	Ellerton	B.J. Marwick	York, WA	1963	BBSA, Vol. 57, 1965	1968
320	Rocky Hills	N.F. Candy & Son	Cuballing, WA	1963	BBSA, Vol. 57, 1965	1969
321	Narrawong	Miss Sandy McI. Purnell	Moriac, VIC	1965	BBSA, Vol. 58, 1966	
322	Tallengower	N. McDonnell	Chetwynd, VIC	1966	BBSA, Vol. 59, 1967	1968
323	Wimbledon	M.K. Reynolds	Meckering, WA	1966	BBSA, Vol. 60, 1968	1970
324	Subiaco	Robert W. Queale	Horsham, VIC	1966	BBSA, Vol. 60, 1968	1970
325	Franketti	Mrs F.J. Groom	Ulverstone, TAS	1966	BBSA, Vol. 60, 1968	1968
326	Geerak	K.D. & M.I. Luhrs	Cavendish, VIC	1966	BBSA, Vol. 60, 1968	1969
327	Wilyungulup	J.C. Brown	Boyup Brook, WA	1966	BBSA, Vol. 60, 1968	1970
328	Mt. St. Clare	Mousley Bros.	Lowden, WA	1968	BBSA, Vol. 62, 1970	1970
329	Fosterine Park	A.J. Foster	Stoneville, WA	1969	BBSA, Vol. 63, 1971	1972
330	Bara-Simbil	Robin Harwood	Lilydale, VIC	1971	BBSA, Vol. 64, 1972	

#	Name	Owner	Location	Year	Reference	Year
331	Brams	E. & M. Bramich	Elizabeth Town, TAS	1973	BBSA, Vol. 66, 1974	1979
332	Yu-Yana	E.O. Walter & Son	Hensley Park, VIC	1973	BBSA, Vol. 67, 1975	1982
333	Sunbury Farm	K.R. Toovey	Tenterden, WA	1973	BBSA, Vol. 67, 1975	1978
334	Severina	R.S. & M.L. Coppin	Timboon, VIC	1975	BBSA, Vol. 69, 1977	1978
335	Koenarl	Colin R. Taylor	Timboon, VIC	1977	BBSA, Vol. 71, 1978	
336	Weemilah	Mrs Dianne M. MacFarlane	Jindivick, VIC	1978	BBSA, Vol. 72, 1979	
337	Wallingford	D. & L. Picken	Healesville, VIC	1978	BBSA, Vol. 72, 1979	1979
338		N.V. & L.F. Henderson	Kangaroo Ground, VIC	1979	BBSA, Vol. 72, 1979	1980
339	Yalamurra	Dr W.J. Granger	Timboon, VIC	1979	BBSA, Vol. 72, 1979	1982
340	The Glen	M.S. Gadd	Wodonga, VIC	1979	BBSA, Vol. 73, 1980	1983
341	Koorana	S.W. & C.R. Collings	Kapunda, SA	1979	BBSA, Vol. 73, 1980	1982
342	Neelloc	J.K. & C.A. Nagel	West Pingelly, WA	1979	BBSA, Vol. 73, 1980	
343	Chintillga	J. & J. Hobill	Freshwater Creek, VIC	1980	BBSA, Vol. 74, 1981	1984
344	Arracoola	Beverley B. Scott	Dunkeld, VIC	1980	BBSA, Vol. 74, 1981	1984
345	Parkside	Miss Virginia M. Cooke	Lue, NSW	1981	BBSA, Vol. 74, 1981	
346	Sicol	B.L. & D.M. Coles	Bungendore, NSW	1981	BBSA, Vol. 75, 1982	1982
347	Marengo	Bellbrook Partnership	Richmond, TAS	1981	BBSA, Vol. 75, 1982	1997
348	Littlewood	Mrs Carol Stuart	Euroa, VIC	1981	BBSA, Vol. 75, 1982	1988
349	Dishley Hall	D.A. & M. Woodhouse	Northam, WA	1984	BBSA, Vol. 77, 1984	1985
350	Kevley Hill	K.T. Rorke & L.A. Johnson	Balhannah, SA	1983	BBSA, Vol. 77, 1984	
351	The Delta	John Stone	Laanecoorie, VIC	1983	BBSA, Vol. 77, 1984	1996
352	Ostlers Hill	P. & E. Stephenson	Flinders, VIC	1984	BBSA, Vol. 77, 1984	2015
353	Cottage Hill	R.M. Thompson	Pimpama, QLD	1983	BBSA, Vol. 77, 1984	
354	Jarmon	Mrs M. Kendall	Mornington, VIC	1985	BBSA, Vol. 79, 1986	1986
355	Castle Lea	Robert C. Yavion	Lawsons Creek, NSW	1985	BBSA, Vol. 79, 1986	
356	Razorridge Farm	P. & S. DiBona	Kangarilla, SA	1985	BBSA, Vol. 79, 1986	
357	Towamba	R. Cohen & N. Clarke	Balwyn, VIC	1983	BBSA, Vol. 80, 1987	
358	Karlee Park	P.J. & G.D. Foureur	Macclesfield, SA	1985	BBSA, Vol. 80, 1987	1988

THE HISTORY OF ENGLISH LEICESTER SHEEP IN AUSTRALIA

359	Piccadilly	David A.S. & Ines Parker	Burnside, SA	1986	BBSA, Vol. 80, 1987	1997
360	Tay Brig	Jane G. Seater	Oakbank, SA	1985	BBSA, Vol. 80, 1987	
361	Green Hills	Gliksten Pastoral Co.	Adelong, NSW	1986	BBSA, Vol. 80, 1987	1988
362	Ewenme	M.J. Chigros	Littlehampton, SA	1986	BBSA, Vol. 80, 1987	
363	Dee Pee Jay	D.P. Johnson	Balhannah, SA	1986	BBSA, Vol. 80, 1987	1988
364	Nulla Vale	T.A. & P.A. Chapman	Port Fairy, VIC	1984	BBSA, Vol. 80, 1987	
365	Mingara Heights	H.C. & C.B. Wilkinson	Happy Valley, SA	1987	BBSA, Vol. 81, 1988	1988
366	Kookaburra Springs	P.J. & L.J. McVee	Mt Barker, SA	1987	BBSA, Vol. 81, 1988	1988
367	Cirrus	A. & R. Hughes	St. Arnaud, VIC	1987	BBSA, Vol. 81, 1988	1996
368	Sunnylea	G. & P.S. Henry	Poowong, VIC	1988	BBSA, Vol. 81, 1988	
369	Shan-Taurus	F.J. & M.W. Lawson	Spring Ridge, NSW	1988	BBSA, Vol. 82, 1989	
370	Cherry Tree Farm	Mrs Sue Flynn	Sebastopol, VIC	1988	BBSA, Vol. 82, 1989	
371	Tranquility	B.P. & P.J. Sullivan	Woodend, VIC	1988	BBSA, Vol. 82, 1989	
372	Kuitpo Valley	H.S. & R.W. Dixon	Meadows, SA	1988	BBSA, Vol. 82, 1989	
373	Old Dalmore	Mrs D.M. McDonell	Romsey, VIC	1990	BBSA, Vol. 83, 1991	
374	Karrama	I.N. & M.E. Hay	Grassmere, VIC	1990	BBSA, Vol. 83, 1991	
375	Greenmount	J.E. Eddington	Selbourne, TAS	1990	BBSA, Vol. 83, 1991	
376	The Hermitage	Mrs J. Sloan	Pakenham, VIC	1990	BBSA, Vol. 83, 1991	
377	Boonah Springs	Mr & Mrs L.C. Glare	Springvale, VIC	1990	BBSA, Vol. 83, 1991	1997
378	Willow Farm	G. & S. Oliver	Hoddles Creek, VIC	1990	BBSA, Vol. 83, 1991	1998
379	Moira	M. Starritt	Tatura, VIC	1991	BBSA, Vol. 83, 1991	
380	Yarn Time	Mrs M. Kingman	Narre Warren, VIC	1992	BBSA, Vol. 84, 1992	2013
381	Coolibaa	W. & L. Middleton	Deer Park, VIC	1992	BBSA, Vol. 84, 1992	
382	Coricancha	Mrs P. Earnst	Ballan, VIC	1992	BBSA, Vol. 84, 1992	1996
383	Lonarch	S.L. & C.D. Lyons	Tooma, NSW	1992	BBSA, Vol. 85, 1993	
384	Bracken Hill	D. & I. Simpson	Glengarry, TAS	1993	BBSA, Vol. 85, 1993	1996
385	Hope Springs	I.D. Flint	Mount Pleasant	1993	BBSA, Vol. 86, 1994	
386	Gembrooke	Mrs C. Gillham	Bannockburn, VIC	1994	BBSA, Vol. 86, 1994	1996

387	Jarob	R. & J. Brown	Mount Lonarch, VIC	1994	BBSA, Vol. 87, 1995	
388	Shingle Hill 2	Mrs B.C. Teniswood	Triabunna, TAS	1995	BBSA, Vol. 88, 1996	
389	Hadfield Park	V. & M. Bell	Cooma, NSW	1995	BBSA, Vol. 88, 1996	
390	Aries One	R. & S. Wade-Ferrall	Glaziers Bay, TAS	1995	BBSA, Vol. 88, 1996	
391	Whistletop	Whistletop Ptn.	Horsham, VIC	1996	BBSA, Vol. 88, 1996	2000
392	Carramar 2	J.H.A. & L. Beard	Koroit, VIC	1996	BBSA, Vol. 88, 1996	
393	Claybon	P. Harkness & B. Savage	Timboon, VIC	1996	BBSA, Vol. 88, 1996	
394	Lizziwell	M.R. Walker	Frankston, VIC	1996	AFR, Vol. 89, 1997	
395	Wanstead	R.M. & P. Baird	Campbell Town, TAS	1997	AFR, Vol. 90, 1998	
396	Torwood 2	Ms E. & L. Long	Richmond, TAS	1997	AFR, Vol. 89, 1997	1996
397	Peaceful Gardens	Peaceful Gardens Pty Ltd	Nerrena, VIC	1997	AFR, Vol. 90, 1998	
398	Elonview	E.K. Rayner	Mowbray Heights, TAS	1997	AFR, Vol. 90, 1998	2001
399	Klicitat	C. & B. Tatterson	Lima, VIC	1997	AFR, Vol. 90, 1998	2001
400	Nant	I.K. & J.D. Campbell	Evandale, TAS	1997	AFR, Vol. 91, 1999	
400	Nant	G. Willows	Evandale, TAS	1997	AFR, Vol. 109, 2017	
401	Somerset	J. & J. McCarthy	Neerim East, VIC	1999	AFR, Vol. 91, 1999	
402	Brudari Hills	B. & R. Foster	Goomalibee, VIC	2000	AFR, Vol. 93, 2001	
403	Kaehlou	Kirsty Harker	Goorambat, VIC	2000	AFR, Vol. 93, 2001	
404	Kings	Kings College	Warrnambool, VIC	2002	AFR, Vol. 95, 2003	
405	Pointck	Joe La Greca	Point Cook, VIC	2002	AFR, Vol. 95, 2003	
406	Collfarm	Collingwood Children's Farm	Abbotsford, VIC	1997	AFR, Vol. 95, 2003	
407	Charpanya	Mrs R. Rogers	Yallourn North, VIC	2002	AFR, Vol. 96, 2004	
408	Rose of Kerry	M.C. Culloch & K. Gawley	Welshman's Reef, VIC	2003	AFR, Vol. 96, 2004	2006
409	Silent Dale	E. Barry & A. Cassar	Schofields, NSW	2004	AFR, Vol. 97, 2005	
410		Gradia International Pty Ltd	Warrandyte, VIC			2006
411	Bo-Peep	R.H. & M.E. Cowburn	Meredith, VIC	1999	AFR, Vol. 98, 2006	
412	Wingul	E. Hogarth	Pipers River, TAS	2007	AFR, Vol. 100, 2008	2016
413	Moseley Park	I.F & J.L. Vandenbroek	Kimba, SA	2008	AFR, Vol. 100, 2008	2011

414	Snowtop	Ms S. Williams	Bairnsdale, VIC	2008	AFR, Vol. 100, 2008	
415	Epona Park	G. & J. Burgun	Benalla, VIC	2008	AFR, Vol. 100, 2008	
416	Ginabba	Miss G. Loveridge	Longwarry, VIC	2011	AFR, Vol. 104, 2011	2012
417	Ethelgrae	G.L.T. Clarke	Invermay, VIC	2011	AFR, Vol. 106, 2014	
418	Dragon Point	N. Bray	Oatlands, TAS	2011	AFR, Vol. 104, 2011	2013
419	Upland	N. & V. Wootton	Main Ridge, VIC	2012	AFR, Vol. 106, 2014	
420	Langs Crossing	J.T. & C.E. Southwell	Mullion Creek, NSW	2013	AFR, Vol. 106, 2014	
421	Caesia	P. & J. Gelmi	Burekup, WA	2014	AFR, Vol. 106, 2014	
422	Torran	K. Wheeler	Lavington, NSW	2015	AFR, Vol. 108, 2016	2019
423	Beersheba	W. Beer	Moorngag, VIC	2015	AFR, Vol. 108, 2016	
424	Bellevue Park	J.F. & J.W. Fletcher	Cooma, NSW	2016	AFR, Vol. 108, 2016	2022
425	Tarleea	V.W. Gorring	Wellcamp, QLD	2015	AFR, Vol. 109, 2017	2019
426	Sprgvel	Jennifer Shields	Dookie, VIC	2018	AFR, Vol. 110, 2018	
427	Nerrena	Claire Crocker	Leongatha, VIC	2011	AFR, Vol. 111, 2019	
428	Omazel	B. Cardinal & T. Urek	Weerite, VIC	2020	AFR, Vol. 112, 2020	
430	Windsong Valley	G. Chenik	Bedfordale, WA	2021	AFR, Vol. 113, 2021	
431	Junction Lodge	P. Talbot & P. Ottey	Carisbrook, VIC	2022	AFR, Vol. 114, 2022	
432	Dalmally	I. & M. Morris	Doreen, VIC	2022	AFR, Vol. 114, 2022	
433	Church Hill Park	L. Church	Wonga Park, VIC	2022	AFR, Vol. 114, 2022	
434	Gracies Lane	M. & R. Mamers	Main Ridge, VIC	2022	AFR, Vol. 114, 2022	
435	Kiah Killibinbin	V. Robinson	Bairnsdale, VIC	2022	AFR, Vol. 114, 2022	
436	Rocky River	K. Stewart	Rocky River, NSW	2022	AFR, Vol. 114, 2022	
437	Willow Drive	B. Shalders & S. & T. Holmes	Grassmere, VIC	2022	AFR, Vol. 115, 2023	
438	Beartach	A. & L. Stewart	Armidale, NSW	2021	AFR, Vol. 115, 2023	

Notes:

1. Flock No.9 prior to Vol.27.

2. Became Flock No.8 in Vol.27.

3. Sold to Cole Bros. in 1912.

4. Sold to Bradshaw, March 1945.

5. Sold to I.D. Macdonald, January 1946.

6. Sold to Nicholas Cole, November 1946.

7. Prefix change and transfer to B. Heazlewood.

8. First book published by ASBBS.

9. Vol. 19 first to include a Prefix.

10. J. MacGregor & Son in Vol. 29.

Appendix 2: Sheep Numbers

The Australian Longwool Sheepbreeders' Association

Volume	Year	Total Flocks	Total Ewes	TAS Flocks/Ewes	VIC Flocks/Ewes	NSW Flocks/Ewes	SA Flocks/Ewes	QLD Flocks/Ewes	WA Flocks/Ewes
Vol. 1	1909	23		18/	1/	2/	2/		
Vol. 2	1910	21	2643	15/1043	1/900	2/119	2/461	1/120	
Vol. 3	1911	20	2052	14/1170		2/191	2/548	1/143	
Vol. 4	1912	23	1936	15/1215		3/226	4/330	1/165	
Vol. 5	1913	23	2265	15/1076		6/1093	1/90		
Vol. 6	1914	26	1523	17/924	1/73	8/526			
Vol. 7	1915	37	3516	19/1029	4/976	13/1421	1/90		
Vol. 8	1916	34	4317	17/1324	4/1084	11/1856	2/53		
Vol. 9	1917	34	5823	18/1408	4/1490	10/78	2/147		
Vol. 10	1918	27	3480	17/1537	3/670	5/1090	2/183	1/	
Vol. 11	1919	33	2595	19/764	3/183	7/1458	2/190		
Vol. 12	1920	29	3031	17/1198	3/636	6/1030	1/102		2/65
Vol. 13	1921	30	2704	15/1007	4/516	7/981	1/82		3/118
Vol. 14	1922	27	2166	14/840	3/166	6/1009	1/90		3/61
Vol. 14	1923	27	1845	14/774	3/118	6/800	1/82		3/71
Vol. 15	1924	23	1971	12/700	3/154	4/845	1/80		3/192
Vol. 16	1925	19	1797	10/559	3/160	5/994	1/84		
Vol. 17	1927	15	1762	8/398	2/232	5/1132			
Vol. 18	1928	19	1874	9/465	2/199	8/1210			
Vol. 19	1929	16	2116	8/555	2/230	6/1331			
Vol. 19	1930	18	2362	9/615	2/270	7/1477			
Vol. 20	1931	17	2570	9/665	2/266	6/1639			
Vol. 21	1932	17	2157	9/517	2/310	6/1330			

British Breeds of Sheep in Victoria (Flock Book published by RASV)

Volume	Year	Total Flocks	Total Ewes	TAS Flocks/Ewes	VIC Flocks/Ewes	NSW Flocks/Ewes	SA Flocks/Ewes	QLD Flocks/Ewes	WA Flocks/Ewes
Vol. 1	1898	2	600		2/600				
Vol. 2	1899	2	683		2/683				
	1900	2	656		2/656				
	1901	2	636		2/636				

British Breeds of Sheep in Australia (Flock Book published by RASV)

Volume	Year	Total Flocks	Total Ewes	TAS Flocks/Ewes	VIC Flocks/Ewes	NSW Flocks/Ewes	SA Flocks/Ewes	QLD Flocks/Ewes	WA Flocks/Ewes
Vol. 3	1902	2	488		2/488				
	1903	2	543		2/543				
	1904	2	714		2/714				
	1905	2	713		2/713				
	1906	4	935		4/935				
Vol. 4	1907	2	125		1/31			1/94	
	1908	4	190		3/84			1/106	
	1909	9	354		8/234			1/120	
Vol. 5	1911	10	323		10/323				
	1912	11	403		11/403				
	1913	19	554		19/554				
Vol. 6	1914	21	2017		20/1982	1/35			
Vol. 7	1915	23	2282		21/2234	1/36	1/12		
Vol. 8	1916	25	2591		25/2591				
Vol. 9	1917	31	2856		31/2856				
Vol. 10	1918	35	3118		34/3099			1/19	
Vol. 11	1919	47	3916		46/3869			1/47	
Vol. 12	1920	44	4009		43/3966			1/43	
Vol. 13	1921	37	3549		35/3464			2/85	
Vol. 14	1922	31	2773		30/2692			1/81	
Vol. 15	1923	30	2477		30/2477				
Vol. 16	1924	24	1642		24/1642				

THE HISTORY OF ENGLISH LEICESTER SHEEP IN AUSTRALIA

1925, Australian Society of Breeders of British Sheep formed and published the Flock Book from Vol. 17

Volume	Year	Total Flocks	Total Ewes	TAS Flocks/Ewes	VIC Flocks/Ewes	NSW Flocks/Ewes	SA Flocks/Ewes	QLD Flocks/Ewes	WA Flocks/Ewes
Vol. 17	1925	25	1907		25/1907				
Vol. 18	1926	24	2140		20/1918	1/27			3/195
Vol. 19	1927	31	2601		25/2315	1/24	3/135		2/127
Vol. 20	1928	35	2627		27/2180	1/28	4/163		3/256
Vol. 21	1929	38	3264		27/2716	1/33	6/242		4/273
Vol. 22	1930	35	3420		25/2862	1/8	5/227		4/323
Vol. 23	1931	39	3846		26/3214	1/22	7/276		5/335
Vol. 24	1932	40	4009		25/3297	1/24	9/317		5/371
Vol. 25	1933	41	4793	1/6	25/3640	2/451	9/356		4/340
Vol. 26	1934	40	4725	1/4	26/3681	2/446	8/392		3/202

First combined (ASBBS and ALSA) Flock Book published in 1935, Vol. 27

Volume	Year	Total Flocks	Total Ewes	TAS Flocks/Ewes	VIC Flocks/Ewes	NSW Flocks/Ewes	SA Flocks/Ewes	QLD Flocks/Ewes	WA Flocks/Ewes
Vol. 27	1935	48	5990	7/440	24/3135	7/1824	7/400		3/191
Vol. 28	1936	55	6414	10/538	28/4513	6/743	8/406		3/214
Vol. 29	1937	58	6567	11/645	24/3461	8/1708	11/474		4/279
Vol. 30	1938	60	6693	12/750	27/3564	9/1650	9/410		3/319
Vol. 31	1939	58	6725	15/918	25/3658	8/1354	7/428		3/428
Vol. 32	1940	59	7009	13/1006	27/3847	7/1399	6/337		6/420
Vol. 33	1941	78	8564	24/1369	31/4505	8/1627	8/479		7/584
Vol. 34	1942	90	8479	29/1561	37/4673	8/1095	9/562		7/588
Vol. 35	1943	99	9576	33/1973	42/5334	9/1124	8/531		7/614
Vol. 36	1944	122	9971	44/2219	54/5557	8/1076	9/478		7/641
Vol. 37	1945	122	9843	42/2072	56/5532	10/1141	8/446		6/654
Vol. 38	1946	125	9087	42/2191	53/4552	13/1175	9/509		8/660
Vol. 39	1947	130	9141	41/2066	61/5115	13/930	8/450		7/580
Vol. 40	1948	133	9981	43/2411	58/5286	15/1103	7/517		10/664
Vol. 41	1949	136	10368	42/2348	63/5773	14/1054	5/473		12/720
Vol. 42	1950	109	7265	34/1721	49/3603	11/952	5/336		10/653
Vol. 43	1951	86	4972	24/1265	39/2264	8/598	5/285		10/560
Vol. 44	1952	75	4263	19/1078	33/2018	6/341	7/291		10/535
Vol. 45	1953	66	3787	17/859	31/1922	3/268	5/194		10/544
Vol. 46	1954	59	3579	13/597	29/1984	3/256	5/183		9/559
Vol. 47	1955	54	3411	11/580	26/1770	5/336	3/186		9/539
Vol. 48	1956	50	3217	9/543	24/1696	3/256	4/125		10/597
Vol. 49	1957	49	3468	8/451	24/1938	4/313	4/149		9/617
Vol. 50	1958	54	3917	9/715	27/2093	3/229	4/203		11/677

Volume	Year	Total Flocks	Total Ewes	TAS Flocks/Ewes	VIC Flocks/Ewes	NSW Flocks/Ewes	SA Flocks/Ewes	QLD Flocks/Ewes	WA Flocks/Ewes
Vol. 51	1959	52	3843	8/674	26/2035	3/269	3/192		12/673
Vol. 52	1960	48	3745	8/579	23/2024	3/242	2/208		12/692
Vol. 53	1961	49	3414	8/540	22/1849	3/237	3/204		13/584
Vol. 54	1962	50	3391	8/526	23/1851	3/198	3/218		13/598
Vol. 55	1963	45	3292	9/604	21/1787	2/83	2/222		11/596
Vol. 56	1964	46	3026	9/553	21/1542	2/74	3/250		11/607
Vol. 57	1965	46	2689	9/528	20/1461	1/41	2/7		14/652
Vol. 58	1966	40	2588	9/594	18/1377	1/38	1/4		11/575
Vol. 59	1967	39	2631	9/514	19/1422	1/37			10/658
Vol. 60	1968	40	2617	10/528	18/1394	1/33			11/662
Vol. 61	1969	34	1982	7/257	16/1043	1/38			10/644
Vol. 62	1970	31	1875	7/348	14/937	2/71			8/518
Vol. 63	1971	29	1478	7/284	14/742	2/57			6/395
Vol. 64	1972	26	1258	5/141	14/705	1/30			6/382
Vol. 65	1973	26	1204	6/171	14/652	1/30			5/351
Vol. 66	1974	22	1074	7/185	11/695				4/194
Vol. 67	1975	22	1048	7/197	11/674				4/177
Vol. 68	1976	21	900	7/186	10/525				4/189
Vol. 69	1977	21	886	7/205	11/491				3/190
Vol. 70	1977	20	894	7/242	10/461				3/191
Vol. 71	1978	22	953	7/239	12/514				3/200
Vol. 72	1979	24	1064	7/316	15/583				2/165
Vol. 73	1980	25	1096	6/268	15/656		1/3		3/169
Vol. 74	1981	26	1081	6/278	15/626	1/3	1/3		3/171
Vol. 75	1982	28	1114	7/262	15/668	2/7	1/8		3/168
Vol. 76	1983	24	1006	5/184	15/660	1/2			3/160
Vol. 77	1984	28	1058	7/256	13/500	2/11	2/112	1/3	3/167
Vol. 78	1985	23	846	7/259	11/464	2/11	2/54	1/4	3/54
Vol. 79	1986	26	918	7/242	11/493	3/19	2/54	1/4	2/106
Vol. 80	1987	31	955	6/219	11/455	4/119		1/5	2/39
Vol. 81	1988	34	924	6/202	15/500	4/119	7/63	1/5	1/35
Vol. 82	1989	31	959	5/195	15/595	4/45	4/35	1/10	2/79
Vol. 83	1991	30	940	5/199	19/652	2/49	3/32	1/8	
Vol. 84	1992	30	998	4/170	21/766	1/26	3/28	1/8	
Vol. 85	1993	32	1085	5/176	21/837	2/38	3/26	1/8	
Vol. 86	1994	30	1003	5/204	20/741	2/27	3/31		
Vol. 87	1995	25	958	5/211	16/653	2/43	2/51		
Vol. 88	1996	28	969	6/208	18/636	3/80	1/45		

THE HISTORY OF ENGLISH LEICESTER SHEEP IN AUSTRALIA

From Vol. 89 onwards the Flock Book name changed to Australian Flock Register

Volume	Year	Total Flocks	Total Ewes	TAS Flocks/Ewes	VIC Flocks/Ewes	NSW Flocks/Ewes	SA Flocks/Ewes	QLD Flocks/Ewes	WA Flocks/Ewes
Vol. 89	1997	26	888	6/199	17/612	1/30	2/47		
Vol. 90	1998	28	886	5/118	19/583	3/133	1/52		
Vol. 91	1999	26	744	7/147	18/557	1/40			
Vol. 92	2000	21	656	6/129	15/527				
Vol. 93	2001	21	603	5/100	15/483	1/20			

Federal name change from Australian Society of Breeders of British Sheep to Australian Stud Sheep Breeders Association

Volume	Year	Total Flocks	Total Ewes	TAS Flocks/Ewes	VIC Flocks/Ewes	NSW Flocks/Ewes	SA Flocks/Ewes	QLD Flocks/Ewes	WA Flocks/Ewes
Vol. 94	2002	19	585	4/76	13/449	2/60			
Vol. 95	2003	21	460	4/77	16/363	1/20			
Vol. 96	2004	19	472	3/65	15/392	1/15			
Vol. 97	2005	19	472	3/65	15/395	1/12			
Vol. 98	2006	18	452	3/68	14/357	1/27			
Vol. 99	2007	15	441	3/91	12/350				
Vol. 100	2008	17	470	3/51	13/412		1/7		
Vol. 101	2009	16	384	4/105	11/272		1/7		
Vol. 102	2010	16	383	4/95	11/278		1/10		
Vol. 103	2011	16	423	4/101	11/312		1/10		
Vol. 104	2012	17	461	5/124	12/319				
Vol. 105	2013	15	460	5/120	10/340				
Vol. 106	2014	17	480	4/115	11/339	1/22			1/4
Vol. 107	2015	16	510	4/119	10/368	1/19			1/4
Vol. 108	2016	18	460	4/135	10/277	3/40			1/8
Vol. 109	2017	18	431	4/119	9/241	3/52		1/11	1/8
Vol. 110	2018	18	466	4/125	10/277	2/37		1/18	1/9
Vol. 111	2019	20	508	4/144	11/286	3/46		1/24	1/8
Vol. 112	2020	18	507	4/160	12/309	1/30			1/8
Vol. 113	2021	17	695	3/139	11/304	1/45			2/7
Vol. 114	2022	21	514	3/181	15/279	2/51			1/3
Vol. 115	2023	18	503	3/208	13/279	1/13			1/3

About the Author

Brenton Heazlewood

Brenton is a 5th generation English Leicester breeder farming at Whitemore, northern Tasmania. He is proud to be looking after this breed on the same property that his great grandfather registered the first stud in 1871. Each generation has run registered English Leicesters and exhibited at Tasmanian as well as mainland shows. Brenton takes a team of English Leicester's to the Australian Sheep & Wool Show in Bendigo, Victoria each year as well as exhibiting at several local shows.

Brenton is a federal longwool sheep judge, having judged nationally, as well as in the United States and Sweden.

He is currently the Federal President and Tasmanian State Chairman of the Australian Stud Sheep Breeders Association as well as President of the English Leicester Association of Australia.

www.ingramcontent.com/pod-product-compliance
Lightning Source LLC
Chambersburg PA
CBHW052005070526
44584CB00016B/1625